The Seven States of
CALIFORNIA

Crescent City

River

Klamath

Humbolt Bay
Eureka

Cape
Mendocino

Tule Lake
▲ Mount Shasta

MODOC
PLATEAU

III

IV

Redding

CASCADE RANGE

▲ Lassen Peak

II

Donner Pass

GREAT

Point Arena

Sacramento

COAST
RANGES

SIERRA NEVADA

Lake
Tahoe

Mono Lake

Point Reyes

San Francisco

San Jose

VI

Turlock

V

Bishop

Monterey

San Andreas Fault

Fresno

Death
Valley

COAST
RANGES

VALLEY

Ancient
Owens
River Bed

Bakersfield

TRANSVERSE RANGES

Barstow

I

Point Concepcion

Santa
Barbara

MOJAVE DESERT

Los Angeles

Los Angeles
River

Palm Springs

VII

PENINSULAR
RANGES

Salton
Sea

San Diego

The Seven States of
CALIFORNIA

A Natural and Human History

■

PHILIP L. FRADKIN
with Photographs by the Author

A JOHN MACRAE BOOK
Henry Holt and Company ■ New York

Henry Holt and Company, Inc.
Publishers since 1866
115 West 18th Street
New York, New York 10011

Henry Holt® is a registered
trademark of Henry Holt and Company, Inc.

Published in Canada by Fitzhenry & Whiteside Ltd.,
195 Allstate Parkway, Markham, Ontario L3R 4T8.

Library of Congress Cataloging-in-Publication Data
Fradkin, Philip L.
The seven states of California:
a natural and human history/
Philip L. Fradkin.—1st ed.
p.   cm.
"A John Macrae book."
Includes bibliographical references and index.
1. California—Description and travel.
2. California—Geography.
I. Title.
F861.F75   1995   94-39984
917.94—dc20   CIP

ISBN 0-8050-1947-2

Henry Holt books are available for special promotions
and premiums. For details contact:
Director, Special Markets.

First Edition—1995

Designed by Paula R. Szafranski

Maps by Jeffrey L. Ward

Printed in the United States of America
All first editions are printed on acid-free paper. ∞

1   3   5   7   9   10   8   6   4   2

Grateful acknowledgment is made to reprint an excerpt from *Twilight* by Anna
Deavere Smith. Copyright © 1994 by Anna Deavere Smith. Used by permission
of Doubleday, a division of Bantam Doubleday Dell Publishing Group, Inc.

For my wife,
Dianne Caccioli Fradkin

Contents

"You got to live here to express this point, you got to live here to see what's goin' on.
You gotta look at history, baby,
    you gotta look at history."

—Big Al, as quoted by Anna Deavere Smith
in *Twilight: Los Angeles, 1992*

Preface

What explains California? The landscape, to a great extent, has shaped its history and destiny. There are many landscapes; there is no one California. The state contains distinctive parts of the entire continent: the hot deserts of the Southwest, the wet rain forests of the Northwest, the high peaks of the Rocky Mountain states, the cold steppes and waving grasslands of the Plains states, the meandering bayous of the Gulf states, the wooded valleys and mountains of Appalachia, the sandy coasts of the Mid-Atlantic states, and the rocky coast of Maine. California is all those places, yet none of them. It is its own diverse, transcendent self.

I came to the conclusion that landscape was a determining factor of great importance when I reflected on my own California experience. Like others, I came from elsewhere and have moved frequently within the state. I count fourteen changes in street addresses during the thirty-five years that I have lived in three Californias: the north and south coasts and the Central Valley. I dressed differently, had different friends, and thought and lived differently in each place. The differences were matters of degree that were derived, I believe, in large part from the landscapes that shaped the prevalent cultures to which I sought to adapt.

I am not alone in my belief. Other writers, scholars, architects, and geographers have remarked for almost a century upon the intimate relationship between landscape and culture in California.

The author Mary Austin was a perceptive mystic. Raised in the Midwest, she lived around the turn of the century in the California desert, about which she wrote in *Land of Little Rain*, "Not the law, but the land sets the limits." She returned to this theme in another book: "There can be no adequate discussion of a country, any more than there can be of a woman, which leaves out this inexplicable effect produced by it on the people who live there."

More recently, when Tony Hiss, a staff writer for *The New Yorker* who specialized in the emerging science of place, walked through the main concourse of Grand Central Terminal in New York City, he experienced "a spontaneous and quiet change in perception." (I ask: Is there not a similar adjustment in the desert, among the redwoods, or along the San Andreas Fault?) Hiss wrote in his book *The Experience of Place*, "We all react, consciously or unconsciously, to the places where we live and work in ways we scarcely notice or that are only now becoming known to us." He attributed the unconscious ability to absorb the nature of a place to a sixth sense that combined all other senses. He called it "simultaneous perception."

Josiah Royce, who was born in the Gold Country and spent most of his working life as a professor of philosophy at Harvard University, wrote, "But after all, I think that in California literature, in the customary expressions of Californians in speaking to one another, and, to a very limited degree, in the inner consciousness of any one who has grown up in California, we have evidence of certain ways in which the condition of such a region must influence the life and, I suppose in the end, the character of the whole community."

Architects are particularly attuned to place. John Brinkerhoff Jackson, a specialist in landscapes who taught environmental design at the University of California and elsewhere, cited the "slow adaptation to place" and called for expanding our knowledge of landscapes. Jackson wrote in his influential book *Discovering the Vernacular Landscape*, "Only very rarely is there a glimpse of the history of the landscape itself, how it was formed, how it has changed, and who it was who changed it, and even more rarely does landscape research produce any speculation about the *nature* of the American landscape."

Geographers have also cited the relationship between place and culture. A school of geography referred to variously as geographic determinism, environmentalism, or human geography flourished at the turn of the century but fell into disrepute around 1930. Its practitioners put too much emphasis on "nature-man relations," said the critics. One such critic, Jan O. M. Broek of the University of Minnesota, wrote of Ellen Churchill Semple of the University of Chicago that her books were "classics of their kind, but distress the modern geographer in their overemphasis on the earth's effects on man."

Miss Semple was also accused by her male colleagues of being naive and guilty of that most unpardonable of academic offenses—having popularized her theory in a book. She wrote in 1911, "Man has been so noisy about the way he has 'conquered Nature,' and Nature has been so silent in her persistent influence over man, that the geographic factor in the equation of human development has been overlooked." She cited "this physical basis of history" and added, " 'Which was there first, geography or history?' asks [Immanuel] Kant. And then comes his answer: 'Geography lies at the basis of history.' "

More recently James J. Parsons, a cultural geographer at the University of California at Berkeley, made the case for such an approach on a regional basis. He wrote, "The regional consciousness of California, remarkably strong for so restless and rootless a population, has had its origins in the common problems and interests imposed by geography." Parsons noted the cultural and physical diversity of the state. "Although in one sense California can properly be thought of as approximating a physical, economic and cultural unit, it probably contains as many aberrant cultural groups and distinctive landscape types as any like area on earth. There is not one California, but many."

I have divided California into seven geographic/cultural provinces that have similar appearances, economies, customs, and heritages. They are, in order of presentation: Deserts (southeast), The Sierra (east), Land of Fire (northeast), Land of Water (north coast), The Great Valley (central interior), The Fractured Province (central coast), and The Profligate Province (south coast). The boundaries are not the artificial limits of local political jurisdictions, but rather the cohesiveness of natural features and cultures. In each province I have selected an emblematic landscape feature—in the same order: a chain of dry lakes, a mountain pass, a lava flow, an ocean inlet, a flat piece of ground, an earthquake fault, and a former river—upon which to hang my narrative.

I see California's history unreeling in a number of filmic scenes. The rhythm of its history is spasmodic; the recording device is fitful. The camera zooms in for a close-up of a person or a detail of nature and then retreats for a wide-angle shot that establishes context. Between the two extremes of focal length is everyday California life.

The strands of natural and human history are interwoven throughout the book. Indeed, I do not see how it could be otherwise. Thus, I profile aspects of the natural environment and people and their institutions that mirror the seven landscape provinces. My selections were guided by my experiences as a newspaper reporter, state bureaucrat, magazine writer, university lecturer, book author, and constant wanderer throughout California. I see this book as part history, natural guide, travelogue, memoir, and elegy.

The impressions of William H. Brewer, who crossed and recrossed California as part of the first scientific survey of the state, are the baseline for the narrative. He wrote a series of letters, later preserved in the book *Up and Down California in 1860–1864,* that are an excellent record of the past. My journeys of rediscovery, which began one hundred years later, are contemporary impressions. The bulk of the American experience in California separates the two of us.

Much has been written about the California dream, a concept embraced by those people who have achieved certain advantages. Little has been written about the California nightmare. I have sought to correct that imbalance; for I believe the future will be a dark, chaotic time—a time whose origins can be traced back through the natural and human histories of the state.

Volcanoes, tsunamis, floods, earth slides, droughts, wildfires, glaciers, and earthquakes have periodically shifted or scoured the land throughout geologic time. These wild displacements punctuated and shaped the human history of California. Transient peoples with different skin colors, languages, and customs constantly warred with one another from the time, some thousands of years ago, when the second tribe of ancient Indians displaced the first grouping from a watered valley. This legacy of violence and conflict has filtered down through the years to the most recent Los Angeles riot and earthquake.

There are other common attributes that cross provincial lines and bind the state together. California was conceived of as an island myth, and according to geologists, it will end as an island. Separateness has always been one of its basic characteristics. Geologically and historically the state is young. Ever since it was first represented accurately on a map, in 1846, California has been a known center of incessant activity. The land and people are still in constant motion.

Finally, it is a richly textured land of great extremes and extreme changes: the highest mountain, the lowest valley, the oldest life-forms, the youngest population, great wealth, grinding poverty, the tallest trees, dwarf forests, abundant water, widespread aridity, startling fecundity, great beauty, and violent death. The landscape is deceptive. Great pleasure and pain ripple across its surface. This region of moderate climate and gentle, flower-dappled hills can beguile or, alternatively, burst into deadly flames. Long, long after the end, the fossil remains of this once-great civilization will consist of a thin layer of ashes embedded in the floor of a desert playa.

The Seven States of
CALIFORNIA

The Approach

The name begot the myth of quick riches and easy living and set the tone for the false perception of its history. California was conceived as a fantasy before it was discovered by the Europeans. Most probably it will survive in memory as a myth, much like Atlantis, or better yet the lost continent of Lemuria, the Pacific Ocean's equivalent of the Atlantis legend. Lemuria has been located by some at Mount Shasta, whose snowy peak stands alone like a gigantic, sparkling eye at the center of the northern end of the state.

The myth attached to California by the Spanish involved gold and the exotic; and from it the state derived its name, its nickname (the Golden State), and its repute. But it took an easterner to discover the source of the fantasy. Edward Everett Hale, a Boston Unitarian minister who was the author of *The Man Without a Country,* came across a forgotten Spanish novel in 1862. The novel, *Las Sergas de Esplanadián* (*The Exploits of Esplanadián*), was written in 1510 by Garcí Ordóñez de Montalvo.

One of the most persistent themes of the Judeo-Christian heritage is of an earthly paradise. Montalvo wrote at a time when the Old World was

1

aflame with stories of limitless riches and exotic peoples in the New World. The myths foretold the existence of paradise on earth, variously designated as El Dorado, Quivira, the Seven Cities of Cibola, and California.

In Montalvo's book, there were several references to an island named California that was inhabited by a race of black Amazons ruled by a queen named Califia. California, Montalvo explained, was an island with steep cliffs and a rocky shoreline on the right hand of the Indies and very close to the Terrestrial Paradise. The weapons of the Amazons were made entirely of gold, there being no other metal on the island.

It also happened that on this rugged isle there were many griffins who bore the heads, wings, and talons of eagles and the torsos of lions. In the ancient world the griffin symbolized the sun, the sky, and the golden light of dawn. Then came Christianity, the perspective from which Montalvo wrote, and the griffin came to represent evil in the form of the Devil, who fled with the souls of Christians. "Griffin" was also the common name for a type of Spanish vulture, and in California at that time there were wild condors—giant, black vultures.

The women warriors were adept at trapping the griffins. The men the Amazons took prisoners in their wars, and the boys they subsequently gave birth to, were fed to these hybrid beasts. The griffins repaid their female captors by snatching males from the ground, flying off, and releasing them from great heights.

Queen Califia participated in the siege of Constantinople. After she was tamed by Christianity, the queen married a relative of Esplanadián, and they returned to the island of California. "What happened after that, I must be excused from telling," wrote the author.

No one knows for sure where Montalvo found the name California, or whether he invented it. There were vague precedents. A member of the Calpurnia family was one of Caesar's wives. Calefurnia (spelled variously) was the medieval equivalent of our Jane Doe. Two archaeological sites in Sicily were known as Calaforno and Calaforninu. The people of Cali-ferne, meaning the Arabs or the subjects of the Caliphs, were mentioned in the *Chanson de Roland*. In medieval Latin, *Calidus furnus* is said to mean cauldron. Marco Polo had mentioned a fabulous isle just off the coast of Asia.

Nor does anyone know who first applied the name to the southern tip of the Baja (Lower) California peninsula sometime between 1533 and 1542, when it was depicted on a map as a group of islands. Cabo California was a bleak place, raising the possibility that the name was applied derisively. California was then enlarged to take in the whole peninsula. It became a single island on a map that vaguely resembled the state and the lower peninsula in 1648. The English designation *Nova Albion* had a fleeting existence when the pirate Francis Drake claimed the land north of San Francisco for his queen in 1579. But the Latin appellation drifted off maps when the English did nothing to enforce their claim.

When the Spanish sailed north in the late eighteenth century, the more resilient name of California was stretched to include all the territory between the cape, now known as Cabo San Lucas, and Alaska. Reflecting geopolitical reality, the name later shrank on maps to encompass Baja and Alta (Upper) California during the mission and Mexican periods. When the Anglos took over in 1848, the territory became known simply as California, with Baja California remaining a part of Mexico.

The earliest inhabitants, the Native Americans, had no single name for the region. They did not conceive of it as a whole, but rather as parts—their individual parts. The diverse landscape divided the Indians into more than one hundred tribes with a multitude of cultures. The California Indians may have had the greatest linguistic diversity in the world. The five main language stocks were subdivided into twenty-one language families that were further splintered into more than one hundred dialects. The Indians who spoke one dialect were unintelligible to Indians who spoke another. Just as California is now the most populous state in the nation, so in prehistoric times there were more Indians within its borders (a number variously estimated at 300,000) than in any other comparable area in North America.

I

Deserts

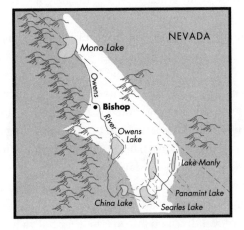

I felt my way into California in 1960 via the desert and arrived on the crest of the wave of immigration that would establish it as the most populous state in the union and an exceedingly rich and well-endowed nation in its own right. Those were proud and naive times.

My route crossed the midsection of California. Inadvertently, I had chosen the perfect introduction to the landscape.

I had not known real heat until I entered Death Valley in early July, nor had I ever encountered such physical extremes in close proximity as when I drove from below sea level to the ten-thousand-foot heights of the Sierra Nevada on that same day.

I dallied only for gas at Furnace Creek, then proceeded across the valley floor as the temperature rose to over one hundred degrees. I pushed my Volkswagen Bug, loaded with all my worldly possessions, as fast as I could up the alluvial fan that led out of that inferno and into the next valley. I stopped for nothing in that fearful landscape.

The descent was into the Panamint Valley and past more history I was ignorant of: a history that bespoke other transients crossing this violent, convoluted land and making their way via foot or horse or wagon or train or car or plane across deserts, over mountain barriers, and into the luscious heartland of California.

The route lay across yet another pass and into the Owens Valley. It wound past the blinding, white salt flats of a desiccated Owens Lake, and at last reached the visual safety of a row of green cottonwood trees at Olancha. Far above, the granite heights of the towering Sierra Nevada formed a seemingly impenetrable barrier.

North went the road, and still does, but it is widened now into an occasional four-lane expressway through Owens Valley, a tectonic trough framed by the White Mountains to the east and the Sierra Nevada to the west. At another dying lake, this one named Mono, I turned left onto the narrow road over Tioga Pass and entered Yosemite National Park. I spent that night in Tuolumne Meadows, and the next morning ice had formed around the rim of my cooking pan. On the Fourth of July I headed for Yosemite Valley. Even in those days the valley was crowded on holiday weekends, but a family made room for me at their campsite.

The next day I departed, as I was anxious to reach the Pacific Ocean and end my journey. I dropped into yet another hot valley, this one called the San Joaquin, where I would return to live in six months' time. I descended into an unfamiliar landscape of green, irrigated lands clasped between the dry thighs of foothills. I pushed the small, uncomplaining car over the Coast Range, where the lustrous gold of summer grasses flowing across the soft flanks of hills was both dazzling and sensuous— almost too much for this easterner to bear.

A hint of cool ocean breezes filtered across the Salinas Valley. I headed south to Big Sur where, after asking directions, I took a dirt road to the right and drove to Pfeiffer Beach. It was dusk and the fog had descended to isolate presences. A small boy darted out of a ranch house where a lamp was lit in a window. It was his job to collect the twenty-five-cent toll. By making it necessary to stop and back up to negotiate a sharp curve, the family cannily forced me, and other motorists, to pause just long enough for someone to run out and collect the money.

I drove the remaining short distance to the beach, walked the dark path between the coastal scrub, and ran across the white sand to the water. I dove straightaway into the surf, as I had done thousands of times before in summertime on the other shore. I was shocked by the frigid water. I surfaced gasping and stared directly into two black holes that held the eyes of an alien beast. It stared unblinkingly at me and then slowly sank into the water. Something slimy brushed against my leg.

I swam toward shore against an unseen force, but lost ground; the riptide dragged me out. I lunged with desperation verging on panic against the current, then swam sideways and finally struggled to shore. I threw myself upon the sand that still retained a hint of the sun that had blazed earlier in the day, and the warm land slowly revived me.

I had barely survived my entry into California. I learned a few months later that I had encountered a harbor seal and giant kelp, both of which were benign. By that time I had been initiated into other California mysteries and was swimming toward the mainstream of the California experience.

It is difficult to fix time in the desert. A tiny fraction of the last one billion years, which is only one-fifth the age of the earth, has been given over to long glacial periods alternating with warm, dry periods. We live in an exceedingly short interval within one of the periods of glaciation called the Quarternary Period, which began about two million years ago. The Quarternary Period is divided into two epochs, the Pleistocene and the Holocene. The Holocene extends into present time. The dividing line between the two epochs was about 10,000 years ago, the point of transition between the last full-blown ice age, known as the Wisconsin, and the present, warmer epoch.

Change is a constant. There are variations within epochs. The warmest time within the Holocene occurred not long after the changeover and lasted 3,000 to 4,000 years. Then came a time of renewed glaciation. That was followed by a warming trend and then the Little Ice Age that extended into the last century. Another warming trend lasted until the 1940s, then it cooled and was wetter for thirty years. The last twenty-five years have been warmer and drier.

The Owens River ran its full course from Mono Lake through a chain of lakes to Death Valley in the late Pleistocene Epoch. There was a different landscape then. A cooler, wetter climate produced woodlands and grasslands where there is now desert. As the glaciers that covered the Sierra Nevada to the four-thousand-foot elevation mark during the late Wisconsin disgorged their contents into the Owens River, marshes formed around the edges of the lakes and became rich food sources for roving bands of ancient Californians. Dark coniferous forests and thick aspen groves descended much lower than they do today. Horses, camels, bison, deer, and the saber-toothed cat prowled these savannas.

When the Pleistocene gave way to the Holocene Epoch, there was a wave of extinctions. The principal victims were the larger animals. Gone were the five species of Pleistocene horses, bison, western camels, large cats, giant beavers, and the Shasta ground sloth. They may have been hunted to death by the early humans or were unable to adapt to the increasing aridity.

The flowing rivers became intermittent, and the deep lakes evaporated. The dry lakes, or *playas* as they are also known (from the Spanish word for beach), are ultimate desert. There are more than fifty playas in the California desert. Heat rises off their white surfaces in shimmering waves that cause mirages. An early desert traveler noted, "Everywhere you meet with the dry lake-bed—its flat surface devoid of life and often glimmering white with salt."

The playas are both ancient lake beds and contemporary landing fields for drug traffic from Central America, experimental aircraft, and flights from space. The dry lakes have received a fair amount of scientific attention, since the runways at Edwards Air Force Base in the Mojave Desert are laid out across a playa and test pilots like Chuck Yeager have landed exotic craft, such as the first jets, space shuttles, and stealth bombers, on the flat terrain.

The California desert has a wide assortment of landscapes. There are mountains, valleys, alluvial fans, badlands, canyons, oases, dry washes, and sand dunes. The playas are the only flat component of the desert. Their surfaces vary one inch or less per mile. To qualify as a playa, evaporation on this flat surface must exceed precipitation.

Dry lakes may seem like extremely stable landforms, but they are dynamic. They can be under a few inches of water one week and be dry with a new surface coating a few weeks later. Over longer periods of time, say a few thousand years, the variations are more extreme. Some lake beds that are dry now were once seven hundred feet under water.

Playas were formed when the runoff from surrounding mountains accumulated in large depressions. When the water evaporated, mineral deposits left a white frosting on the surface. Dry lakes are seemingly devoid of life. When it rains, however, brine shrimp, flies, mosquitoes, toads, and tiger salamanders emerge from the ancient lake beds. Where the saline, sulfate, and carbonate minerals are greater, pickleweed pre-

dominates. Where these minerals are less, mesquite grows in isolated patches.

The glittering surfaces, which can be seen from satellites, are expanding as the climate grows drier. This raises interesting questions: What if they should expand greatly, and what if another form of intelligent life should catch sight of them from space? Such observers could be fooled. Sheet wash stains, giant contraction polygons, evaporate-pressure polygons, contraction stripes, parallel vegetation stripes, phreatophyte mounds, and giant fissures might indicate that these places of near lifelessness were created for a purpose by intelligent beings, like the "canals" on Mars. They could also be mistaken for the ruins of an ancient civilization.

To confuse such observers even further, each playa has a distinctive surface, and there are long, arcing lines across the surfaces that end in an object, such as a rock, stick, or burro dropping. There is an unconfirmed explanation for such phenomena. Two researchers, one from Caltech and the other from the University of California at Los Angeles, wrote:

> No authenticated record has been discovered of anyone seeing a stone actually make a track by natural means on Racetrack or any other *playa*. Some immutable law of nature probably prescribes that movements occur in the darkness of stormy, moonless nights, so that even a resident observer would see newly made tracks only in the dawn of a new day.

The guess is that when rain causes a fine layer of slippery clay to form, and a strong wind comes along, the rocks and other objects glide over the surface of the playa, much like a sailboat before the wind. They leave tracks in their wake for as long as two miles. But there are some puzzling circumstances, such as what caused two rocks on opposite sides of one dry lake to move toward each other?

Six of these playas were once large lakes connected by the Owens River, which extended from Mono Lake to Death Valley. The fossilized course of that ancient riverbed can be traced today, and along it is found a revealing slice of the natural and human history of the desert.

The birth and continuing growth of the desert are geologic processes that are rife with violence. The California desert and the Sierra Nevada formed in tandem. As the young, dynamic mountain range rose, the deserts to the east dropped. Sometimes the movement was sudden. The earth shook, adobe buildings collapsed, and twenty-seven people were killed in Lone Pine in 1872 when an earthquake—perhaps the strongest in California's recorded history—rippled across the sparsely settled eastern slope of the Sierra Nevada.

The newspaper that served the area, the *Inyo Independent*, dropped the usual ads that crowded the front page and ran the following headline decks in descending order: HORRORS!!/APPALLING TIMES!/EARTHQUAKES/ AWFUL LOSS OF LIFE/25 PERSONS KILLED/EARTH OPENS/HOUSES PROS-TRATED/LONE PINE! ITS TERRIBLE CONDITION/MOST HEART RENDING SCENES!/MIRACULOUS ESCAPES!/INDIVIDUAL HEROISM!/A DEMORALIZED PRINTING OFFICE. The story began:

> Between 2 and 3 o'clock Tuesday morning last, the inhabi-
> tants of this region experienced one of the most terror striking
> awe inspiring sensations that ever falls to the lot of mortal
> man—an earthquake—an earthquake in all its mighty power!
> The solid earth was loosened from its very foundations, and
> heaved and tossed as if in the throes of a terrible agony. It was
> a terrible scene when all were so rudely awakened from deep
> slumber to face death in its most terrifying form. Strong
> wooden houses bounded up and down and rolled to and fro
> like ships in a heavy sea way. Crockery smashed and furniture
> danced about the floors, chimneys dropped instantly to the
> ground, stone and adobe houses crumbled and went to earth
> like piles of sand, burying the miserable occupants in the
> ruins, and the whole world was in its last convulsions!

A young John Muir in Yosemite Valley on the opposite side of the Sierra Nevada thought when the earthquake struck that his beloved mountains were falling down. He wrote, "The shocks were so violent and varied, and succeeded one another so closely, that I had to balance myself carefully in walking as if on the deck of a ship among waves, and it seemed impossible that the high cliffs of the Valley could escape being altered."

The earth rumbled and heaved. Birds fled in panic. "A cloud of dust particles, lighted by the moon, floated out across the whole breadth of the Valley, forming a ceiling that lasted until after sunrise, and the air was filled with the odor of crushed Douglas spruces from a grove that had been mowed down and mashed like weeds."

The day following the earthquake a mass grave was dug and fifteen coffins containing sixteen bodies of "the foreign born," meaning mostly those of Mexican extraction, were buried outside the ruined town of Lone Pine. The Anglos were buried individually in the regular cemetery. Dust obscured the Sierra Nevada for two days following the quake. The town quickly rebuilt.

The mountains rose twenty-three feet within a few moments during the 1872 Lone Pine earthquake. Such violent earth movements are not unusual in California. The San Gabriel Mountains north of Los Angeles were tossed upward six feet during the moderate San Fernando earthquake in 1971, and the earth slipped horizontally twenty feet during the 1906 San Francisco earthquake. Like the seismographs spread throughout the state, the human psyche also registered such jolts.

The California desert is both distant past and near future; it enfolds the California experience. How long humans have been inhabiting the North American continent, and the California desert in particular, is a subject fraught with much emotion, contention, and bitterness. In other words, there is an intense academic debate on the subject. Careers, reputations, and money are at stake.

There are those, including the late Louis S. B. Leaky of African archaeological fame, who believe that human occupation of the California desert extends far back into the past—as far back as 200,000 years or more—and that the oldest remnants of a human presence in the Western Hemisphere can be found at the Calico Early Man Site in the Mojave Desert. Here campsites and workshops were scattered around the shores of an ancient lake, and thousands of stone tools and flakes were found during nineteen years of digging. Calico is the most completely excavated Early Man site in North America. According to reliable scientists, there is evidence of a pre–*Homo sapiens* population.

Nonsense, say the critics. The rocks these scientists have found and dated are not artifacts but just rocks—chipped and flaked by such natural processes as rolling down steep alluvial fans or streambeds. More likely, ancient humans appeared during a window of time when there was

still a land bridge over the Bering Strait, and the ice from the late Pleistocene Epoch was retreating—perhaps some 12,500 years ago. There is a
middle ground between the archconservatives and the radicals that falls
between 20,000 and 50,000 years ago, when land bridges may have come
and gone. Best to extend the time line back slowly, say these moderates.

As the last ice age waned, humans certainly were present in California; they began, at first tentatively and then with boldness and artistry, to
leave their marks on the fragile desert pavement from Mono Lake to Baja
California. With remarkable prescience, they did not aim their markings
at themselves, or at others like them, but rather directly at the vast sky
and all that it might hold. No vaulted ceilings came between them and
their gods and spirits.

Modern man discovered the symbols from the sky. In the summer of
1932, an ex–U.S. Army pilot flying from Las Vegas to Blythe, just across the
Colorado River from Arizona, happened to look down and saw an enormous human figure with sticklike limbs and a representation of an animal
etched on the desert below. As George Palmer circled a few miles north of
the desert town, more human figures (some exceeding one hundred feet in
length) along with abstract shapes of animals and squiggles and circles
became visible. The pilot landed at Blythe and returned with his box camera. He took the photographs to the Los Angeles County Museum, and the
museum staff was amazed at the discovery of the intaglios.

But, of course, it was no discovery, just a rediscovery. The ancient Indians, and those who immediately followed them, knew of the figures;
although contemporary Indians living in the region did not. In 1851, Anglo
explorers had come across these scrapings in the desert and reported them.
Army Captain Lorenzo Sitgreaves, exploring along the Colorado River,
wrote in his report:

> Nov. 7, Camp No. 33—A well worn trail leads down the river,
> by the side of which in several places were found traced on
> the ground Indian hieroglyphics, which Mr. Leroux and a
> Mexican of the party, who had passed many years among the
> Comanches, interpreted into warnings to us to turn back, and
> threats against our penetrating farther into the country.

The Comanches knew nothing of these markings; and Antoine Leroux,
the guide, had never passed this way before. They interpreted the signs to

suit their purposes, which is not unusual. The expedition was stalked by fear. Leroux had been wounded a few days earlier in one of the many Indian ambushes they encountered. Now the expedition was in the territory of the fierce Mojave Indians, who were taking their toll. The expedition's doctor had just been wounded, and a trooper had been killed. Their food was low. The landscape was bearing down on them. "The most perfect picture of desolation I have ever beheld, as if some sirocco had passed over the land withering and scorching everything to crispness," said Sitgreaves. They fled to the protection of Fort Yuma.

As on the plains of Nazca in Peru, where there are similar figures, modern minds have not been able to comprehend with any degree of certainty how these etchings were created, what they represented, or for whom they were intended. Geometry, hallucinogens, and shaman-induced out-of-body experiences have been suggested as aids in the process of creation. As to what they represent, the quadruped figure near Blythe has been variously described as a horse or a jaguar. Horses, which had become extinct before the Spanish reintroduced them, and jaguars inhabited these lands long before the coming of the Europeans. A serpentlike figure is readily identified as a snake, a common symbol in creation myths.

To some the figures and lines are abstract: messages for the deities, ceremonial pathways, homages to fertility or water. To others they are realistic: landing strips for UFOs, signs for extraterrestrial people, racetracks that may have been used for some ancient sport. The figures, lines, circles, and mazes remain silent, as silent as the artifacts of our civilization will be thousands of years from now.

Space science, the development and testing of weapons, and other military activities are the dominant presences in the contemporary desert. There are nine military bases and testing grounds totaling 3.1 million acres, with huge blocks of reserved airspace. Scientists are preparing for the future—probing the outer reaches of the universe and devising ingenious and horrible means of death for wars yet to come—in close proximity to the markings of the ancients on the rocks and floor of the desert.

Tucked into fifty-two square miles of the 1,000-square-mile Fort Irwin Military Reservation in the Mojave Desert, the Goldstone Deep Space Communications Complex lies adjacent to a dry lake. The gleaming white parabolic antennas that reach out billions of miles into deep space sit amid common creosote bushes, the oldest known living organisms. Some of these desert plants date back ten thousand years to the end of the last ice age.

At dusk the softly glowing lights on the space center's gigantic dishes, whose interior parts resemble the probes of insects, add a touch of science fiction to the prehistoric landscape. Not surprisingly, many books and films of this genre, not the least of which were *Dune* and *The Martian Chronicles,* used the stark reality of the Mojave Desert to suggest future worlds.

The white parabolas located around Goldstone Dry Lake are oriented upward, as are the intaglios. They serve a variety of purposes. They are in contact with unmanned spacecraft and satellites transmitting weather data and other observations as they orbit Earth or pass by Mercury, Venus, Mars, Jupiter, Saturn, Uranus, and Neptune. The strength of the detectable signals shrinks to a billionth of a trillionth of a watt as the spacecraft hurtle through the heliosphere and enter interstellar space. The signals will cease entirely when their power supplies fail, and the spacecraft will coast to an end somewhere out among the stars.

Goldstone is also the listening post and transmitting station for the National Aeronautics and Space Administration's search for intelligent life elsewhere. "Are we alone?" is the question. Imagine when the answer comes back, as it eventually must. What better place to receive such news than the desert, which may be the mirror image of that distant planet whose inhabitants will send the message. Many of the photographs of the surfaces of planets transmitted back to Earth by these interplanetary spacecraft closely resemble the floor of the California desert.

The Goldstone site was selected in 1958 for its silence, since any nearby electrical signals would have to be quieter than the weakest signal from space. The bowl-shaped terrain formed by the desert mountains and the sunken playa acts as a shield against interference. There are five antennas. The largest is Mars, whose six-million-pound bulk rotates on a pressurized film of oil the thickness of a single sheet of paper. In order to avoid losses in signal power, the parabolic antenna rotates to point within 0.006 degree of a moving spacecraft.

At nearby China Lake, a scientific community is diligently working on another scenario—the extinction of our species. Deep within the labyrinthine corridors of a laboratory, around an oak conference table that

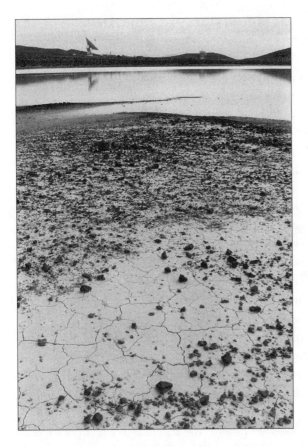

The antennas and the lake

sits upon a brown carpet in a windowless vault with thick walls and sealed ducts, men and women discuss the design of the next generation of weapons. They are aided by a supercomputer that can make two hundred million computations in the time it takes to tell a second-grader that two plus two equal four.

Thirty years after I made my first crossing of California I drove back into the desert from the opposite direction to begin a counterclockwise tour of the state. I threaded my way down the steep eastern slope of the Sierra Nevada on the narrow road from Monitor Pass. The feeling was one of being suspended over a restless ocean. Coming at me from the east, from the direction of Nevada, were waves upon waves of brown mountain ranges whose frothy white crests, which were about to break against California, bore the remnants of winter snow.

I stopped, got out, and looked back at the stacks of thick clouds that towered above the Sierra Nevada. Caught in fluid motion halfway between the mountains and the desert was a skateboarder. His lightness of being was heightened by the backdrop of dark clouds. With consummate grace and skill he slalomed down the long, black asphalt road. His long blond hair was a small flag of defiance set against gravity and the approaching storm.

Briefly, I wondered if he was an apparition, and then I knew he was another California image that would be indelibly etched on my memory. He typified the fluidity, health, and youthfulness of surfers, skiers, and, more recently, snowboarders whom I had watched and, in some cases, sought to emulate over the years. The journey of rediscovery that I would undertake over the next three years would be rich in new images and older ones recalled from previous California peregrinations.

I got back in my Volkswagen camper and descended to the floor of the desert. Technically, there are three deserts in California. The Great Basin Desert stretches from the Oregon border to the three-thousand-foot elevation mark in the south. The middle desert is the Mojave, and the most southern desert is the Colorado, a subdivision of the Sonoran Desert.

The three deserts, together roughly the size of Ohio, account for one-fourth the total land area of California (one hundred million acres) and are its single largest province. Two percent of the state's population lives in the deserts. They include the rich of Palm Springs, the migrant laborers of the Imperial Valley, scattered desert rats, a few ranchers, a growing number of retirees, military personnel, civilian scientists, and the people to serve all of the above, plus the tourists. Nine out of ten inhabitants were not born in the desert. Most migrated there because of job transfers, military assignments, or for health reasons. Besides the military bases and testing grounds, there are over one hundred communities of varying small sizes.

The deserts are crisscrossed by the arteries that connect California to the rest of the nation and give it life. There are railroads and interstate highways. More than three thousand miles of high-voltage lines carry one-quarter of Southern California's electricity, and twelve thousand miles of oil and natural gas pipelines transport fuel and energy into the state whose economy regularly ranks among the top ten wealthiest nations in the world. No rivers traverse the desert, but water for the nation's number-one agricultural producer (and consumer of water) and for domestic consumption snakes across the desert via irrigation canals and aqueducts.

Nearly six million acres of the deserts have been identified as having wilderness values. There are 2,000 species of plants. As far as anyone knows—and the counts will never be precise—there are 42 species of fish, 64 species of reptiles, 94 species of mammals, and 419 species of birds. A few of these plants and animals have been declared rare, threatened, or endangered by the federal or state government, and sometimes by both.

By the end of the century, perhaps seventy thousand acres in the desert will have been disturbed by mining activities. Grazing is on the decline while off-road vehicle use keeps increasing. Four-wheel-drive vehicles and air-conditioning opened up the desert to visitors and inhabitants after World War II, and solitude and silence are about as rare now as some of those endangered plants and animals. Visibility is poor. A light gray pall from the Los Angeles Basin slithers through the mountain passes and spreads across the desert on most days.

It wasn't always like that. John C. Van Dyke, an art historian who wandered across the Mojave and Sonoran deserts on foot and horse nearly one hundred years ago, thought of the desert in terms of fired colors. The mountains were "roasted to a dark wine-red and the foot-hills burnt to a terra-cotta orange."

He felt the transience of the human presence in the desert. "Nothing human is of long duration," Van Dyke wrote in his classic work *The Desert*. "Men and their deeds are obliterated, the race itself fades."

I drove south and camped beside Mono Lake, the first stop on my journey down the ancient course of the Owens River that ended in Death Valley, the ultimate sink. Change has distinguished the geologic history of Mono Lake. The lake may have spilled once or several times between 22,000 and 13,000 years ago into Adobe Valley and thence into the Owens River. Each time the river formed, it was for a sufficiently long enough period of time for many generations of human inhabitants to believe they had a permanent source of water at their disposal.

Volcanoes erupted around and within the lake. Nineteen separate layers of volcanic ash have been uncovered in the basin. One consolidated ash layer, known as the Bishop Tuff, lies between 1,300 and 1,600 feet below the surface. The ash from that eruption, far greater than the 1980 eruption of Mount St. Helens, fell over most of western North America some 700,000 years ago. A "rain of fire" fell on the immediate region. The temperature of the incandescent rock was well over 1,200 degrees Fahrenheit. Life was eradicated. Some 220 years ago hot magma from the bowels of the earth burst forth in the midst of the lake, and Paoha Island was formed almost instantaneously in a massive, boiling cloud of steam and ash.

To this day the region is exceptionally unstable, even by California standards. A U.S. Geological Survey scientist, writing in 1976, noted, "The young age and frequency of eruptions along the Inyo-Mono volcanic chain indicate an active volcanic zone; eruptions of similar kind and magnitude could occur in the future." A bulge in the earth south of Mono Lake was detected and swarms of earthquakes, including six moderate jolts within forty-eight hours, rippled across Mono County during the last decade. Officials readied evacuation plans for the towns. The introduction to the Long Valley Caldera–Mono Crater Contingency Plan, drawn up by the Forest Service, reads, "The increased earthquake activities in the Long Valley Caldera, starting in May 1980, gave significant evidence of a strong possibility that a volcanic eruption could occur. Continued episodes since May 1980, have strengthened the possibility of an eruption."

Mono Lake

Scientists have been attracted to the strange landscape over the years. William H. Brewer, who was Josiah Dwight Whitney's chief assistant in the California State Geological Survey, spent a few days at Mono Lake in July of 1863. It was the weirdest lake he had ever seen. From its surface rose steam and vapors. The water had a nauseating taste, but a silk handkerchief washed in its alkaline waters emerged "with a luster like new." Brewer sampled the brine flies offered by the Indians and commented, "It does not taste bad, and if one were ignorant of its origins, it would make fine soup." He noted the mushroom-shaped tufa formations and then departed to continue his remarkable four-year journey around the state.

Brewer was a full-bearded Victorian gentleman who was an enthusiastic traveler, a careful observer, and a talented writer. Educated at Yale University, he continued his studies in the agricultural sciences, such as chemistry and botany, in Europe. There Brewer undertook a six-hundred-mile walking tour of Switzerland. Returning to this country, he took a teaching position at a Pennsylvania college. His wife and infant son died, and shortly thereafter he was offered the California job. He accepted to escape his despondency, obtain valuable field experience, and advance his career. Historian Francis P. Farquhar termed his journal, which is still in print, "an unabridged, undecorated record of the times." Brewer spent the last forty years of his life at the Sheffield Scientific School at Yale University.

Eighteen years after Brewer departed from Mono Lake, another scientist entered the region. When Israel C. Russell, a geologist with the U.S. Geological Survey, rode into the Mono Lake Basin in 1881, he noted the "stern and wild" and the "silent and lifeless" landscape that had been formed by earthquakes, glaciers, volcanoes, wind, and water. He remarked that "the earth's crust over this entire region is far from being in a state of stable equilibrium."

Of the abrupt conjunction of desert and mountain landscapes, Russell wrote in his subsequent report, "One has the desolation and solitude of the Sahara, the other the rugged grandeur of the Pyrenees." Of the latter, Russell said:

> No prosaic description, however, can portray the grandeur of
> fifty miles of rugged mountains, rising above a placid lake in
> which each sharply-cut peak, each shadowy precipice, and
> each purple gorge is reflected. Nor is it possible to bring to

these pages the wonderful transparency of the dry atmosphere of California or the gorgeous coloring of the background against which the mountains repose at sunset.

Russell rode on a mule into Mono Basin at a time of transition. When
he camped at Warm Spring on the eastern shore of the lake, the only evidence of previous occupation was a trail leading to a spring. When he
returned early the next year, there was a railroad crossing the valley and
a station where he had previously camped. There were also a few white
settlers around the southern and western edges of the lake. Russell noted
that livestock had depleted most of the natural pastureland.

Indians still camped around the lake during autumn months. The
women gathered fly larvae and separated the soft kernels from the hard
cases. "Such scenes are not only novel, but add a bit of life and color to a
landscape apt to impress one as somewhat dreary and somber," Russell
wrote.

He named the two islands in the lake: *Paoha* was the Paiute name for
"hot spring" and also for the ethereal spirits with long, coiling hair who
were sometimes sighted in the wisps of vapor that escaped from the
fumaroles and vents. *Negit* was understood by Russell to be the Indian
word for the blue-winged goose. There were steam vents, noxious
vapors, and strange upwellings of water. The surface of the lake was "literally darkened" in the fall and early winter with ducks, geese, swans,
gulls, grebes, and other migrating birds. They fed on the brine shrimp
and larvae.

Taken altogether, Mono Lake was an unusual place, and Russell
attempted to make scientific sense out of it. He noted the terraces 672
feet above the lake level, and deduced that they indicated the highest
level of an ancient lake, whose surface he set at 7,060 feet above sea level.
Others would later place the maximum height of the lake at 7,200 feet. At
that point the ancient lake was seven times deeper and had five times the
surface area of the contemporary lake.

When Russell first visited Mono Lake in 1881, a time roughly coin-
ciding with the end of the Little Ice Age, the level of the lake had fallen
820 feet from its greatest height and was 6,380 feet above sea level. A
century later the level had fallen eight feet from Russell's time and 56
feet from its historic high of 6,428 feet in 1919. A drier climate and the
diversion since 1941 of water south to Los Angeles accounted for the

decrease. Russell spent portions of three years at the lake compiling material for his report. For his efforts the ancient lake, as distinguished from the contemporary one, was named Lake Russell.

While Russell was poking around the shoreline of Mono Lake, another scene was unfolding just outside the basin at the booming mining camps of Bodie and Aurora. To the north of the lake was frontier America par excellence, and, in the case of Bodie, now a state historic park, its fossilized remains are still visible near the Nevada state line.

The desert preserves its past, and the past has been mythologized on the printed page and in television and movies. But, as is often the case, the myths do not fit the realities. For example, take frontier violence, usually depicted as being generalized and random. It was more specialized, aimed more at specific targets than the random type of terrorism exercised today. The young, the old, women (with the exception of prostitutes), children, the weak, and the unwilling were generally not harmed. Gunfighters were poor shots. Swaggering under the influence of alcohol caused more fights than malevolence.

Historian Roger D. McGrath studied frontier and present-day violence. Based on a close reading of the histories of Bodie and Aurora, McGrath concluded:

> The violence and lawlessness that visited the trans-Sierra frontier most frequently and affected it most deeply, then, took special forms: warfare between Indians and whites, stagecoach robbery, vigilantism, and gunfights. These activities bear little or no relation to the violence and lawlessness that pervade American society today. Serious juvenile offenses, crimes against the elderly and weak, rape, robbery, burglary, and theft were either nonexistent or of little significance on the trans-Sierra frontier. If the trans-Sierran frontier was at all representative of frontiers in general, then there seems to be little justification for blaming contemporary violence and lawlessness on a frontier heritage.

Mark Twain lived for a short time in Aurora, and *The WPA Guide to California* noted in 1939:

The lake has achieved a certain measure of fame in literature and legend. Mark Twain tells in *Roughing It* of a dog that attained a running speed of 250 miles an hour after taking a swim in the lake. A more modern tale is of a long-haired dog that emerged from a swim with nothing more than its bark. But the water that feeds Mono Lake will not go to waste much longer, for, soon purified, it will go into the new Mono Basin Aqueduct and help slake the thirst of metropolitan Los Angeles.

Over the years the causes of contention shifted. Environmentalists from outside the basin and some inhabitants from within took umbrage at the draining of the lake by the Los Angeles Department of Water and Power. In the late 1970s they initiated a water war, using the modern-day tools of publicity, lawsuits, and lobbying in the state legislature and Congress. One of the leaders of the war was David Gaines, who assumed Russell's mantle.

A founding member of the Mono Lake Committee, Gaines was a great admirer of Russell and attempted to follow in his footsteps. "No white man before or since has come to know this dynamic land more intimately," wrote Gaines in the introduction to the reprint of Russell's report published by Gaines in 1984. It wasn't the first time the report had been reprinted. It was first reissued by early Mono Basin residents who wanted to attract tourists. Gaines reprinted it to draw attention to the beauty and uniqueness of the region.

With a great deal of zeal and a deft hand at publicity, the Mono Lake Committee succeeded in thwarting, at least for a time, the diversion of water south to Los Angeles that was partially responsible for drying up the lake. This remote place at the apex of the California desert became nationally known. "Save Mono Lake" became the public rallying cry. Eleven scientists from the National Research Council, aided by a staff of three, pondered the issue and eventually produced a report in 1987 that said the pumping of water south had reduced the lake level by forty feet since 1941, and there was now danger to the birds.

Gaines wrote a guidebook for the lake in which he sketched its human history. The Indians were followed by white miners, ranchers, aqueduct builders, and tourists. At interludes a steamer plied the lake, health addicts sought relief in its waters, and homesteaders attempted to raise goats, rabbits, other livestock, and vegetables on Paoha Island. Gaines became part of that history. He died in a winter automobile accident in 1988 and was buried beside the lake.

I awakened before dawn at Black Point, a volcanic extrusion that had been launched upward through the lake some 13,500 years ago. As the sky turned from gray to rose, I climbed through the sagebrush to the five-hundred-foot summit and watched the sun rise. It was like witnessing the dawn of the world.

A sliver of orange rose slowly over a slight depression in the low ridge to the east, its undulating outline punctured by volcanic cones. The surface of the lake glowed like burnished pewter. The dark, serrated shafts of tufa pinnacles pierced the still water. The tufa formations were created by the precipitation of minerals from underwater springs. The fossilized springs resemble stalagmites (to whom they bear a close chemical relationship), concrete cauliflowers, or the fanciful droppings of wet sand.

The air was still enough to hear the gulls crying incessantly on the two islands. Late spring is the time of year when 40,000 to 50,000 California gulls nest at the lake. While on their migration to South America, as many as 100,000 Wilson's phalaropes might stop at Mono Lake to complete their molt and put on fat by eating the plentiful supplies of brine flies and shrimp. The night before, I had heard their nemesis, the yips of coyotes; this morning I saw coyote tracks in the black volcanic soil. Made uncomfortable by my continued presence, a pair of the world's most efficient predators and scavengers—the common raven—rose with a series of *kraaak*s from the summit of Black Point.

The wily raven is *the* bird of the desert. After we are gone, the raven will be here to tell our story, just as many cultures have related the story of the raven. The raven was associated with ancient deluge myths that predated the Bible. From a four-thousand-year-old Arab myth to Jewish folklore and Estonian traditions and on to the myths of the Arctic and sub-Arctic people of Siberia and Alaska, the raven was the bird dispatched from the ark to find land but instead dallied to eat carrion. For this indiscretion, the raven was punished. He was colored black or condemned to eat carrion forevermore, depending on the particular tale. It was the faithful dove that returned with the olive branch from the land.

In North America raven myths flourished among the Indians of the Pacific Northwest, where Raven assumed the role of both creator and trickster. Raven created the earth, the moon, the sun, the stars, and people. Since he was the creator, Raven could do as he pleased. Raven created mosquitoes because people had it too easy. At first Raven made humans out of rocks. When they proved to be too durable, he used dust.

When the Europeans entered the continent, Raven took on another form. Edgar Allan Poe depicted the raven as the bird of ill omen in his famous poem about the death of a beautiful woman. And the same metaphor was used in a recent book titled *The Ravens,* about the pilots who flew in the secret war in Laos. Barry Lopez returned to the trickster image in his book about the desert. He wrote, "The raven is cautious, but he is thorough. He will sense your peaceful intentions. Let him have the first word. Be careful: he will tell you he knows nothing."

The raven has a number of human traits. The black bird with iridescent plumage mimics humans, has a large vocabulary that some researchers claim to understand, sings, plays, uses tools, shows off during courtship, and mates for life. Ravens travel in pairs. They are intelligent. "It is at the top of the most species-rich and rapidly evolving line of birds," wrote Bernd Heinrich, a zoologist who studied ravens in Maine. Their strong, hooked beaks and grasping talons make them efficient predators. They have one known enemy: humans.

The raven population has paralleled the growth of the human population in the California desert. Prior to 1940, ravens were almost unknown. When the curator of the San Bernardino County Museum went looking for a raven to add to the museum's collection in the 1940s, it took him two years to locate one. Then people and their detritus began to occupy the desert in growing numbers. There was a parallel explosion in the number of ravens. Between 1968 and 1988 there was a 1,528 percent increase in the number of ravens in the Mojave Desert, 474 percent in the Colorado Desert, and 168 percent in the Great Basin Desert.

Ravens nest in abandoned shacks, cliffs, railroad trestles, transmission towers, highway billboards, freeway overpasses, caves, tamarisks, and Joshua trees. Flocks of one hundred to four hundred ravens were reported at Harper Dry Lake near Barstow, and at the garbage dumps at Edwards Air Force Base and the Marine Corps Air Ground Combat Center at Twentynine Palms. They feed on animal carcasses and scavenge at Dumpsters outside McDonald's and Taco Bell, at roadside rest areas, campgrounds, garbage dumps, and sewage treatment plants. Ravens prey on small mammals, insects, birds, bird eggs, reptiles, fish, and the desert tortoise, an endangered species.

The tortoise depredations got them into trouble. The Bureau of Land Management drew up a plan for the "control" of the burgeoning raven population. The bureau's solution: shoot and poison the ravens. A pilot program was instituted in 1989. The poison that was placed in hard-boiled eggs was DRC-1339, known as Starlicide. It was developed by the U.S. Fish and Wildlife Service for use on "pest birds." It is also lethal to doves, crows, game birds, and owls. The bird dies in one to two days from kidney failure or central nervous system depression. In California, the marines have been particularly active in using the poison on birds, not only at their desert base but also at Camp Pendleton on the coast.

The sky grew lighter, and the surrounding countryside came into full view from the summit of Black Point. I was standing at nearly seven thousand feet above sea level, and still the mountains towered above me. A slight breeze flitted across the surface of Mono Lake. In concentric rings that rose above the lake, like the successive tiers of a giant amphitheater, were the various terraces that represented different lake levels. Each level signified an era in one of the oldest lakes on the North American continent.

To the south, extending from Negit Island to Mammoth Mountain, I made out the Mono Craters, the youngest mountain range on the continent. The eruptions that formed this range—actually a chain of domes—date back some 200,000 years to the first uplifting of Mammoth Mountain and forward to some two hundred years ago and the rise of Negit Island above the waters of the lake. Russell found the Mono Craters to be on the scale of Vesuvius and Stromboli, but they have received less attention than the Italian volcanoes because of their remoteness and the fact that they are dwarfed by the adjacent Sierra Nevada range.

This was not the first time I had been here. Years ago I had launched my kayak into the lake. As I approached Negit Island the gulls rose in a white, flapping sheet. Their cries echoed in the thin air. I remembered the red rock that I scrambled across, the stench of gull droppings, and the miniature mountain built by a movie company in the 1950s.

I had skied nearby Mammoth Mountain many times in the late 1960s and early 1970s. Mammoth, the mountain, and Mammoth Lakes, the town, were a California phenomenon—a recreational bonanza in the middle of nowhere into which more than twenty thousand people were stuffed for short periods of time. The condominiums and factory outlet stores now lie cheek by jowl within the bowl of the Long Valley Caldera. I could understand why people continued to invest in real estate at Mammoth. In California there is no such thing as a totally safe environment, but the odds always appear to favor eluding the many different forms of natural disasters.

To the west from where I stood and visible as a notch in the Sierra massif was Tioga Pass, which I had traversed on my first journey into California. Just south of the automobile route into Yosemite National Park was Mono Pass; from it, Bloody Canyon drops down the east side of the Sierra Nevada. The descent is exceedingly steep. Early travelers remarked on the stench of rotting flesh. The name of the pass may have been derived from the many horses that plunged to their death there, or the following incident.

Lt. Tredwell Moore passed eastward over this pass with a company of soldiers in 1852 in hot pursuit of some Miwok Indians from Yosemite Valley who had supposedly killed two or three prospectors. Along with the actual number of those killed, there was some dispute as to the instigators of the crime. One miner, hoping to eliminate competition, may have put the Indians up to it. In any case, Moore caught six Indians and summarily executed them near the pass. Chief Teneiya eluded him and hid with the Paiutes of Mono Lake, who eventually killed him.

The lieutenant and his soldiers were the first white men to see Mono Lake, and, like other invaders, they arrived with hate in their hearts and greed on their minds. The lieutenant returned with samples of gold-bearing ore. There was a gold rush into the region. In 1853, one year after the discovery of gold and eight years before the outbreak of the Civil War, Mono Lake appeared on a published map for the first time.

Mono Lake provides a foretaste of the dark side of California history. Actions against the Chinese followed the near extinction of the Indians. The miners of Bodie, as was their practice elsewhere, went on an anti-Chinese binge. The Chinese laborers who were building the Bodie and Benton Railroad took refuge on Paoha Island until the drunken miners had spent themselves. The Anglo residents of Bodie voted 1,144 to 2

against any further Chinese presence in 1880, and immediately thereafter the Chinese departed. Three Mexicans were shot by gringos. The
killings were ruled as self-defense.

To the east over Sagehen Summit was Adobe Valley. The streams that
flow into this valley are the northernmost tributaries of the Owens River.
It was this route that the ancient river followed. I descended from the
summit and drove farther south into the desert.

Below Mono Lake, the river was there, then it wasn't. When it wasn't
there—when it was being diverted through tunnels and aqueducts—and
even when it was there, it was on its way to Los Angeles, but that was the
fate of most rivers in the desert Southwest. At the small town of Independence, in the heart of the Owens Valley, some water was being put
back into the river, the idea being to restore it to a semblance of its previous existence. The city, however, never willingly gave water away. Wells
dug in the Owens Valley replaced the water released from the aqueduct.

In 1903, Mary Austin foresaw the fate of the valley. She wrote in her
first and best-known book, *The Land of Little Rain*: "It is the proper destiny of every considerable stream in the west to become an irrigating
ditch." She was thinking of the valley; others were thinking of Los Angeles
to the south. Three years later, after Los Angeles began acquiring water
rights in the valley, Austin departed to follow her muse. She had written,
"Is all this worthwhile in order that Los Angeles should be just so big?"

Austin's fourteen years in that deep desert trough between towering
mountain ranges were not happy ones, but they were pivotal. She began
them as a wife and mother and ended them as a published writer. Her
daughter, whom she wished to be beautiful and intelligent, was born
retarded, and Austin blamed "a tainted inheritance." Part of the time
she was separated from her husband, a schoolteacher and minor government official. Austin, who had wanted to write since she was seven
years old, strapped her screaming daughter in a chair and paced the
floor as she composed her stories. The townspeople felt sorry for the
child and the husband. Austin befriended Mexicans and Indians, and
labored on.

An acquaintance later recalled: "To me she was one of the most color-
less humans I have ever known. She was a mark for ridicule and, I believe,
had few friends." This woman marveled at what Austin had accomplished.

The Owens Valley

A female physician who treated Austin's daughter and later became Austin's friend described the writer as a "rather short, somewhat dumpy looking woman with a homely, heavy lower face, a sullen mouth, fine eyes, a high forehead and abundant, beautiful gold-brown hair." The doctor at first abhorred Austin's neglect of her child. She then came to sense her sorrow, frustration, and loneliness.

Austin described herself thusly during these years:

> You are to figure to yourself a small, plain, brown woman,
> with too much hair, always a little sick, and always busy about
> the fields and mesas in a manner, so they say in the village, as
> if I should like to see anybody try to stop me.

The essays she wrote of the natural and human life in the valley became her first book. Austin celebrated the landscape of a place where her life was "long dull months of living interspersed between the few fruitful occasions when I actually came into contact with the Land."

It was "a land of lost rivers" that "begins with the creosote" and is inhabited by the raven, "the least objectionable of the inland scavengers." Myths suffused the place: "The palpable sense of mystery in the desert air breeds fables, chiefly of lost treasures." The people were a bit off: "There are many strange sorts of humans bred in a mining country, each sort despising the queerness of the other." And finally, the theme that Austin emphasized, perhaps as an explanation for her behavior: "The manner of the country makes the usage of life there, and the land will not be lived in except in its own fashion."

Once she departed, Austin never returned to the place that was to give her lasting fame. Nor did she ever visit her daughter, whom she deposited in a private sanatorium. Her husband remained in the valley to fight the unsuccessful water war against Los Angeles. They were eventually divorced. She was now free to write. Austin wrote more books—none as successful as her first one. She consorted with the leading writers of her time in the literary salons of San Francisco, Carmel, Pasadena, New York City, Santa Fe, and London.

Austin's house—"the brown house under the willow-tree at the end of the village street"—still stands on Market Street in Independence. Down the street is Austin's General Store, where liquor, ice, groceries, and bait for fishing in the reconstituted Owens River can be purchased.

When the wind blew from the direction of Owens Lake to the south, Austin noted the "smell of the bitter dust that comes up from the alkali flats." Owens Lake, the second in the chain of playas, is a dry, dusty plain. Not too long before Austin's time, steamships navigated the coffee-colored waters that lay in the shallow depression between the soft tan-and-rust-colored desert mountains to the east and the gray granite of the knifelike Sierra Nevada to the west. The lake covered an area of 240 square miles following the last ice age. It had shrunk to 112 square miles by 1872, when the first steamboat was launched.

With the help of Owens Valley ranchers drawing off extensive amounts of irrigation water around the turn of the century, and the City of Los Angeles diverting the remaining water in 1913, the lake went dry in 1926. Since then, slight amounts of water have periodically appeared in the lake bed when Los Angeles dumps surplus water from its aqueduct or when the spring runoff from the Sierra Nevada is excessive.

Dust and death hover over this dry lake.

On August 3, 1878, Anna Mills, a cripple since childhood, set off for the summit of 14,495-foot Mount Whitney, the highest mountain in the contiguous states. Three local fishermen had made the first recorded climb of Mount Whitney five years earlier. (Very probably Native Americans had reached the summit sometime before 1873, but there was no record of those ascents.) The indefatigable Mills injured her back, but continued on. She later wrote:

> Climbing over those rocks was no easy task—and how my back did smart and burn! But I didn't mind such trifles when there was so much at stake; my heart was set on something higher, and nothing short of the highest point would satisfy me. I would reach that and die if need be.

The party, which included three other women, successfully navigated the steep, narrow "Devils Ladder" and emerged on the summit ridge, from where it was an easy climb in the cold, thin air to the summit. They were the first women to climb over 14,000 feet in the country, it was thought. Mills, a schoolteacher, gazed back toward Owens Lake and commented:

> This desert, or vast plain of sand, called by some an extinct or dry lake, is locked in on all sides by rock-ribbed mountains whose peaks mount upward among the clouds. One could imagine himself descending into the valley of death and having the gates closed after him.

It truly was the Valley of Death for the Indians. As the white settlers entered the valley, Indians and whites traded atrocities. There were more whites, and they were better armed. The water of Owens Lake was stained red with the blood of the Paiutes. William Brewer wrote of the atrocity that led to the 1863 massacre:

> We camped by the river and took a cooling bath. Our camp was the scene of a fearful tragedy a year ago. The Indians attacked a party of one man, a nigger, two women, and a child. The nigger was on horseback and fought well, killing several. In attempting to cross the river the team of horses was drowned. He gave his horse to the women—both of whom got on it, and they and the white man escaped to Camp Independence. The Indians caught the negro and afterward said that he was tortured for three days.

Charley Tyler, the black, was known as a very effective Indian fighter. He had been in the party that had gunned down a chief and participated in two battles against the Paiutes. After the three days of torture, the Indians bound Tyler with willow thongs and slowly roasted him to death. Tyler was memorialized by the naming of Charley's Butte, near the intake to the Los Angeles Aqueduct.

U.S. Army troops and civilian volunteers who took up the chase of the offending band of Paiutes were ambushed near Owens Lake; the casualties consisted of a clean shot through the hat of one trooper. The whites deployed and forced the Indians slowly down the alluvial fan toward the water. Their marksmanship left something to be desired. They inflicted more casualties on themselves than the Indians did.

Nevertheless, they managed to kill sixteen Indians on land. The remainder jumped into the lake and attempted to swim away. A strong east wind kept forcing them back to shore. As a full moon rose over the desert, the white men lined the shoreline and methodically picked off the Indians. Twenty Indians were killed in the water. One escaped to curse the whites. Tyler's pistol was recovered from one of the bodies.

Two years later the more bloodthirsty local militia was chasing the Indians by itself, the federal troops having departed. A militiaman wrote to an army colonel, "I think things have reached the point, that either the white people must leave this Valley or the Indians killed off, or in some other way got rid of." As the whites were not about to leave, a policy of extermination was commenced.

A white woman and her son were killed and their cabin burned to the ground by the Indians. The family had done much to aid the Indians. The two white men who had been entrusted by the husband to guard the woman and child fled. Rescuers located the cowards and ordered them to leave the country immediately, or be killed. The whites then decided to exterminate the Indians camped on the shoreline of Owens Lake. They attacked at daybreak on January 6, 1865. Three dozen men, women, and children were slaughtered. Six Indians jumped into the icy water of the lake. At least two were shot in the water. Two girls and a boy or two women and two boys, depending on the account, were spared, although fully one-third of the company favored shooting them.

That massacre pretty well broke the back of Indian resistance. Mary Austin wrote, "The Paiutes had made their last stand at the border of the

Bitter Lake; battle driven they died in its waters, and the land filled with cattle-men and adventurers for gold."

The adventurers found gold and silver in the mountains to the east. Discovered by Mexicans, who were either quickly killed or driven from the region, the Cerro Gordo mines soon boomed under the management of an Anglo mining syndicate. By 1872 there were eleven producing mines and the silver ore was being smelted in furnaces at the mountain camp and by the lakeside. The problem was how to get all that bullion south to Los Angeles.

On the sultry Fourth of July morning of 1872, just a few months after the disastrous Lone Pine earthquake had set the valley's economic hopes reeling backward, the steamer *Bessie Brady* was christened on the shoreline of Owens Lake. Twenty wagons decorated with flags and streamers, with a brass band in the lead, trundled down to Ferguson's Landing. At 10:15 A.M. the steamer arrived at the makeshift landing, and the vessel was christened with a bottle of wine by the young daughter of the owner. The poet of the day, W. H. Creighton, rose and intoned the "Ode to the *Bessie Brady*," which read, in part:

> *A noble work before thee lies,*
> *The opening of a country rich beyond compute,*
> *Removal of the stigma "want of enterprise,"*
> *Which hither to we've born in shame and mute.*

That day the *Bessie Brady* carried one hundred and thirty excursionists on the lake. The local newspaper put the best face on the hostile environment: "The almost intolerable heat of the sun glowered unobstructed upon the deck, and the atmosphere was not made cooler by the added pulsations of heat from the boiler and furnace. But 'the ball went on.' "

The steamer plied the lake from north to south with shipments of bullion, but it was too efficient. The teamsters who hauled the silver south over the muddy roads in winter could not keep up, and, much like the Sorcerer's Apprentice, the little steamer kept arriving at the south end of the lake with more and more ingots that unemployed miners stacked for use as shelters. The smelters were forced to close down, and within a year of its baptism, the *Bessie Brady* was laid up.

Another steamer, somewhat smaller, was constructed on the shoreline in 1877 and christened the *Mollie Stevens,* after the name of the daughter of the owner. But mining was on the wane. The *Mollie Stevens* was soon dismantled and her engine put in the *Bessie Brady*. The work was nearly completed when the *Bessie Brady* caught fire and was destroyed in 1882. The railroad, which would also disappear in time, entered the north end of the valley that year and the next year there was a stop at Keeler, a lakeside hamlet at the foot of the grade from the Cerro Gordo.

In the early years of this century, Keeler was the richest industrial community in Inyo County, where there were a number of mining operations. Three plants annually processed forty-seven thousand tons of soda ash and bicarbonate, which represented one-half of all the soda products consumed in the United States. Three dwindled to one as the lake became a playa, and then there were none.

Keeler became one more vestige of the past. Its dreams of wealth evaporated and became salt-encrusted dust. The roiling clouds of deadly dust now silently eat away at the health of the seventy inhabitants of Keeler, the self-proclaimed gateway to the defunct Cerro Gordo mines. In addition to the native alkali chlorides, carbonates, and sulfates that are poisonous and carcinogenic, imported toxic wastes were dumped in the dry lake bed in the 1970s. Visibility is sometimes close to zero. Only the readings taken around forest fires exceed the number of particles generated by Owens Lake during a dust storm.

I drove slowly along the rutted streets of Keeler on a day when the dust blew away from the hamlet. A small blond girl who was riding a bike spotted my vehicle from a distance and fell to the ground. She ran like a startled rabbit into a nearby house. I passed the rotting two-story train station. There were no tracks. The concrete side wall of a store advertised Famous ABC Beer, a beer that is now unknown. The store's false front had collapsed.

A Southern Pacific freight car was someone's home. Abandoned vehicles lay about: old dune buggies, ancient Jeeps, painted school buses from the hippie era. Television satellite dishes pulled the world into the small bungalows that were tucked under vegetation grown to ward off the incessant sun. Signs reading NO TRESPASSING and KEEP OUT barred curious outsiders. The desert overflowed into most yards.

The wind blew and the dust rose in vertical columns and horizontal sheets over the dry lake bed. Satellite photos have shown the dust from

Owens Lake blanketing a 54,000-square-mile area. When precipitated out by rain over Orange County to the south, the dust has turned to acid mud that destroyed the finish of vehicles. The dust has halted military operations at nearby China Lake Naval Weapons Center and Edwards Air Force Base. One dust storm from Owens Lake, it is estimated, will produce 80,000 metric tons of suspended matter.

The scientists at China Lake examined the situation. Quoting W. C. Fields ("The game is rigged; but if we don't bet, we can't win"), they advanced a solution similar to a technique used by the Dutch to rid the soil of the Netherlands of sea salt. The cost would probably be excessive.

When the dust in which the bones of Indians are mixed blows through the weed-choked yards of Keeler or over the well-manicured lawns of Southern California, it affects people with lung ailments. The dust from the Valley of Death seeps into houses through tiny cracks. It coats furniture. People cough, sneeze, rub their eyes, or become anxious because it is difficult to breathe. And cats behave strangely, according to one scientific report.

At 3,880 feet above sea level and 328 feet above the level of the present lake bed, Owens Lake began overflowing to the south some 15,000 years ago, eroding the outlet 120 feet. The overflow into Rose Valley and on through Little Lake into China Lake ceased about 10,000 years ago.

The undulating blue waters of Haiwee Reservoir, set against the soft flanks of the Coso Range, follow a portion of that route before being siphoned off by the aqueduct. Farther on down the dry riverbed is Little Lake, a mile-long remnant of past wetness fed by springs that are immune to dry cycles and the public works of humans: they are dependent upon tectonic activity deep within the earth. Because of the vagaries of the fault system, Little Lake is an anomaly. There is more water in the present-day lake than there was five thousand years ago.

Between the reservoir and the lake is Fossil Falls, the best place to sense the ancient river. Fossil Falls is just that, a dry falls where the river once flowed over rocks, polishing them smooth where the rock was hardest or digging potholes where whirlpools found weakness in the basalt. Across smooth shelves and into these holes the river swirled, boiled, and tumbled. There was life and noise and spray where now there is silence and dry, black boulders within the canyon.

Along this moist corridor the Pinto Basin Culture flourished some four thousand years ago. Seven homesites are outlined by postholes, and numerous projectile points, scrapers, manos, and metatates have been found in the area that is rich in petroglyphs. Mixed in with these indications of an older Indian culture are more recent signs of the Paiute and Shoshone. The Shoshone Indians called the two-mile-long lava cliff to the east of the lake the Rattlesnake, because of its undulating form and the basaltic rocks that resembled scales.

Who really understands the rattlesnake? Not many people do. Laurence M. Klauber studied them for nearly fifty years and published a two-volume book on the subject. His basement was full of rattlesnakes. Over a twelve-year period he milked 5,171 rattlesnakes as part of a venom project for the San Diego Zoo. Klauber became the international authority on rattlesnakes. The prime message in his monumental work was that the rattlesnake had been the victim of bad press.

The problem, as Klauber saw it, was that the serpent, and the rattlesnake in particular, was the symbol of evil. That biblical concept was imported to America when Captain John Smith warned New England colonists of "the danger of the rattell snake" in 1631. As the population moved west, rattlesnakes were discovered in California (an Eden, some claimed). One species of rattlesnake, the Mojave Desert Sidewinder, was identified and named in 1854.

The sidewinder has a distinctive crawl and leaves tracks that can be mistaken for no other snake's. I have seen those tracks near China Lake. They were parallel slanted lines—one-half of a chevron—laid down when the sidewinder pushed against the sand, threw out its head, pushed against the sand, lay down its head, and so on. The side-flowing or looping motion, using vertical force to anchor itself before the thrust, is a more efficient method of locomotion through sand than the sinuous track of other snakes who use transverse force to push off from more solid footing. Down through the millennia, the sidewinder has adapted to the drier climate.

The other distinguishing characteristic of the sidewinder is the two horns over its eyes, giving it a devilish cast. The horns, or what appear to be horns, fold down over the eyes and protect them from abrasion when the snake buries itself in cool sand, its preferred habitat during the heat of the day.

The sidewinder is a creature of the night. As the desert cools, it bestirs itself and sets off at a gait of some 0.3 miles per hour, searching for lizards and small mammals. In short bursts it can move ten times as fast. It prefers lying in ambush near a burrow. As a pit viper, the sidewinder identifies its prey by an infrared device that functions through two pits, one on each side of its head. Klauber wrote, "Finally, there is the pit, a high-temperature differential-receptor that gives the rattlesnake a knowledge of the direction and distance of objects whose temperature is higher than the rest of the surroundings, a valuable organ to a creature living largely on warm-blooded prey."

While Klauber saw the thermal detectors as offensive devices, more recent research suggests they are also used for defensive purposes. In this scenario, the information is used to determine if the sidewinder should rattle to warn off an enemy or slink away if, for instance, a raven gets too aggressive. The remaining option is to kill. The sidewinder strikes without coiling, and its strike is so fast (averaging 8.12 feet per second) that the eye cannot see it. The strike is actually more of a snap, and the snake's body is yanked off the ground by the force of the movement. It can strike repeatedly.

Because of its nocturnal habits and crooked motion, the sidewinder has acquired a reputation for deadliness far beyond fact. Klauber said it was "an average rattler in disposition, neither especially pugnacious nor tranquil." Within the spectrum of rattlesnakes, the sidewinder is not particularly deadly to humans. Its venom yield is low and not too toxic.

However, an aura of fear hovers about all rattlesnakes. There is this account of a hunting trip into the Sacramento Valley in 1855:

> The dread of rattlesnakes destroyed in great measure the pleasures of our sport, for we lost many a good shot from looking on the ground—which men are apt to do occasionally when once satisfied of the existence of a venomous reptile, the bite of which is by all accounts fatal.

Fear begets extermination. Whole communities organized "rattlesnake bees," a social event for which people armed themselves with clubs, rakes, pistols, shotguns, and rifles and struck out in every direction. The next year, or the year after, they wondered why the populations of rabbits, ground squirrels, and gophers had multiplied and were destroying crops.

The Sidewinder, a heat-seeking missile, is the most successful weapon among the many developed and tested at the China Lake Naval Weapons Center during the last fifty years. Promotional literature put out by China Lake describes the Sidewinder as "the Free World's premier dogfight missile" and "the fighter pilot's weapon of choice." Numerous versions of the missile have been in use since 1956.

The Saudi Arabian pilot who knocked two Iraqi jets out of the air during the Gulf War used Sidewinders (the AIM-9L model), and the cheers began at the Sidewinder Technical Office at the north end of Michelson Laboratory and echoed up and down the long, doeskin-colored corridors of the eleven-acre facility. Of such a desert laboratory, the British author Aldous Huxley wrote:

> From their air-conditioned laboratories and machine shops there flows a steady stream of marvels, each one more expensive and each more fiendish than the last. The desert silence is still there; but so, ever more noisily, are the scientific irrelevancies. Give the boys in the Reservations a few more years and another hundred billion dollars, and they will succeed (for with technology all things are possible) in abolishing the silence, in transforming what are now irrelevancies into the desert's fundamental meaning.

The laboratories are the most recent phase in the evolution of the development of destructive technology that dates back thousands of years along the shoreline of China Lake.

That conclusion is implicit in the work of Emma Lou Davis. Known as Davy to her friends, she was born in the Midwest and graduated from Vassar College in New York State in 1927. She lived for a time in a commune in Russia, and helped dig the Moscow subway. Davis went to China and remained long enough to learn the language. Returning to the United States, she got her master's degree in Fine Arts. Davis was a successful sculptor, art teacher, and furniture designer. During World War II, she was an aircraft designer and union organizer at the Douglas Aircraft plant in Long Beach.

After the war, Davis went to Taos, New Mexico, to cure herself of cigarettes, alcohol, and one of a number of bad marriages. It was there that she caught the anthropology bug. She got her master's and then her doctorate in 1964 from the University of California at Los Angeles, persisting after her original doctoral thesis was destroyed in a fire. Controversial, scrappy, colorful, a feminist before such was fashionable, Davis stormed the staid, male-dominated field of anthropology with the same zest that allowed her to become one of the first women to gain admission to the Explorers' Club in New York. She died in 1988.

Twenty years earlier two college students had brought a large grocery box full of artifacts to Davis, who was then the curator of archaeology at the San Diego Museum of Man. Two weeks later Davis, accompanied by others, was plodding across China Lake in the boiling heat. She had found her lifework. From 1969 to 1974, with an infusion of her own money and grants from the National Geographic Society and the National Science Foundation, Davis, dressed in black and accompanied by her dog and a coterie of acolytes, combed the playa and its ancient shoreline.

There were difficulties. There was the heat, the reluctant navy brass who had to be convinced of the worth of the project, and ravens. She wrote, "Ravens are ubiquitous on the bombing ranges where we work and, due to their mischievous curiosity, account for the disappearance or displacement of small bone, stone and wind test samples." Eventually Davis collected some twenty thousand artifacts from which she drew her conclusions.

The level of China Lake, which was never deeper than ninety feet, fluctuated widely and rapidly during the time that humans wandered along its shoreline. The present is the driest period. Davis thought that fifteen thousand years ago, when the climate was wetter, the environment was most ideal for a complex human culture. The people moved with the hunting seasons from Panamint, to Searles, to China Lakes and then up into the Coso and Argus Ranges, all within or bordering the weapons center. Around the forested edges of China Lake there were shallow sloughs, ponds, and marshes. Davis mocked "the male myth" of full-time big-game hunters. Small game and plants, she said, had furnished the bulk of the diet. The fossils Davis discovered at China Lake were similar to those preserved in the La Brea tar pits of Los Angeles, where the environment was comparable during prehistoric times.

From the weapons and tools collected at China Lake, Davis and her colleagues age-tested the artifacts and came up with dates that spanned some forty-five thousand years, a period of time that fits within the mainstream of scientific estimates for human occupation of the desert. That span of time took in such well-defined cultures as the Pinto, Clovis, and Lake Mohave; hinted at other barely defined cultures; and, because of the hit-and-miss nature of artifact collecting over a large area, skipped other human bloomings that have no names.

What is clear from Davis's work at China Lake is that different civilizations used this land when it was cooler and wetter. Down through thousands upon thousands of years their weapons grew in sophistication, from the discovery that obsidian made a better point for a projectile than wood or plain stone, to a heat-seeking missile named after a snake. Davis used the term *technological innovations* to describe a new knife form and a new knapping technique that were quite deadly in their times.

Like the test site for the atomic bomb in Nevada, China Lake was advantageous for its remoteness. Early in World War II, the scientists at the California Institute of Technology (Caltech) began working on rockets and their volatile propulsion systems. The prototypes were tested in a canyon above Pasadena. Three scientists were killed and two were horribly burned in explosions. In August of 1943, Charles C. Lauritsen, who headed the rocket program, flew over the Mojave Desert looking for a more isolated test site. He saw an airstrip, some shacks, and a few mercury miners, and, much like Brigham Young was supposed to have said when he saw the Salt Lake Valley, Lauritsen declared, "This is it." The official account of the founding of the laboratory goes on to state:

> In the area surrounding the lake bed the desert sands were dotted with creosote bushes. Among the jagged peaks of volcanic rock to the north, the airbourne pathfinders could see the inlet by which the water from the Owens Basin had in centuries past flowed into China Lake.

When Caltech scientists and navy personnel arrived at the playa that same year, they found a dry lake bed within a gently sloping bowl named Indian Wells Valley. The 425-square-mile valley was named for some trickles of water utilized by its previous occupants. When William Brewer passed through the valley in 1863, he commented, "A more god-forsaken, cheerless place I have seldom seen—a spring of water—nothing else."

The valley is surrounded by mountains: the Sierra Nevada to the west, the El Paso Mountains to the south, the Argus Range to the east, and the Cosos on the north. It is a very active seismic area. In 1946 a severe earthquake was centered twenty-five miles to the west in the Sierra Nevada. There have been a number of smaller quakes since then. The Garlock Fault, a major landscape feature, slashes horizontally across the center's most secretive ranges. The average annual rainfall is three inches.

Before the arrival of the military, the adjacent settlement consisting of a general store and a few desert residences was known as Crumville. That name had to go. A renaming contest was held. A native of Missouri submitted the name of Ridgecrest, his hometown. Crumville became Ridgecrest, although it neither encompassed a ridge nor a crest. The name Rattlesnake Gulch was rejected, although the name of a species of rattlesnake would bring China Lake (the military base) and Ridgecrest (the town) fame within the military-industrial-academic complex, which was so dominant in California. The lake had previously been named for Chinese laborers who briefly harvested borax from the playa.

Some 650 square miles of public-domain lands were sealed off, and a bulldozer operator, who was shown a mountain in the distance and told to proceed toward it in a straight line, scratched the first flight line for a rocket test into the surface of the desert. The first test was conducted on December 3, 1943. The pilot narrowly missed two civilian scientists crossing the desert in a weapons carrier.

Morale was low. A Quonset hut served as the first officers club, and two Filipino sailors were the mess attendants. The local sheriff took pity on the lack of recreational facilities and supplied the officers with five slot machines. The first China Lake insignia was a cross-eyed jackrabbit riding a rocket. A masked bandit robbed the mail truck. It was a fifty-mile round-trip in a battered touring car to the nearest whorehouse at Red Mountain.

More land in the mountains to the north of the base was taken over to test the nonnuclear detonators for atomic bombs being designed at Los Alamos, New Mexico—another secret desert laboratory under the aegis of a California university. A number of mining families were ordered to leave their claims to make room for Project Camel. They were elderly desert folks who had no place to go and were inadequately compensated for their losses in a time of national emergency. Bitterness over this issue is evident to this day among locals with long memories.

On August 21, 1944, Navy Lieutenant John M. Armitage took off from the county airstrip and flew over the playa at an altitude of 1,500 feet. He fired a Tiny Tim rocket; and just as it ignited, his plane suddenly nosed over and augered into the floor of the desert. "The scene was appalling. No semblance of an aircraft could be detected in the twisted fragments bestrewing the desert sand," stated the official history of the weapons center. The young test pilot had been popular at the base. When the new airfield was completed, it was named for him. Armitage was the first death. Four persons died in accidents during the first eight months.

As the war played out, the center looked around for other projects. William B. McLean, a product of Caltech, arrived at China Lake in 1945. During World War II, the physicist had worked at the National Bureau of Standards on a guidance system for a glide bomb. He believed that missile work had become too complicated. A missile, he said, should be treated as mere ordinance, not as a separate aircraft. The innovative McLean believed in simplicity, which was not the dominant thinking. The air-to-air missiles being developed were small planes flown by robotic mechanisms. They were bulky, complex, expensive weapons.

McLean was a tinkerer out of necessity. The son of a Protestant minister, he had to build what he wanted. That included surfboards, canoes, rafts, and a photographic enlarger. At Caltech he studied under Lauritsen and built a half-million-volt Van de Graaff generator. When he arrived at China Lake at age thirty-one, McLean was recognized as a leader and technical innovator. His trademarks were his bow ties, quiet demeanor, and half-smile. A simple query would produce almost instantaneous results.

The problem with missiles was that the target kept shifting after the rocket was launched. The trajectory of the rocket needed to be corrected, not the trajectory of the aircraft that launched it. The solution was to find some type of seeker for the missile.

Following the war, there were lean times at China Lake. Because there was a restriction on guided missile work, funds from other sources within the China Lake budget were siphoned off to further McLean's tinkerings. They were disguised under such budget code names as "Local

Fuze Project 602" and "Feasibility Study 567." A recent publication of the weapons center noted:

> There are stories about the legendary Technical Director, Dr. William B. McLean, building the seeker in his garage. Some say these stories need to be taken with the proverbial grain of salt, given the magnitude of the program and the many talented people who worked on it, but people who are still here remember noting the presence of a seeker in Dr. McLean's garage.

McLean proposed adding an infrared sensor to a simple ballistics rocket. Congress balked, thinking the concept was too Buck Rogers–ish. To impress visiting dignitaries, McLean and his small crew rigged up a prototype heat seeker on a surplus radar van that tracked a distant person who held a cigarette. McLean persisted in his garage and in a small work space behind Michelson Lab with spare parts and whatever funding he could corral until the program was first officially funded in 1951.

A crash program quickly got under way. There was an advantage to living in the desert and working on a revolutionary weapons system. McLean recalled, "People could and did communicate with each other all day, through the cocktail hour, and for as long as the parties lasted at night. This isolation in a location where the job could be performed provided large measures of the intimate communication that is so essential for getting any major job completed." There were family strains, however, as the missile work monopolized conversations and the men regularly labored into the early morning hours at the height of the cold war.

The first Sidewinders, named by a colleague of McLean's for the snake, were failures. Finally Navy Lieutenant Wally Schirra, who went on to fame as an astronaut, fired the first hit on a drone. There was a shoot-off at Holloman Air Force Base in New Mexico with the Air Force's Falcon missile that the Sidewinder won handily. The missile was declared operational in 1956 and was first used in combat over the Straits of Formosa in 1958. There were four confirmed kills of Chinese Communist jets in the first dogfight, and the Sidewinder was credited with forestalling an invasion of the island.

The Sidewinder was the first air-to-air guided missile successfully used in combat, and went on to become the most widely used air-to-air guided missile in the world. In Vietnam, the missile had a "kill ratio" of 95 percent. From models A to R, the missile's "launch envelope"—the position from which it had to be fired—expanded from having to look up the tailpipe of an enemy plane to almost any direction.

There are some 100,000 of the arrowlike Sidewinders scattered around the world in various arsenals. The parent missile spawned a number of progeny that went under such names as Bombwinder, Bullwinder, Sparrowinder, Spacewinder, and Subwinder. The basic missile can also be fired from land and from ships, in which case they are called, respectively, the Chaparral and the Sea Chaparral. The Soviet Union copied a Sidewinder right down to the color of its wiring, but substituted the Cyrillic alphabet.

McLean came to symbolize the "China Lake way"—the freedom from military rigidity to employ "innovation, change, and responsible risk taking." Original thinking, however, is not always appreciated, nor is competition. McLean and subsequent civilian technical directors have private contractors, naval brass, and Congress to contend with; and some directors have been beaten down or defeated by a combination of these forces. When McLean died in 1976, he was out of favor; he had emphasized the vulnerability of such large-ticket items as giant aircraft carriers to small missiles.

What the civilian scientists have going for them is longevity. They also outnumber the military by a five-to-one ratio. Base commanders come and go, usually in two-to-three-year hitches, but the civilians remain. These scientists fit a pattern. They are a blend of corporate consciousness and academic casualness; their work clothes range from suits to shorts. A few are still recruited from Caltech, and some are "lab brats," the sons and daughters of alumni. But most of the scientists and technicians now come from universities in the Midwest and the interior West. Weapons work in the desert at government pay scales is no longer the most desirable of scientific jobs.

The scientists do think about the morality of working on the machinery of death and destruction. Some say that the purer types of research keep them isolated from the unpleasant realities of what their products might eventually be used for. Others cite the civilian spin-offs, such as the instant replay of televised sports that comes from video techniques developed for immediate bomb and missile scoring. Yet another group believes in the deterrent factor. Most believe that weapons work is necessary for the survival of our particular segment of the species.

Yes, I did meet a Dr. Strangelove type. Across his computer screen flashed an occasional bolt of lightning, then blackness was followed by a scream from the machine at regular intervals—his choice for a screen-saving program. The bulky, bearded man smoked incessantly in a darkened, cave-like office. Stacks of documents threatened to topple over, and what I hoped were nonfunctioning weapons—hand grenades and a Stinger anti-aircraft launcher—lay about. There were copies of the *Oxford English Dictionary* and *Jane's Rocket Ballistic Data*. This man was a lab brat. He was dressed in a sport shirt with a jungle motif, black pants, and cowboy boots.

The digitized screaming was unnerving. He screamed at me about "Claiborne Pell and other Soviet dupes" in Congress who had halted a weather modification program devised at China Lake during the Vietnam War. He boasted that they had flooded the Philippines and made it rain in southern Laos. They had done nothing that the Los Angeles Department of Water and Power was not doing over the Sierra Nevada to produce precipitation, he said.

Another scream. The bearded man continued talking at a pace that verged on the incoherent about "cooking off" people and the "bugs and drugs" program. More screams echoed through that darkened room.

This man was an historian. He recounted the successes at China Lake and gave me copies of reports and articles he had written in specialized publications. Weapons for such conflicts as the Formosa Strait, Korea, the Cuban Missile Crisis, Vietnam, the Falklands War, Libya, the Gulf of Sidra, the Bekka Valley, and the Gulf War had emerged from the 1,100 buildings, 1.1 million acres of grounds and test ranges, and 17,000 square miles of restricted airspace (12 percent of California's airspace) over the desert.

These weapons had such names as Tiny Tim, Holy Moses, Mickey Mouse, Walleye, Bigeye, Skipper, Shrike, Sparrow, HARM, SLAM, Sidewinder, Sidearm, Tomahawk, Zuni, Polaris, and Harpoon. Nuclear weapons components, the most powerful organic explosives, chemical weapons, cluster bombs, fuel-air explosives, propellants, fuses, fire-control systems, parachutes, weather modification schemes, submersibles, an abortive answer to Sputnik, lunar landers, the Space Shuttle escape system, Stealth bomber technology, laser optics for the Strategic Defense Initiative, cold fusion and superconductivity experiments, and the first computer that works as an organic neural network have been perfected in China Lake's air-conditioned laboratories.

What was easy to forget was that all this lethal activity took place on the bed of an ancient lake. But nature has a way of reasserting itself, of reminding humans of their fragility. The naval weapons center does not seem like a fragile place, but it is. On August 15, 1984, three inches of rain fell in a six-hour period, an amount equal to the average annual rainfall. It soon became apparent that the weapons center was located on an ancient floodplain.

The water rose quickly in the dry washes and the streets that fed into the weapons center. The churning, brown mass overturned cars and washed them away like toys. The water ran four feet deep in places, and the more daring youths hitched rides in their inner tubes on the twisting course it took through Ridgecrest.

The water flowed through the heart of the weapons center seeking the flat playa, the lowest point in the valley. The cascade was inexorable and terrifying in its suddenness. Nothing in this center of high technology could stanch the flow. Sheets of mud and water poured into Michelson Laboratory, the nerve center of the weapons complex. Mud coated the first floor, and the basement filled with twelve feet of water. On a sweltering summer day the power failed, the air-conditioning ceased, the telephones went dead, and eight thousand computer screens went blank. The laboratory was hastily evacuated at noon and abandoned to the elements.

It was as if a bomb had struck: Chairs and tables were overturned. Glass was shattered. Nineteen thousand classified documents in basement vaults swirled about in the floodwaters. In the terms of those who calculate such things, the storm was rated a one-hundred-year or possibly even a two-hundred-year event, meaning that no such similar flood had been recorded in the valley during historic times—a very short time span, indeed.

The fire department pumped water out of the building for the next three days. The Los Angeles Department of Water and Power loaned transformers. The soggy documents posed a special problem for the postgraduates, doctorates, technicians, staff, and management who pitched in to clean up the mess. They needed the proper security clearance to handle the sensitive papers.

The laboratory contracted with the firm that had treated water-damaged books following a fire at the main Los Angeles library. Four thousand document boxes, four hundred at a time, were freeze-dried to halt mildew. Ninety percent of the documents were eventually saved. Two weeks and $32 million worth of damage later, Michelson Laboratory reopened for business. A plaque outside the building lists the names of 717 "Mich [pronounced Mike] Lab Muckers." Congress appropriated $5 million for a flood-control project. The water that had ponded in the dry lake bed slowly retreated again under the desert sun.

Three-quarters of a century earlier Dr. Raymond G. Taylor arrived in Sand Canyon at the western end of Indian Wells Valley to work on the gigantic project dedicated to the control and transportation of water. The canyon was a division headquarters for the building of the Los Angeles Aqueduct in 1908. Tramways stretched up both canyon walls to tunnels numbered 16 and 17 that bracketed the largest siphon on the project. Rectangular buildings with peaked tin roofs were located with military precision on the floor of the canyon. The tunnel builders were mostly Irish and Cornish miners. The open ditch workers (using the designations of nationality of the time) were Greeks, Bulgarians, Serbians, Montenegrins, and Mexicans. The gigantic project followed the California tradition of having foreign workers do the manual labor.

Each division headquarters had a field hospital equipped with a doctor, a nurse, and six beds for patients. Taylor was the doctor at Sand Canyon before becoming chief physician of the project. A native of Illinois, Taylor had graduated from the University of Southern California Medical School. At the age of thirty-five he became the youngest president in the short history of the Los Angeles County Medical Association.

Taylor toured Indian Wells and Owens Valleys in a six-cylinder Franklin racing car that broke down frequently and had to be towed by horses. The doctor carried a .22-caliber rifle to shoot coyotes and rabbits and a shotgun for ducks. He recalled:

> An odd thing about the ducks in the Owens Valley was that if
> they got down to the lake where the water was very highly

mineralized and if they stayed any length of time, the salts crystallized on their feathers and they got so heavy that it was difficult for them to get up off the water. As a result, some of them—thousands of them in fact—starved because there was little feed in the salt water and they got so heavy with impregnated salts in their feathers that they couldn't get up to fly away to feed.

During the five years of construction of the aqueduct, there were forty-three accidental deaths, one permanent injury, and 1,282 miscellaneous injuries, according to Taylor's count. He remembered a shift boss at Sand Canyon who walked into a tunnel and was electrocuted. "Nobody knows just how it came about, but the lights went out and somebody went back to see if he needed any help getting out and found him dead."

Taylor hired a doctor for the Sand Canyon hospital who had done research for a major pharmaceutical company. He got disquieting reports on the new doctor's behavior, and was finally notified that the man had disappeared. "I hotfooted it up there and by the time I got there somebody had found his body down on the desert about five miles from camp, deader'n a doornail."

It turned out that the strange doctor had become addicted to heroin while researching the manufacture of the drug for the pharmaceutical company. "We had to bury him on the desert on account of the condition of his body, and his grave was properly marked," the doctor said. When Taylor died, in his eighties, the *Los Angeles Times* obituary noted that he had been "the physician to the 10,000 men who built the Los Angeles Aqueduct."

I camped in Sand Canyon, somewhat higher and cooler than the valley floor, and commuted daily to the weapons center. The spring wind rustled the green leaves of the cottonwood trees and wildflowers grew among the concrete slabs and rotting walls that marked the ruins of the division headquarters. The telltale tracks of a sidewinder were visible in the sand.

From Sand Canyon and China Lake I followed the dry riverbed to the fourth in the chain of ancient lakes. It was a sparkling desert night, and I stepped out of my van for some fresh air. A wedge shape blotted out the stars. A roar followed. I ducked. It was difficult to hide from the military presence in the California desert.

The experimental Stealth bomber flew over my camping spot among the tufa towers on Searles Lake. The bomber was on its way to the super-secret Echo Range, where its penetration abilities, or lack of them, could be tested against mock-ups of Russian land and sea radar systems and what other possible enemies had to offer in terms of detection devices.

Low-flying aircraft are common in the desert, and road signs warn of them. Another time rotating red lights—I thought perhaps it was a pair of helicopters or slow, prop-driven aircraft—slid past me and then dropped out of sight into the Panamint Valley. I heard nothing, possibly because I was listening to a book on tape as I drove down the steep, winding mountain road. It could have been drug smugglers or gunrunners who occasionally land on the playa. Their silent glide favored that theory, but the lights indicated military activity.

The tufa slabs that rose from Searles Lake, and the ravens that perched upon them, were no strangers to such noise. Nor was I completely unprepared for low-flying aircraft in that primeval place. I had been told at China Lake that the navy made an aircraft and pilot available to moviemakers who filmed multiple sequences of the jet flying at an altitude of thirty feet off the deck as it weaved between the towers that jut from the ancient lake bed.

The monoliths attract filmmakers of science-fiction bent, such as those who made the *Star Wars* films, and others who need exotic locales to hawk such products as Budweiser beer and Toyota cars. "Otherworldly" and "moonscape" are the descriptions used by the Bureau of Land Management, which administers Pinnacles National Natural Landmark.

The five hundred or so spires have been classified by scientists into four forms: tower, tombstone, cone, and ridgeline. Most of the pinnacles are between ten and forty feet high; but some, particularly the northern grouping, reach one hundred to one hundred forty feet in height. They are between ten thousand and one hundred thousand years old, and are a good indication of where the surface of the lake stood at various times.

The formation of the pinnacles was a function of tectonic fissures. The earth-warmed, calcium-rich water rose beneath the ancient lake through the various interstices of the Garlock Fault, precipitating out as various minerals when it came in contact with the carbonate-rich lake water. The varied shapes of the pinnacles reflected the composition and

The tufa towers

spacing of the springs, the depth of water, and the rate at which the lake's surface rose and fell. At its maximum, Searles Lake reached a depth of 860 feet, at which point it overflowed into the adjacent Panamint Valley.

To stand in the midst of the tallest towers was like being present at the center of a desert Stonehenge. As the sun set, the long shadows of the monoliths began to interweave. That night the desert wind exploded in strong gusts as it was funneled down on my camper parked in the midst of the columns. I walked carefully (in order to avoid night-hunting sidewinders) to the low ridgeline formed by the talus slopes that connected the towers. In the distance the orange glow of streetlights in the mining town of Trona was spread in a thin line across the northern end of the playa.

Like Ridgecrest, Trona was a company town dependent on a dry lake. There was a difference. Trona was a throwback to the era of smokestack industries. I first came across it in a heavy rainstorm, the smoke and steam of its stacks mixing with the dark clouds to form a Stygian darkness. The industrial buildings were walled off from the remainder of the town by a well-maintained barbed-wire fence. The guarded entrance to the plant commanded a field of fire. A mother and child carried a plastic wreath into the cemetery that lacked the comfort of grass.

Another youth, some one hundred years earlier, had returned to Trona from boarding school in the San Francisco Bay Area via Poison Canyon. Dennis Searles was the son of John W. Searles, the founder of Trona and the first to attempt to extract minerals from the surface of the ancient lake. Young Searles wrote in his journal:

> Now we turn a rocky point and see a large dry lake, as white as snow, in front of us. At the upper end we see the works but no smoke is coming from the smoke stacks and you have the impression that things are shut down but they are not. They burn petrollium.

The 250-square-mile valley took getting used to. In 1946 an English war bride traveled to Los Angeles, where her husband met her and drove her home to Trona. She wrote:

> California was sunny, green, and beautiful along the coast. After two days spent in sight-seeing, we set out across the desert for Trona. The scenery became gradually bleaker until finally we reached Poison Canyon and the view of Searles Lake. My husband then asked me if I had ever seen anything like this before, and I answered, "only in pictures of the moon."

Failure dogged the efforts to mine the dry lake bed. Five years after Dennis Searles's visit, his father sold the operation to another mining company. One set of owners followed another. The American Potash & Chemical Corporation took over in 1926 and merged with Kerr-McGee Chemical Corporation of Oklahoma in 1967. Another plant beside the lake went through two owners before also winding up in Kerr-McGee's hands.

In March of 1970, the three unions at Kerr-McGee's plants in Trona struck. It wasn't the first strike in Trona. In 1943 there had been a three-month strike, and the plant had shut down. This time Kerr-McGee, of Karen Silkwood fame, was not about to shut its plants down.

It fired striking employees and hired scabs, and supervisors worked twelve-hour shifts.

Fear and its offshoots—hate, rage, and violence—washed through the town like a flash flood. The most valuable commodities that an isolated community has in times of trouble are self-sufficiency and cohesiveness. Trona lost its glue. Where once a foreman could warn a worker about the behavior of his errant son, where men dressed in tutus for the annual Firemen's Ball, where the elected constable was also the chief of security for the company, and where the community swimming pool at Valley Wells was the living room for the town, now neighbor turned against neighbor, friend against friend, and families were split asunder.

The company fired more than 75 percent of its workforce, and some people lost their homes and cars. The strikers and their families retaliated. Women cursed and spat at the workers who entered the plant. Men were more furtive, burning homes to the ground and setting a bunkhouse across the street from the plant on fire. A scab's car burst into flames as he was driving it. Company cars were stoned and overturned. All the windows in the administration building were broken. A bomb exploded under the bed of the small daughter of a couple who was working at the plant; no one was at home. A bomb went off in a local garage. The sewer lines were dynamited. In all, there were twenty bombing incidents.

Kerr-McGee requested the help of the San Bernardino County Sheriff's Department. Twenty uniformed sheriff's deputies, a half-dozen detectives, and part of the narcotics division worked around the clock, aided by the plant's security force. They had little luck. Three-quarters of the way through the strike there were fifty-four arrests, but few convictions resulted. The problem in that small community of divided loyalties was hung juries.

Finally there were negotiations. Among those negotiating for the unions was Harry Bridges, the militant president of the International Longshoremen and Warehousemen's Union. Bridges urged the workers to accept the new contract. The one-hundred-fifteen-day strike ended in July, but the settlement did not provide for amnesty. A few years later there were no unions in Trona. "The Okies won," said one old-timer. "They had strikes before, and they came out on top of the heap on this one."

What was more a family than a town became just another community of transients. Many who still worked in Trona moved to Ridgecrest, where there were better schools and a McDonald's. One by one the stores closed. The mobile homes were pulled off their concrete pads and driven elsewhere. The wind blew the tumbleweed across the town's vacant streets, and some lodged in the concertina-type barbed wire that ringed the exercise yard of the jail at the sheriff's substation.

Kerr-McGee of Oklahoma sold out to North American Chemical Company of Kansas in 1990, and the new owner pledged prosperity and no layoffs. When I was in Trona the most recent owner announced a "restructuring," a polite term for layoffs.

The geologic history of the lake extended to a great depth. On July 5, 1968, Kerr-McGee, in an effort to determine what it had purchased, began drilling an exploratory drill hole in the bed of Searles Lake. On September 6, the company ceased drilling 50 feet into bedrock at the bottom of the lake. It had drilled a total of 3,051 feet from the surface of the playa. The slightly fractured bedrock was overlain by 740 feet of sand and gravel deposited by ancient streams.

A geologic portrait was constructed from the findings. The basin was formed some 3.2 million years ago, and the water began to rise and fall. Eventually 2,310 feet of mud and evaporates were deposited on top of the sand and gravel. The average rate of sedimentation was 8.7 inches per 1,000 years. The economic minerals are in the upper 500 feet.

Searles Lake has been the receptacle of the upstream lakes, and that is why the minerals coalesced there. Mono, Owens, and China Lakes collected silt; and the minerals passed downstream. Dry lakes alternated with medium- to large-size wet lakes that covered the surface three-fourths of the time. When the upstream lakes ceased to flow, the wet lakes quickly evaporated.

A cabinet in the office of the company's geology department contained artifacts recovered from various depths. They were stops on the descending elevator of time. At a depth of 68 feet an Ice Age fossil of a fish called a chub was found. At 108 feet a piece of ordinary-looking driftwood had been dated at 27,000 years. A piece of gray wood found at the 130-foot level was 35,000 to 45,000 years old.

China and Searles Lakes needed to coalesce before there was enough water to overflow northward into the Panamint Valley. When the water retreated, a small spring bubbled to the surface at the northeast end of the fifth playa in the chain. Insects attracted shorebirds and waterfowl to the small area of marsh supported by the spring. Around the edges of the marsh grew pickleweed, inkweed, rabbit brush, and alkali sacaton. It was a swatch of soft colors amid the blinding glare of the dry lake. Just to the east were the angled slopes of creosote-dotted *bajadas,* the alluvial skirts that swirled about the towering Panamint Mountains.

Three families of Panamint Shoshone Indians, the same tribe that lived in Death Valley, were living at Warm Springs when the botanist Frederick Vernon Coville passed by in 1891. Coville was part of the Department of Agriculture's Death Valley Expedition, the first scientific venture to enter that desert sink. Although heralded as "ten celebrated scientists," the members were all fresh out of college. They entered Death Valley via Wingate Pass in January of 1891 and spent the next six months in and around the valley. The expedition searched for strange new species in that strange new land. The youths shot sidewinders, caught desert pupfish, and dug up hundreds of plants. One of the plants they noted was the creosote bush. They had some close calls in that land of infernal heat, but all except a pet dog escaped Death Valley alive.

The Indian settlement at Warm Springs was known as Hauta. The people, known as Hautans, irrigated two or three acres and raised corn, potatoes, squash, and watermelons. Salt was obtained from Searles Lake. The Indians and the Chinese did the heavy manual labor for the whites in the Panamint and Searles Valleys.

By the 1890s the Indians used both guns and bows and arrows to hunt, repairing the latter with a strong glue made from gum deposited on creosote bushes by an insect. Coville noted that the gum, when mixed with pulverized rock and heated, made a strong bonding material that could anchor stone arrowheads to willow shafts and mend the broken covers of sugar bowls.

Creosote is the ultimate survivor. It is the oldest known single organism. Creosote bushes have outlived the more dramatic bristlecone pine trees in the White Mountains above the Owens River by some 6,000 years. Extrapolated growth rates reach 11,700 years. The most reliable radio-carbon date for a creosote bush is 10,850 years. Others have been estimated to be 9,400 years old. When desert Indians tell the story of the first thing that grew, they relate the tale of the creosote bush. The plants have survived huge amounts of radioactive fallout at the Nevada Test Site.

The bush clones itself outward from a single seed into what is called a fairy ring, an elliptical circle that can reach a diameter of sixty-five feet. Each daughter shrub that grows outward retains the same genetic makeup of the original seed. As the ring widens, the older shrubs die and rot back into the desert soil.

The creosote is an inhospitable plant. Each clonal ring, or single bush if it is a younger plant, competes with other plants for space in the desert. The regular spacing of these brassy-green bushes indicates the intense competition for water and the toxicity of their poisonous resins, which also protect them from grazing animals.

An expert on creosote bushes wrote, "Often, few other plant species are represented in this habitat, and the only major food source available to a herbivore is creosote bush itself, a plant renowned for the number and type of toxic chemicals contained in its stems and leaves." The low profile of the creosote bush provides little shade and protection from wind for other plants and animals.

Because of these protective characteristics, the lowly creosote bush is the most common plant in the desert. More than one hundred natural products have been derived from it. The Indians and the Spanish used creosote to treat stomachaches, wounds, body odor, dandruff, poisons, syphilis, and poor circulation. It was also used for menstrual cramps, as a contraceptive, and to mend engine blocks by later generations of Indians. The medicinal qualities of creosote survive today as chaparral tea.

I encountered the descendants of the Hautans while camping at Warm Springs.

I had been walking among the creosote bushes on the *bajada* and left my mark, the ribbed soles of running shoes. The fragile desert varnish had been disturbed many years before by the placing of rocks in circles by ancient Indian tribes. Nearby were the freshly cut tracks of off-road vehicles.

It was midday when I returned to the spring. Waves of heat rose from the adjacent playa. There was a cacophony of crickets. The contorted branches of a dead bush formed a pattern resembling shattered glass that framed the hazy Argus Range to the west. Where the Panamints dipped to the valley floor, the eroded hillside resembled an elephant's foot. It was difficult to see the landscape as it actually was.

I felt pressed, and retreated to the shade of my camper where I took up a copy of Coville's monograph on the Panamint Indians. Just then a 1960s-vintage Detroit car pulled up and parked behind me on the same flat space adjacent to the marsh. Two elderly Indian women, faces barely discernible over the dashboard, got out. I could not believe the timing. It took me a few astonished minutes to put the report down and walk over to join them.

Pauline Esteves and Grace Good divided their time between Death Valley in the winter and the Owens Valley in the summer. They were looking for yerba santa plants, but they didn't quite know what to look for.

Pauline explained, "They used it for bathing wounds or for bathing muscular pains and then they drank it for arthritis, and what was that latest thing they were talking about? Cancer or some kind of stuff like that. I don't know which type of cancer they were talking about. Some people said they had success with it."

"But they had to boil the root, not the leaves, just the root," said Grace.

Their mothers had told them about the beneficial effects of the plant, and the two women were looking for samples to take to some Indian children who were on an educational outing at a nearby ranch.

"These children are from Lake Isabella. They are different from us," said Pauline.

"I think they are a lost tribe," Grace said.

Pauline nodded. "They are. They are."

We got to talking about Coville's description of Warm Springs one hundred years ago. I read them a portion of the report. Coville began, "To a traveler passing by rail across our southwestern desert region, it is a matter of great wonder how the Indians of that country contrive to subsist." He then proceeded to describe their various means of subsistence.

The two women listened intently. They used the long stalks of a plant to swat the giant flies that circled about. They seemed fascinated. I promised to send them a photocopy of the botanist's report, and subsequently did so.

After they left, I watched the moon rise and give substance to the landscape. The temperature cooled off just about the time the frogs began croaking in earnest.

Not long after Coville's visit, the desert town of Ballarat briefly bloomed on the eastern side of the playa, a few miles to the south of Warm Springs. Ballarat was created to serve the Ratcliff Mine in Pleasant Canyon, it being adjudged that the nearby community of Post Office Spring was on too marshy a footing. Post Office Spring began as a mail drop for outlaws who hid in the tall grasses and willow trees surrounding the marsh.

Ballarat, located a half-mile away on higher ground, gained official status when it was awarded a postmaster in 1897. A number of adobe, wood, and tent structures sprang up in the instant town, which had a school and a schoolteacher for all of one year. There were six saloons in town that served warm beer and strong whiskey. The only two-story structure on the treeless desert was a hotel. A doctor moved to town and advertised medical, dental, and real estate services. There was a red-light district and a jail, but no church.

The Indians stayed away from Ballarat. "Lots shooting and fights. No place for Indian," said one.

Law enforcement was a hit-and-miss matter. A bitter feud between the justice of the peace and the town constable resulted in the town's only homicide: The justice, Richard Decker, was an alcoholic sot. The constable, Henry Peitsch, was known as a fiery Dutchman, which meant he was of German descent. Peitsch entered the judge's office shortly after noon on an early fall day in 1905. There was a shot. The judge crawled on his hands and knees out onto the adjacent wooden boardwalk, trailing blood. By the time the doctor arrived, he was dead. The constable said the shooting was accidental. An eleven-man jury found Peitsch not guilty.

Ballarat passed into official ghost-town status in 1917, when it lost its postmaster. An historical monument was dedicated at the town site in 1963. By then the place was down to its last resident, Seldom Seen Slim, a shy prospector who tended the few graves in the cemetery and then was buried amid the creosote bushes in 1968. On his gravestone was carved his motto: "Me lonely? Hell no! I'm half coyote and half wild burro." When I visited Ballarat, two vases of plastic flowers placed on Slim's grave had been tipped over by the desert wind.

Ballarat was an incongruous sight. Along with the soft adobe ruins of a true ghost town were the sharply rectangular shapes of mobile homes and a concrete-block building with blue trim that looked like a convenience store. It was the headquarters for a group of Southern Californians who had bought the town site and hoped to turn it into a tourist attraction. The caretaker for this group—Slim's linear descendant—was a young man who shaved his head and lifted weights in the desert sun. He talked fast and cracked his knuckles incessantly. The caretaker told me of the Charles Manson mementos that lay hereabouts.

The word *helter* was scratched on the inside of the adobe ruins, and right in front of me on blocks was a surplus army vehicle used by the Manson clan to get around in the desert, he said. On the playa to the north, not too far from the rock circles constructed by the Indians, straight rock alignments guided drug planes to their landings. To the south, on the outskirts of Trona, was a wrecking yard where the Manson bus slowly rusted back into the earth.

As he talked, a large ore truck passed on its way from the Keystone Mine to a mill where gold would be extracted. Mining was undergoing something of a revival in the Panamints. The mine was located in the Goler Wash, and farther up the wash was the Manson hideout, which has gained the status of a minor shrine.

The Bureau of Land Management office in Ridgecrest receives telephone calls from across the country asking for directions to the Barker

The cemetery and Slim's gravestone

Ranch. Someone recently tried to stake the Barker Ranch as a mining claim, but the bureau disallowed it. The ranch is part of the nation's public domain and is maintained by volunteers under the bureau's "adopt a cabin" program. It gets a fair amount of visitors, which bureau rangers find unsettling. The name of Manson and the images of brutal murders are inextricably mixed.

I walked up the rutted road past coiled rattlesnakes and vigilant ravens. The Manson cabin was just north of the Garlock Fault and Wingate Pass, the ancient passage for water from Searles and Panamint Lakes to Lake Manly, which once covered the floor of Death Valley and was the last in the chain of lakes linked by the ancient Owens River.

California has had more than its share of gruesome murderers, mass slayers, serial killers, assassins, and chroniclers of such deeds. Of the latter, Erle Stanley Gardner, Raymond Chandler, Dashiell Hammett, James Ellroy, and Joseph Wambaugh, among others, come immediately to mind. Dark souls are periodically set adrift in this island paradise; the depravity of their deeds is echoed in the violence of the landscape.

Charles Manson and his "family" rode out of the desert, slaughtered haphazardly and with abandon, and then returned to the remote cabin in Death Valley country where they were eventually apprehended for car theft. They were a whirlwind of death whose precise message was lost in a babble of tongues loosened by drugs, sex, and rock and roll, but no one said, How unusual for California.

Manson, who went by the name of Charlie, was born in Ohio, the son of a teenage mother who ran around a lot and a father he never knew. He bounced around West Virginia, Kentucky, and Ohio with neighbors, relatives, and his mother before she went to jail for robbery. By the age of thirteen, Charlie was committing his own armed robberies. He was in and out of boys' schools, juvenile halls, honor camps, jails, and eventually state and federal prisons. Charlie was both cruel and captivating by the time he arrived in California in 1955, where he was soon in jail and then out again.

Manson operated in and around the margins of society, the mean fringes of Hollywood, and elsewhere. He absorbed the messages of the dominant cults, to which he added his own litany. When the hippie years descended on California, Manson was the right person in the right place at the right time. With his guitar, the mad Pied Piper attracted the dissatisfied children of the middle class, who brought their parents' credit cards along with them. At the age of thirty-three, Manson loaded his followers onto an old yellow school bus and headed for the Mojave Desert in November 1967.

His message fit the times; in fact, Manson may have been ahead of his time. He foresaw a racial apocalypse, calling it "Helter Skelter," after a Beatles song. It could have just as well been called chaos.

Manson and his brood prepared for the future in the desert. They armed themselves and conducted maneuvers in dune buggies and four-wheel-drive vehicles. They searched incessantly for their version of the California myth. Calling themselves the Hole Patrol, they crisscrossed the desert looking for the entrance to a cave in Death Valley leading to a sea of gold. Through this underground cavity ran a river of milk and honey. Trees with twelve kinds of fruit, one for each month of the year, grew there. The light was golden, the temperature was neither too hot nor too cold, and there were warm springs and cool water.

Charlie Manson and his followers were alien to the desert, yet were of it, too. Their junk food was imported, they were dependent on large quantities of stored gasoline, and they suffered from the heat and the cold. Although they committed some of the most heinous murders in California's long history of murder, dating back to the extermination of the Indians, Manson would not let his people harm a rattlesnake. He had a rapport with the animals of the desert. He said:

> I studied the animals. I know how they get moisture from the sand. I've seen them do it. I'd watch the rodents burrow and exist in places where no one could live, how they'd go down deep, and get the water, even store it. I didn't eat meat. I didn't eat meat, and I could communicate with coyotes.

Manson achieved the status of a godhead and prophet to his own people in an environment that was conducive to the emergence of such gurus. The bearded ex-convict, who was regarded as Jesus Christ by his followers and was booked as "MANSON, CHARLES M., aka JESUS CHRIST, GOD," was hiding under a sink when he was captured by lawmen. He was clad in buckskin, had the eyes of a fanatic, a slight build, a thin face, and shoulder-length hair.

Eventually Manson and some of his followers were linked to the ghoulish Hollywood murders of actress Sharon Tate, wife of director Roman Polanski, and others. They killed with knives and ropes and by repeated bludgeoning. The victims were stabbed and beaten beyond death. Their fresh blood was smeared upon the walls in various apocalyptic messages. The killings seemed vengeful and personal, but it turned out that the victims had been selected at random.

Fear stalked the land. The rich and famous hired bodyguards and bought guns and attack dogs. After his capture, *Rolling Stone* magazine referred to Manson as "the most dangerous man alive." Manson and a few family members were convicted of nine murders, but other followers admitted to between thirty-five and forty killings. Manson-related killings continued for a number of years, and one of his followers, "Squeaky" Fromme, later shot and wounded President Gerald Ford in Sacramento.

Manson lives on in prison, and some of the family members still reside nearby in the desert, working at mundane jobs. The wind picked up, and dark clouds formed on the horizon. I was spooked and turned back before I reached the cabin.

Death Valley, some eighty miles east of Mount Whitney as the crow flies, is the driest, hottest, and lowest place in the Western Hemisphere. It is difficult to envision 618 square miles of one of the world's great salt pans covered with water to a depth of six hundred feet, but that was the case some sixty-five thousand years ago.

It still was no Garden of Eden. Lake Manly resembled the Mono Lake of today, the first lake in the chain and some six thousand feet higher in elevation. Brine shrimp, flies, mollusks, pupfish, and minnows inhabited the saline waters. Willows and cottonwoods dotted the banks of the streams that fed the lake. Along the lowlands greasewood and sagebrush predominated, above were piñon and juniper, and a dense belt of pine laced the upper canyons.

With an evaporation rate of between twelve and fourteen feet a year, no body of water in Death Valley could be a permanent fixture. Lake Manly, named after one of the first white men to cross the valley, fluctuated widely.

At least twelve former shorelines are visibly incised in Shoreline Butte at the south end of the valley, indicating depths varying from three hundred thirty to six hundred feet. Some two thousand years ago the remnant of the lake was thirty feet deep. Then there was an earthquake, the floor of the valley tilted toward the east, and the water evaporated.

When it rains hard and long, say, three inches in a day or so—almost twice the average annual rainfall—or when the snowmelt from the adjacent mountains is excessive, the water will pond in the valley, and a small lake will form. At such times it is possible to canoe across the floor of Death Valley, and people have done so. Such lakes do not last long in the blazing desert sun.

Consider the heat. It comes from both below and above. The mantle of the earth is thinner here, and thus the heat from the molten core has a greater effect on the surface temperature. Ground temperatures can be 200 degrees. The Indian name for the valley was *Tomesha,* meaning "ground afire." An air temperature of 134 degrees was recorded in 1913, making Death Valley the hottest spot on the earth. Considering the limitations of the thermometer, the temperature could easily have been higher. One veteran desert traveler died in that heat wave when his car bogged down in sand.

The Indians and Civilian Conservation Corps workers fled to the mountains in the summer. Chinese laborers mined borax for a pittance, but not in the summer months because the mineral did not crystallize properly in the heat. Death Valley became a national monument in 1933, and the National Park Service began manning the facilities year-round. Some Park Service housing did not have air-conditioning. A swamp cooler lowered the temperature twenty degrees, say from 125 to 105. The cold water faucet became the hot water faucet as the ground heated up the water, and the hot-water heater served as an insulated container for cooler water. Food was cooked on one day and frozen to last for a number of days. Park Service employees drove to Las Vegas for supplies before dawn and returned after dark. They wore cotton gloves so that their hands were not burned by the metal in their vehicles.

The place made the name, and the name made the place. Death Valley was named for the one person who died there in 1850 when the first overland immigrants crossed it. In the oxygen-rich environment of a land below sea level, the hearts of some elderly tourists burst when they become too active. Death Valley is like honey to suicides, would-be suicides, and fake suicides. The name tempts adventurers to pit themselves against the heat, an intangible substance that they think they can beat. Some lose. Their rigid bodies are found with skin cracked, eyes sunken, and tongues swollen.

The name also attracts flocks of foreign tourists in the summer. They are mostly European. The French in particular are searching for a Wild West experience. Three-quarters of them find that less than one day is more than enough time to spend in Death Valley; and on they drive in their air-conditioned, rented cars—some without even stopping. At night drug smugglers and gunrunners land on the remote dry lakes and paved roads. Witches' covens, and others who perform odd rites, gather there. And jet aircraft from China Lake break the sound barrier over the ancient lake bed.

I had not known any of this when I entered California via Death Valley in July 1960. More than thirty years later I drove to the park headquarters and looked at the superintendent's report for that earlier month. The thermometer had registered 129 degrees on one day, and for four days, highs of 124 degrees were recorded. There were 10,264 visitors, some of whom might have been lured by articles on Death Valley in *Holiday* and *Desert* magazines. Two mountain lions were spotted in the monument. Because of heat, work on the road ceased on July 29 and would resume again on September 15.

II

The Sierra

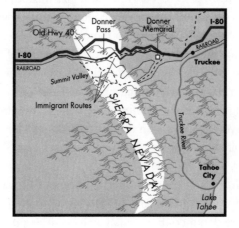

I was led to Donner Pass by Ellen Churchill Semple, the geographic determinist who wrote ninety years ago: "Mountains influence the life of their inhabitants and their neighbors fundamentally and variously, but always reveal their barrier nature." For a migrating, warring, and trading people, such as the Californians, interest in the mountains has centered in the passes; Donner Pass, Miss Semple pointed out, was "the line of least resistance across the barrier mass." It was also a name that came to symbolize horror and dread as emigrants from the east in the mid–nineteenth century poured across the deserts and mountain passes in pursuit of the Califia myth.

Donner Pass lies near the north end of the 430-mile-long mountain range that occupies one-fifth of California's landmass. The Sierra Nevada is the largest and highest mountain range in the contiguous United States. No other mountain range towers so far above its base. The forty-to-eighty-mile-wide range rises 11,000 feet above the Owens Valley at its eastern base and 14,000 feet above the Great Valley at its western base. From 7,000-foot peaks at its junctions with the Cascade Range to the north and the Tehachapi Mountains to the south, the Sierra Nevada rises in a spectacular granite escarpment—called the Giant Scarp—to climax in a serrated rank of 14,000-foot peaks near Mount Whitney.

To pass that scarp, as I did when I entered California, was to encounter either the dramatic end or the beginning of something unusual. It was as if the Sierra Nevada—Spanish for a mountain range covered with snow—had purposely been placed near the continent's western terminus to serve as a bookend, or a gilded entrance into another land.

In an otherwise arid region, the mountains, which capture and hold the precipitation in the form of snow, are the source of the water that is pumped or flows by gravity through huge arteries to the cities and farmlands of California. There are some sixty small glaciers scattered throughout the range. But it is the winter snowpack melting in the spring that spawns the seventeen rivers on the western slope. They supply water to such diverse places as Los Angeles, San Francisco, and Turlock in the San Joaquin Valley.

The search for and claiming and entrapment of water has been an incessant California activity. All but one of the rivers that flow into the Great Valley are dammed. Four smaller rivers descend the eastern slope. The range, dominated for most of its length by a single crest, runs generally north and south. The river canyons are incised along an east-west axis. It is for this reason that it is far more difficult to travel from north to south than through the east-west passes.

The most conspicuous of landscape features, mountains appear to be eternal. As measured by geologic time, however, mountain ranges come and go relatively quickly. The Sierra Nevada is a young, dynamic range; thus its sharp features, unlike the rounded ridges of the Adirondacks.

The range, an uplifted then tilted block of granite, is the product of extensive faulting. Once under water, the Sierra's Giant Scarp, whose size continues to inch upward, is not much more than eight hundred thousand years old. The glacial scouring of Yosemite Valley and Donner Pass took place only during the last few thousand years. Human history, in its recorded form, first took note of the mountains in 1776 when Pedro Font, a Franciscan missionary, saw a snowy range in the distance and recorded, *"Una gran sierra nevada."* What drew Font's attention was the late-spring snowpack.

John Muir called the Sierra Nevada "the Range of Light" because of the reflective qualities of the sparkling snow and granite, which are almost tinsel-like under the right conditions. The Sierra also has its somber side. Dark reds, grays, browns, and blacks indicate extensive volcanic activity, such as pyroclastic explosions and flows of lava. Mary Hill wrote in *Geology of the Sierra Nevada* that "violence—fire, ice, and earthquake" were the hallmarks of the Sierra Nevada.

The mountains are a barrier to the movement of weather, as well as to plants, animals, and humans. Wet Pacific storms dump most of their loads on the lush, western slope; the high peaks intercept most of the precipitation. Life-forms vary accordingly—the yellow pine belt and the spotted owl on the west side need more rain than the sagebrush belt and indigenous inhabitants such as the piñon jay on the east side.

The Indians moved haltingly across the mountains, as did the early Anglos who followed—the first being trapper Jedediah Smith, who made two efforts in 1827 to cross the mountains before finally succeeding that same year. Later, when machines prevailed over all but the most fearsome snowstorms, traffic flowed more fluidly over the eleven paved roads and three railroads extending from Fredonyer Pass in the north to Tehachapi Pass in the south. Still, three of these passes are closed all winter, and the others can be blocked on an intermittent basis.

Nonetheless, it was through these portals and also by sea—yet another formidable barrier to entering the state—that the population took its greatest single surge forward during the first rush into California. A mountain pass called Donner was the chief obstacle, the eye of the needle, the safe passage for some and the burial ground for others. As Oscar Lewis wrote in *High Sierra Country,* "That since the dawn of history the range has wielded a profound influence on the way of life of the inhabitants of a vast area of Western America is undeniable."

Faulting lowered the Sierra crest at Donner Pass, and glaciers carved enough of a notch in the granite to allow wild animals, people, horses, wagons, railroads, and rubber-tired vehicles to pass over the barrier. All the separate paths that led across Donner Pass, be they game trails or an interstate highway, formed an intertwined history of transportation into and out of the state.

The characteristics of the human portion of that history were alluded to by landscape specialist John Brinkerhoff Jackson. He wrote that the national highway systems of such imperial powers as the Incas and the

Romans had three characteristics in common: "first, a vastness of scale; second, a disregard of local landscape features, topographical as well as man-made; and last, a persistent emphasis on military and commercial functions." The railroad and interstate either burrowed through the landscape or pushed it aside to serve military and commercial purposes.

In terms of its physical presence, of the eleven passes bisecting the mountain range, Donner is average in height at an elevation of 7,135 feet above sea level. The pass is by far the busiest and most congested crossing of the Sierra Nevada. In a recent year 5.5 million vehicles crossed the mountains from east to west at this point. (The state does not measure outgoing traffic.) There is a pipeline that carries gas, diesel, or jet fuel, two high-voltage electrical transmission lines, an underground transcontinental telephone cable, overhead telephone lines, microwave relay towers, two fiber-optic cables, an abandoned wagon road, an old coast-to-coast highway, a new interstate freeway, and a transcontinental railway.

Along with wild animals such as deer and mountain lions came the first humans. Successive cultures left their marks in the granite in the forms of petroglyphs just below the summit of the pass. The Martis culture may have been the first, arriving about four thousand years ago. They were distinct from the Washo, who had the misfortune of appearing on the scene a couple hundred years before the whites came tromping across the pass.

Perhaps three thousand Washo roamed the Sierra from Beckwourth Pass in the north to Sonora Pass in the south. Occasionally, Washo hunters pushed over the mountains and into the foothills east of Sacramento. Although based around Lake Tahoe and in the eastern mountain valleys, the Washo were a trans-Sierran people. A few made it all the way to the Pacific Ocean to gather shellfish and bring back abalone and mussel shells with which to trade with Great Basin tribes to the east.

Anthropologist A. L. Kroeber thought the gentle Washo possessed one of the simplest cultures on the North American continent. But the Washo had the physical wherewithal to cover a huge amount of varied territory. Another anthropologist, James F. Downs, wrote:

> Bands of young men are said to have traveled as far south as San Diego to obtain particularly fine obsidian knives from tribes in that region. The Yosemite Valley was well known to

Donner Pass and Donner Lake, with the freeway on the left and the railroad on the right

the Washo who viewed the area from the surrounding mountains. They were afraid to enter the valley because of a belief that the Indians who lived there were sorcerers.

Trouble came from outside the region. Ute Indians and Spaniards from the east raided the Washo, seeking slaves. Spanish-speaking people from the settlements in New Mexico worked mines in the region, then slipped away, leaving no records of their presence. But the Indians remembered in their legends and language. Some Washo words, such as the ones for horse, cow, and money, are similar to the equivalent Spanish terms. Jedediah Smith traveled through Washo territory in the mid-1820s. Twenty years later many more Anglos came.

The Washo did not resist the invasion. They melted back into the forests and watched the approach of the *do ba ah*, whose wagon trains resembled monster snakes. Every once in a while the emigrants would catch sight of a Washo Indian flitting about the woods on circular snowshoes while they and their livestock floundered in the deep snow. They pleaded: please guide us, please feed us. Sometimes the Washo would do this. More often they slipped away. Word had spread about the *do ba ah*.

Near Honey Lake a party of trappers traded with the Indians for furs. A misunderstanding arose. The trappers "let them feel the weight of whips applied by 'Kentucky George,' who understood his business," one of the men recalled. The whites banished the Indians from their camp. When the Indians returned, five were shot and their bodies cremated "to set an example." A tribal history noted that "reports of the cannibalism at Donner Lake convinced the Washo that whites were inhuman." A Washo woman in Carson Valley became sick when she first saw a white person.

In *The Opening of the California Trail,* George R. Stewart, a professor at the University of California at Berkeley during the middle years of this century and the author of a number of popular works on California history, termed the Stevens party of 1844 the "discoverers" of Donner Pass. He took note that an old Paiute Indian whose name the Stevens party understood to be "Truckee" (actually the Indian word for the equivalent of *okay,* which he may have been saying rather than repeating his name) had guided the emigrants toward the pass. Stewart dismissed any claims of discovery or use of the pass by indigenous peoples with the statement "Before 1844, there were only Indians." He defined "discoverer" in the following manner:

> The important question really is, not who first happened to be there, but who first solidly explored, brought back the word, established the route, and annexed the territory to the known world. By all these standards of judgment the men of the Stevens Party should have the credit for the discovery of the Truckee Route and of Donner Pass.

The accomplishments of the Stevens party have been eclipsed by the notoriety of the Donner party, which came stumbling through the area two years later. There are those who believe the pass should be renamed Stevens Pass. The Stevens party was successful under the same conditions that caused the Donner party to make a terrible botch of things. Elisha Stevens was a solitary trapper who knew how to command effectively. He died in obscurity forty years later, when trains were comfortably transporting passengers over the pass.

The Stevens party consisted of twenty-six men, eight women, nine girls, eight boys, and eleven wagons. Nearly half made up the extended Murphy family, who had come to Missouri via Canada and Ireland. There were other Catholics in the party, Stewart pointed out, adding, "From an American point of view this group must be classed as 'foreign.' Nearly all its members had been born in Ireland or Canada." No matter that some had subsequently become citizens of the United States.

The party left the Missouri River in mid-May 1844. Five months later they were at Truckee Meadows, the site of present-day Reno. For thirty-four days the Stevens party bushwhacked up Truckee River Canyon in a tortuous journey that now takes less than a half hour on the interstate. They averaged two miles a day. One day they crossed the winding river in the narrow gorge ten times within one mile. The oxen's feet became so tender from the repeated immersions in the cold water and the sharp rocks that they bled. All night the animals bawled because of the lack of adequate feed in this wasteland. Later emigrants found a detour around the canyon.

As the canyon bottom rose and finally widened into a broad, sagebrush-and-grass-filled valley, the dazzling mountains loomed ahead. Lying perpendicular to the valley, they were a perfect visual barrier.

The party reached Donner Lake, formed at the foot of the pass by a glacial moraine, in mid-November; here, after a brief rest and a scouting trip, they split into three parties. One group, consisting of two women and four men mounted on horses, swung south along the shoreline of Lake Tahoe and then headed west into the heartland of California via the tributaries of the American River. The bulk of the party, including all the children, hurried up Donner Pass in an attempt to beat the heavy snows, there already being two feet of snow on the ground. Hurrying meant the five tortuous days it took to gain twelve hundred feet in altitude from the lake to the pass.

Their route followed the north shore of the windswept lake, whose gray waters reflected the somber color of the low clouds against the backdrop of dark-green pine trees. Here the Indians hid and watched the serpent crawl up the granite rock. When figures climbed out of the back of the monster to lighten its load, it looked as if the beast had given birth to little monsters. The Washo puzzled over the unfamiliar tracks.

The party came to a long, horizontal rock that blocked their passage. There was a narrow rift in the rock just wide enough to allow the passage of an unyoked ox. The oxen were led through one at a time, re-yoked, and hitched to chains that were attached to the wagons. With the beasts of burden pulling and the people pushing, one wagon at a time was hauled over the obstacle.

The last few hundred feet were a gentle grade through stunted trees. The actual summit was but a slight hump, where they paused to gaze back at the hard-won ground and the elongated lake that was partially obscured by snow flurries. The way down the western slope passed a small rock-ribbed lake and then emptied into the broad expanse of Summit Meadow, where the party rested before proceeding on another twenty miles to the headwaters of the Yuba River. Here they were trapped by winter.

The third party, consisting of two men and a youth, the latter a seventeen-year-old named Moses Schallenberger, returned to the eastern end of Donner Lake where they had volunteered to spend the winter guarding the wagons and goods that had to be jettisoned for the crossing. Schallenberger was one of seven children born to an Ohio couple who died when Moses was six years old. He went to live with a sister and brother-in-law, who were members of the Stevens party. Schallenberger's account of his winter ordeal has survived. He wrote:

> The morning after the separation of our party, which we felt was only for a short time, Foster, Montgomery and myself set about making a cabin, for we determined to make ourselves as comfortable as possible, even if it was for a short time. We cut saplings, and yoked up our poor cows and hauled them together. These we formed into a rude house, and covered it with rawhides and pine brush. The size was about twelve by fourteen feet. We made a chimney of logs eight or ten feet high, on the outside, and used some large stones for the jambs and back. We had no windows, neither was the house chinked or daubed, as is usual in log-houses, but we notched the logs down so close that they nearly or quite touched. A hole was cut for a door, which was never closed. We left it open in the day-time to give us light, and as we had plenty of good beds and bedding that had been left with the wagons, and were not afraid of burglars, we left it open at night also. This cabin is thus particularly described because it became historic, as the residence of the ill-fated Donner party in 1846.

As soon as the little cabin was completed, it began to snow. It snowed for one week. They killed two starving cows and hung up the meat, which immediately froze. Soon the cabin was nearly covered "and we began to fear that we would all perish in the snow." They fashioned snowshoes and went hunting. "There was no game. We went out several times but never saw anything. What could we expect to find in ten feet of snow?" As November turned to December, Schallenberger wrote, "Death, the fearful, agonizing death by starvation, literally stared us in the face."

They decided to make a break for California. They barely made it to the top of the pass. The two mature men were stronger than the youth, who was exhausted and doubled up with cramps. He would hold them up. All might die. Schallenberger decided to return to the cabin. The men continued on. "The feelings of loneliness that came over me as the two men turned away I cannot express, though it will never be forgotten, while the 'Good-by, Mose,' so sadly and reluctantly spoken, rings in my ears today." The young man returned to the cabin, so exhausted that he needed both hands to raise his legs over the nine-inch doorsill.

Schallenberger was starving. He set traps, using slices of flesh and fat from the skulls of the cows. He caught a mangy coyote, which tasted horrible no matter whether he roasted or boiled the meat. But it was food. He took three days to consume the coyote. Then Schallenberger caught two foxes. They were delicious. He had just enough coffee for one cup, and that he saved for Christmas. "My life was more miserable than I can describe. The daily struggle for life and the uncertainty under which I labored were very wearing," he lamented.

In the belongings that had been left behind was a plentiful supply of books. He read late into the night beside a roaring fire, and then slept as long as possible the next morning. His reading matter had a decidedly English flavor: *Lord Chesterfield's Letters to His Son*, and the works of Byron. Schallenberger talked aloud to himself. "What I wanted most was enough to eat, and next the thing I tried hardest to do was to kill time. I thought the snow would never leave the ground, and the few months I had been living here seemed years."

Near the end of February, as Schallenberger was standing near the cabin, the youth thought he saw an Indian approaching. No, it was no Indian; the face was vaguely familiar. It was a member of the party. His sister had asked this man to search for him. The next morning they departed and made it safely over the pass and eventually into the Great Valley, as did all other members of the Stevens party. Schallenberger was to become a moderately prosperous farmer near San Jose; he died in his eighties. The ridge south of the small cabin was given his name.

Though the cabin lay deserted, the Indians would not enter the open door. On August 25, 1846, Edwin Bryant, a Kentucky newspaperman, paused at the cabin. He noted:

> Just before we struck the shore of the lake at its lower or east-
> ern end we came to a tolerably well-constructed log-house,
> with one room, which evidently had been erected and occu-
> pied by civilized men. The floor inside of this house was cov-
> ered with feathers, and strewn around it on the outside, were
> pieces of ragged cloth, torn newspapers, and manuscript let-
> ters, the writing in most of which was nearly obliterated.

Bryant's party camped at the western end of the oblong lake that night. Two men climbed a peak. They saw a female grizzly bear and her cubs. They killed a fat deer, hauled it to camp, and ate the succulent meat. "Nothing can exceed the almost awful profoundness of the solitude by which we are surrounded," noted Bryant in his book, *What I Saw in California*.

In the nearly two years since the passage of the Stevens party, a trail had been beaten into the ground by the hundreds of emigrants. Bryant and his party departed on this trail as soon as the sun cleared the ridges to the east. The scenery was less ominous in the daytime; Bryant, as did other easterners who were at a loss for fresh images, compared the Sierra's unique brand of beauty to "Alpine and Elysian scenery." At the bottom of the grade, Bryant paused and contemplated the pass:

> Standing at the bottom and looking upwards at the perpen-
> dicular, and in some places, impending granite cliffs, the
> observer, without any further knowledge on the subject,
> would doubt if man or beast had ever made good a passage
> over them. But we knew that man and horse, oxen and
> wagon, women and children, had crossed this formidable
> and apparently impossible barrier created by Nature
> between the desert and fertile districts on the coast of the
> Pacific.

The party, mounted on horses and with no wagons to hold them back, tackled the pass with great alacrity. The horses leaped from crag to crag, according to Bryant's account, and easily climbed "nearly perpendicular precipices of smooth, granite rocks." There was some momentary difficulty with a mule whose slide backwards was arrested by a rock, but they easily gained the summit in the morning hours of that same late August day.

The pass held different fates for different people. One month later an elderly man died at sunset at the pass. "Strange as it may seem," remarked his son, within an hour his wife gave birth to a grandchild. She was "a lively, little girl," he said. A chronicler of that event noted "all chances of human existence seemed to concentrate in the pass."

Two months after Bryant's easy passage, the Breen family arrived at the foot of the grade along with others in the forward element of the Donner party. Three feet of snow covered the ground on November 1. Two days later, there were five feet of snow. Two men struggled to the summit, then returned. The party decided to wait. It snowed heavily that night. They had lost their chance to escape the vise of winter.

Retreating a few miles, they went into winter quarters at the east end of the lake. The Breens, who arrived first, claimed the cabin. The Kesebergs erected a lean-to against one wall of the cabin, thus saving themselves some effort. It snowed.

Since July 19, when the Donner party took what they thought was a shortcut across Utah, the emigrants were riven by bad decisions, bad luck, and violent tempers. The ill-advised Hastings Cut-Off proved to be a disaster in the making, and, along with lengthy rest stops, greatly prolonged the time it took to reach the mountains. Still, it could have been a season of late or little snowfall instead of being an early winter of mammoth proportions.

There were no experienced mountain men along, as there had been with the Stevens party. They never could catch up with Lansford Hastings, who wrote the book that touted his route. Hastings successfully guided a large group of emigrants over the mountains just ahead of the Donner party.

George Donner, at the age of sixty-two, was a nominal leader. Each looked out for himself, his family, or his look- and talk-alikes. Class and ethnic and religious prejudices were rife. The rich had servants to cook and drive their wagons for them. The poor walked. There were those of Anglo-Protestant, German, and American Indian descent. One-third of the group were Irish Catholics in an age when Know-Nothingism, a political movement opposed to the foreign-born and especially to Catholics, held sway. A survivor later wrote with some bitterness, "I know, there was some noble souls among them, but the majority were a miserable selfish set."

Even the historians who chronicled the Donner party were a splintered lot. They mirrored their eras, as historians tend to do. Stewart, writing in the mid-1930s in *Ordeal by Hunger,* labeled the Irish, even though they were American citizens, as "foreigners." He termed the seven children of the Catholic Breens "true Irish prodigality." The Indians were "skulkers" and "cowardly" and those of mixed parentage were "mongrels" or "half-breeds" or "little Mexicans." Writing in a 1960 supplement to his book, Stewart was disparaging of the first popular history of the Donner party written in 1879 by C. F. McGlashan, a Truckee newspaper editor. The professor said that the editor was not a trained historian and had indulged in "nineteenth-century oratory." In *Winter of Entrapment,* published in 1992, Joseph A. King, a retired California community college teacher with Irish roots, cited Stewart's "strongly racist view" toward the Irish and those of mixed parentage, whom King tended to elevate to hero status.

As the bickering group made its way across the deserts of Utah and Nevada toward the foothills of the Sierra Nevada they lost wagons, horses, oxen, personal possessions, and five lives in a series of incidents that bred suspicions of theft, betrayal, and murder. They desecrated an Indian burial ground and left one of their number to die alone in the desert, refusing to walk back or loan horses for others to search for the ailing man. They were damned before they reached Donner Lake, where they camped as they had journeyed—in separate groups. Eighty-one men, women, and children were sealed off from the warm bosom of California and from retreat by the deepening snow. Only forty-six would live to see the promised land. Women stood a much better chance to survive than men.

The incessant search for food began. In mid-November, William H. Eddy, a carriage maker from Illinois, walked along the shores of Donner Lake. He spotted a large grizzly bear digging for roots. Eddy fired. The bear reared up in pain and charged. Eddy and the bear circled a tree. The wounded bear became confused. Eddy sneaked up behind it, fired a shot at close range, and then clubbed the bear to death with the rifle. This was the last fresh meat the Donner party obtained from hunting.

The fearsome grizzly bear (*Ursus horribilis*) was at the top of the food chain until the Europeans came along with rifles. The bear ranged throughout all regions of the state except the extreme northeast and the deserts. What would become Los Angeles, San Francisco, and Sacramento were then grizzly bear habitats. Because of the great diversity of the landscapes, bears differed between regions in size, color, and skull characteristics. The number of subspecies was not known with precision, since no mammalogists were at work within the state at that time, but by the late nineteenth century it was known that the California variety, named *californicus,* was distinct from grizzly bears found in the American Rockies, Canada, and Alaska.

The Indians feared grizzlies in much the same way they feared the whites: the Washo names for an unpredictable wild beast and a white man were the same. The earliest recorded sighting of a California grizzly was in 1602 by a Spanish explorer who saw bears feeding on a whale carcass on the beach at Monterey. Spaniards' and Mexicans' herds of horses and cattle became wild and spread throughout the state, thus providing an abundant food source for the grizzlies. The bears multiplied accordingly.

For the first time in North America, bears (both black bears and grizzlies) provided sport. The Mexicans hunted the bears, as the Indians had been unable to do, on horseback with guns, lances, and *reatas.* There were organized bear hunts and man-bear and bull-bear contests. The bears were goaded with nails on a stick to attack the bulls, who usually lost. One bear fought three bulls, one at a time, and succumbed only when exhausted by a fourth encounter.

With the coming of the North Americans, what little sport there was went out of bullbaiting, and a commercial element was introduced. Bear-bull fights were taxed by the state. Bears were put on display and made to perform anthropomorphic tricks. And they were shot by bounty hunters, farmers, meat hunters, sportsmen, and just about anyone who could raise a heavy-caliber rifle to his shoulder. Grizzlies became the major four-legged nemesis of the early settlers. They attacked oxen on Donner Pass, devoured livestock in the Great Valley, and killed or maimed the careless hunter no matter where he was.

The hunters retaliated, and the mountains were stained red. One bear hunter boasted of shooting two hundred grizzlies in his lifetime, another said he had killed more than that number in a single year, and three hunters in Southern California were said to have felled one hundred and fifty bears in one year. The bear population collapsed. The grizzly became extinct in the 1920s, the last verified sighting being in the southern Sierra Nevada. It was there, in 1864, that William Brewer had spotted a female grizzly and her two cubs. He wrote of the encounter:

> She was enormous—would weigh as much as a small ox. After we looked at her a few minutes we all set up a shout. She rose on her hind legs, but did not see us, as we sat perfectly still. We continued to shout. She became frightened at the unseen noise, which echoed from the cliffs so that she could not tell where it came from, so she galloped away with the cubs.

The grizzly bear was reduced to an emblem, a stand-in for the real thing. It is the official state animal. Its likeness appears on state monuments, flags, and the state seal. On the seal, the grizzly has been reduced to the size of a puppy dog standing beside the seated figure of Minerva, the Roman goddess of political wisdom and martial prowess. The bear (or bruin) is the mascot of the two principal University of California campuses. More than five hundred landscape features, including seven rivers, carry the name *bear*. In one small Sierra basin alone there are Bear, Little Bear, and Cub Lakes. The day I hiked through that basin, I heard rifle shots echo off the granite walls. It was hunting season.

The bear meat was consumed quickly by the emigrants. As the snow piled up alongside the cabin, Patrick Breen, fifty-one years old and suffering from kidney stones, began to record events in a diary, the only surviving document that was produced at the time and not later to fit the needs of contemporary journalism or the vicissitudes of history. The Breen family operated as a unit. They took care of themselves first, others were next in line. Much of what Breen wrote about concerned the weather, but there are glimmers of less superficial matters on the eight small sheets of letter paper that were folded to make a journal of thirty-two pages.

The diary began with a notation made on Friday, November 20, 1846, that there had been two unsuccessful attempts to cross the pass since the party had settled at the eastern end of the lake. It snowed for eight days straight, then cleared. The snow almost disappeared from the valley, then accumulated again. There was more snow, a thaw, rain, sleet, more snow, and then the comment at the end of the month that "no liveing thing without wings can get about."

On December 1, it was still snowing. The next day the snow was over six feet deep at the cabin and much deeper at the pass. On December 9, the family was in good health, thanks to a steady supply of beef from their cows that had been slaughtered and frozen. Breen recorded the first death in mid-December.

"Saw no strangers today from any of the shantys," Breen noted on December 18. The Irish-Catholic Breens took in an Irish-Protestant family. Four more deaths were noted on December 21. "Offered our prayers to God this Cherimass morning the prospect is apalling but hope in God *Amen.*"

On New Year's Eve, Breen inscribed another prayer in his diary and noted, "Snow Storms are dredful to us." Dogs and animal hides were being consumed. Children began dying. On January 26, 1847, Breen noted: "provisions getting very scant people geting weak liveing on short allowances of hides." One family seized another's goods as payment for the food they had supplied. Milt Elliott died on February 9. A relief party from California arrived with some supplies.

On February 25, the unspoken thought emerged on a "fine and sun-shiny" day at the lake camp. "Mrs. Murphy says the wolves are about to dig up the dead bodies at her shanty, the nights are too cold to watch them, we hear them howl." Mrs. Murphy added that she "thought she would commence on Milt and eat him. I don't [think] that she has done so yet, it is distressing. The Donnos told the California folks that they [would] commence to eat the dead people 4 days ago, if they did not succeed that day or next in finding their cattle then under ten or twelve feet of snow. I suppose they have done so ere this time."

On February 27, an Indian passed by the cabin with a heavy pack on his back. He would not allow Breen to approach him and left his offering on the snow—"5 or 6 roots resembling onions in shape taste some like a sweet potatoe, all full of little tough fibres." The Indian departed silently on snowshoes. This was the first Indian the emigrants saw at the lake.

Thomas Sanchez imagined what the Indians were witnessing in his novel *Rabbit Boss*:

The Washo watched. The Washo watched through the trees. The Washo watched through the trees as *they* ate themselves. His chin lifted, head cocked rigid to one side as he watched through the leaves, the branches, the bark. The waiting winter light fell flat on the trees, on him, on *them.*

Breen ended his diary on March 1, the day the second group of rescuers reached the camp. The day was "fine and pleasant." The Californians noted signs of cannibalism—a mutilated body in the Murphy cabin, body parts scattered about the Donner camp five miles distant, and the pink stains of blood in the snow. There was a foul odor. The Breens departed. They tasted human flesh on their arduous trip across the Sierra, but did not write about it. The family survived and later prospered in San Juan Bautista, a Catholic mission town.

Louis Keseberg, who would be the last to be rescued, moved into the cabin originally built by Schallenberger and subsequently occupied by the Breens. It was Keseberg who had desecrated the Indian burial site back in Sioux territory. From that point on, the wealthy German immigrant became the party's pariah. It didn't help that he was well-educated, could speak four languages, and had a fiery temper. Keseberg, who had a wife and two children, was hampered by an infected foot. Stewart, who wrote between the two world wars, made Keseberg the villain of the tragedy. McGlashan and King took his side. The newspaper editor sought Keseberg out when he was an old man and interviewed him. Keseberg recalled:

> For nearly two months I was alone in that dismal cabin. No one knows what occurred but myself—no living being ever before was told of the occurrences. Life was a burden. The horrors of one day succeeded those of the preceding. Five of my companions had died in my cabin, and their stark and ghastly bodies lay there day and night, seemingly gazing at me with their glazed and staring eyes. I was too weak to move them had I tried. The relief parties had not removed them. These parties had been too hurried, too horror-stricken at the sight, too fearful lest an hour's delay might cause them to share the same fate. I endured a thousand deaths. To have one's suffering prolonged inch by inch, to be deserted, for-

saken, hopeless; to see that loathsome food ever before my eyes, was almost too much for human endurance. I am conversant with four different languages. I speak and write them with equal fluency; yet in all four I do not find words enough to express the horror I experienced during those two months, or what I still feel when memory reverts to the scene.

Keseberg's one-year-old son had died on January 24. His wife and three-year-old daughter, Ada, departed with the first rescue party. Keseberg contemplated suicide, but the thought of his wife and child stayed his hand. Four days after Keseberg's provisions gave out, he began to eat human flesh.

Keseberg was one of the few members of the party who spoke openly about cannibalism. A questionable newspaper account at the time said that he enjoyed eating human flesh. Keseberg told McGlashan, "I can not describe the unutterable repugnance with which I tasted the first mouthful of flesh. This food was never otherwise than loathsome, insipid, and disgusting."

Stewart wrote with tongue in cheek that "with a certain lack of good taste" Keseberg later opened a restaurant in Sacramento. At the time of the tragedy there was a fair amount of joking at the expense of the emigrants among those who participated in the rescue. George McKinstry, the sheriff of Sacramento, wrote a navy captain who was on one of the rescue missions, "I advise you both to look out for those Man eating Women, from what I can learn from Glover they prefer that kind of meat in larger than *Nine inch* pieces too." A fair amount of money was raised to rescue the Donner party. As with any large-scale mercy effort, there were those who were well-intentioned, those who were out to make money, and those who were outright thieves.

Keseberg was extracted from his pit of despond in late April. There was one final horror remaining for Louis Keseberg. The rescue party would have little to do with him, and he followed in its tracks across the Sierra. One night, as he approached a campsite that had been used by others, he stopped. Richard Rhodes described the scene in his novel, *The Ungodly*:

He noticed a piece of calico sticking from the snow. It lay within reach of his hand and waiting for the water to boil in his coffeepot he began to play with it. It could not be a kerchief. He gave it a tug and the snow loosened around it. It was larger than he had thought. He saw the pattern of the calico then and something broke loose in him and he stood and with both hands jerked on the calico and the frozen body of his daughter Ada sprang forth from the snow into his arms.

Keseberg thought he had been born under an evil star. The Kesebergs arrived childless in California. They had eight daughters there; but when McGlashan interviewed Keseberg in 1879, his wife was dead, only four daughters were alive, and none had long to live. Two were subject to epileptic seizures. One after another of Keseberg's business attempts failed. Sixteen years later Keseberg died penniless at the age of eighty-one in the Sacramento County Hospital, where under "Remarks" on his hospital record it was noted, "A last survivor of the Donner party."

In June 1847, Gen. Stephen Kearny, fresh from victory over the Mexicans in California, crossed Donner Pass in deep snow and descended to the lake. The hardened soldiers accompanying Kearny found the spectacle at the cabins "revolting & distressing," according to one member of the party. The general ordered a major to bury the remains. The soldiers put some of the body parts in a pit and set fire to the Schallenberger-Breen cabin. They departed hastily, leaving two other cabins untouched and other bones and human hair lying scattered about.

It was virtually impossible to eradicate this stain upon the landscape, although its retelling at Donner Memorial State Park has been sanitized and turned into a paean to the state's white pioneers. Twenty years later the owner of a nearby tavern hauled out a sack of bones and displayed them for passing travelers. And, of course, there is the usual California story of gold buried at the lakeside, which was given currency by the unearthing of some coins in 1891. To this day, visitors scan the shoreline with metal detectors.

What had gone before was nothing to what would occur in 1849 in terms of sheer numbers, a commodity that California came to excel in. Between 1841 and 1848 approximately 2,700 emigrants made it overland to California. In the gold rush year of 1849, nearly 50,000 men and a few women and children completed the overland journey via northern, central, and southern routes. An additional 41,000 came by water. Gold rush historian J. S. Holliday described this mass movement of treasure seekers:

> Never before had this country, or any other, experienced such an exodus of civilians, all heavily armed or intending to purchase rifles and pistols, mostly young men on the road for the first time, many organized into formal companies, others alone or with a few friends from their neighborhood. Impatient, curious, somewhat fearful of the uncertainties and dangers ahead, yet buoyed by their common expectations.

If the seekers after gold came by land, they crossed a desert first and then a mountain range no matter what route they took. They followed the bleached bones of bodies picked clean by wolves and ravens and then buffed and polished by the sun and wind. The abandoned spoor of past lives and the exhausted tracks of people and livestock marked the trails.

Some who chose the Truckee River route of the California Trail found that when they reached Donner Lake, whether because of its gruesome reputation or overcrowding, they preferred to avoid the lake and take one of two detours just to the south over the "summit of the long dreaded Sierra Nevada," as one emigrant put it. Others went out of their way to gaze upon such relics of the Donner party as women's clothes, long hair, bleached bones, and the stumps of trees from eight to twelve feet high that demonstrated the depth of the snow at the time they had been cut. The curiosity seekers referred to it in their diaries as "Starvation Camp" or the "canible cabins."

Confronting all of the emigrants was the shimmering granite wall of Donner Pass or the black lava flows and volcanic ash that were spewed across the two other nearby passes. The three routes looked impassable from a distance. A phalanx of three mountains over eight thousand feet high linked by the crest of the Sierra Nevada blocked the way. One gold seeker gazed upward on August 21, 1849, and said, "The main ridge of the Sierra now before us seems so high and cloud like that we can have no chance of passing it with our waggon but the morrow will tell." The three routes snaked between Donner Peak, Mount Judah, and Mount Lincoln, as they would come to be called.

Each pass had its disadvantages. If they chose not to mount the most direct route, which was Donner Pass proper and its granite outcrops that had to be circumnavigated, then it was up Coldstream Valley on the south side of Schallenberger Ridge and then a long, steady grade to Coldstream Pass, seven hundred feet higher than Donner and still snowbound later in the season. A half mile farther south at the same elevation was Roller Pass, which had a steep final pitch, up which wagons had to be winched. The actual passage proved less fearsome than expected. Most made it in a single morning.

The summit was not only a physical divide but a psychological one, as well. One traveler wrote:

> Having reached the highest of the last mountain ranges so we could look forward from its summit to the land of our dreams, toil, and hope, we gave three loud cheers. Looking down the steep gorge whence we had come, we bade adieu to its dark avenues, towering cliffs, sequestered shades, bright waters and melancholy scenes.

It was but a short ride downhill to Summit Valley, a lush mountain meadow where there were thick grasses and water for starved animals and flowers blooming in August for those who were weary of desert

The freeway and the immigrant trail that crossed the granite incised by iron-rimmed wagon wheels

travel. Here they were truly in the promised land, and here they held their rites of passage. One emigrant noted:

> We passed through a grove of woods and then emerged into a beautiful valley and encamped. We were all in the most joyous & elated spirits this evening. We have crossed the only part of road that we feared, & that without any breakage, loss or detention. I had but one & only one bottle of "cognac" that was in our camp, & which I had managed to keep since leaving the Old Dominion. This I invited my mess to join me in, & which invitation was most cordially accepted. When lo & behold, upon bringing it out, it was empty—yes positively empty. The cork was bad & with numerous joltings, it had gradually disappeared. This was a disappointment many of us will not soon forget.

The diaries of the emigrants tended to peter out at this point, as if they were overwhelmed by the new landscape, suddenly became too busy, or chose to wipe the slate clean. As they descended from the mountain pass, they dropped out along the way to make quick forays in search of gold at such places as Green Horn Creek. "It deserves its name from a company that worked here last fall said to have taken out 50 lbs of gold just below the crossing," noted an emigrant, who thought he had arrived in "the golden land of gold."

By 1850 there was a "restaurant" in Bear Valley, where on September 19, a traveler found a copy of the July 9 issue of the *New York Tribune* in which there was the story of the death of President Zachary Taylor. This news had most likely beaten the overland emigrant to California via ship to Panama, mule across the isthmus, ship to San Francisco, and thence by a combination of boat and horseback into the mountains. By this time much of the grass had been grazed along the trail, and California was showing the first signs of wear.

History in California now speeded up and was played to the strident beat of Manifest Destiny. Gold was discovered at Sutter's Mill on January 24, 1848. Nine days later (the negotiators of the Treaty of Guadalupe Hidalgo being unaware of the discovery) Mexico ceded Alta California and other territories to the United States. On September 9, 1850, California became a state.

Modern California was founded on greed and violence. One gold seeker wrote, "Money is our only stimulus and the getting of it our only pleasure. Never was any country so well calculated to cultivate the spirit of avarice." Walton Bean and James J. Rawls wrote in *California,* the most comprehensive and relevant of many linear histories of the state, "The gold rush was the product of a kind of mass hysteria, and it set a tone and created a state of mind in which greed predominated and disorder and violence were all too frequent." No gold rush stories better illustrate these twin legacies than the legend of Gold Lake and the violence at Rich Bar.

Gold Lake was thought to exist in the Sierra Nevada at some point near Donner Lake, or possibly to the north. That its exact location was never determined and some are still searching for it is an indication of the enduring quality of the myth. There was a golden lake, *Laguna del Oro,* in the Andean mountains of Colombia that was sacred to the Chibcha Indians at the time of the Spanish conquest. The seed of that legend wafted north and was watered by the fertile imaginations of the California gold seekers, who were not having much luck in the diggings.

The obsession with a lake around whose shores were nuggets of gold and fierce Indians arose in a number of different places in different forms at about the same time. A trapper who accompanied the Stevens party and guided later emigrants related the story, and an expedition was hastily organized to find such a lake. A black man supposedly returned to South America fabulously wealthy from all the gold he had gathered from the lake. An Indian told a miner of the lake, the miner supposedly found it, and was forced to flee when attacked by Indians. A search party led by the miner combed Last Chance and Humbug Valleys, but to no avail.

The stories spread, became more general, and acquired greater variations on the main themes. Not all were taken in. From the gold fields William Swain wrote his family in western New York State on July 19, 1850:

> There is the greatest swindling, wild and visionary buying and selling in these claims and the gold business generally that ever existed. He who comes into the mines runs great risk of being outrageously shaved.

One report in circulation among the miners here is with regard to a certain Gold Lake in this vicinity. One man who has been to these diggings says that he could take three ounces to the pan and average nine out of ten. This man guided out a company of one hundred men, $50 each: but his searches for such diggings have proved a failure, like all other reports of this kind. Hundreds of miners have been running over the mountains in pursuit of these golden lakes.

A miner overheard a conversation about a lake where gold could be scooped up by the handful. This miner related the story to Thomas Stoddard, an Englishman who had been wounded in the leg some ten years earlier. Stoddard had immigrated to Pennsylvania where he had worked as a schoolteacher and a newspaper editor before coming to California during the gold rush. A few weeks later Stoddard appeared in both Marysville and Nevada City with some gold and the story that, while lost in the mountains somewhere to the east, he had come upon a lake that was lined with gold. He was attacked by Indians, wounded in the leg, and fled with what little gold he could easily carry. Some briefly wondered about the wound being old, but cupidity was joined to avarice.

An expedition was organized. It was described by one onlooker as "a motley crowd following a crazy man." The group set off from Nevada City and acquired various hangers-on from the mining camps it passed through on the way toward the crest of the Sierra Nevada. The ragtag army numbered more than five hundred as it traipsed along the eastern slope of the mountains, stopping at one lake after another.

Meanwhile, the stories intensified like a California wildfire feeding on its own flames. Indians supposedly carved seats and couches out of huge gold nuggets at lakeside. There was a lake where gold could be scraped off the ground. Wise men and fools alike succumbed to the mass hysteria. Farmers and businessmen who had previously forsaken the get-rich-quick life for a steadier climb toward prosperity dropped their plows and boarded up their stores. Marysville was depopulated almost instantly, despite skyrocketing prices for mules and mining supplies. "Some thousands" left the Sacramento Valley, and the roads to the northeast were "lined with the infatuated multitude," one newspaper reported.

A timid newspaper editor wrote of the madness that seized whole towns:

When a conviction takes such complete possession of a whole community, who are fully conversant of all the exaggerations that have had their day, it is scarcely prudent to utter a qualified dissent from that which is universally unquestioned and believed.

Back up in the Sierra, Stoddard became lost. He babbled incoherently and was described as being "tattered and torn, barefooted and uncomfortable." He was a pitiful sight. Most of Stoddard's followers favored lynching the Englishman, but a few argued for mercy. They won the day. Stoddard wandered out of the mountains. The expedition disbanded, but smaller parties and individuals fanned out to continue the search, described by one observer thusly:

> The mountains swarmed with men, exhausted and worn out with toil and hunger; mules were starved, or killed by falling from precipices. Still the search was continued 'til the highest ridge of the Sierra was passed. The disappointed crowds began to return, without getting a glimpse of the grand "disideratum," having had their labor for their pains.

As Gold Lake fever subsided, there were spin-offs. A man reported a "silver-spouting volcano" to the northeast, thus setting off a stampede in that direction. Peter Lassen, who gave his name to another volcano, chased that chimera and was killed by Indians. A lake was named Gold Lake, but its most valuable yield proved to be mountain trout and serenity for summer visitors. The northern Sierra was thoroughly explored for the first time by the newest culture to dominate the region, and small deposits of gold were soon found in more likely places, such as Rich Bar.

Rich Bar sprang instantaneously into existence in the summer of 1850 when a prospector found $3,000 worth of gold on the banks of the North Fork of the Feather River. Among those who flocked to the wide gravel bar in the deep canyon were a New England doctor and his wife, Louisa Clapp. She penned twenty-three letters to her sister in the East during the thirteen months she and her husband were in Rich Bar and the neighboring settlement of Indian Bar. After the letters were published in a San Francisco periodical, Bret Harte and Mark Twain helped themselves to material from them for their more famous stories of the era.

Clapp saw almost everything there was to see in a gold-rush mining settlement. On the eve of her departure she wrote, "I certainly fancied that I had a right to brag of having taken a full view of that most piquant specimen of the brute creation, the California 'Elephant.' "

The elephant became the symbol of the gold-rush. "To see the Elephant" was a phrase used in many gold-rush diaries, journals, and letters. Rocks along the overland trails, stationery used in the gold fields, posters containing the Miners' Ten Commandments, and souvenir items produced in San Francisco were emblazoned with a likeness of the giant beast.

From Big Bar on the American River in May 1851, a miner wrote to a friend at home on a piece of stationery whose letterhead bore the likeness of an elephant:

> If you have never seen the Elephant take a peep at the above or else amuse with a trip across the desert wilds that lay between the rocky and sierra mountains. Travel in the midst of an epidemic that is slaying its hundreds. Stand guard all night when all the elements seem to be at war while it not only rains but pours down. And you will be better able to realize what seeing the "Elephant" means.

The phrase had originated in the East during the previous decade when a farmer longed to see an elephant at a time when they were still a novelty. The story continued:

> When a circus, complete with elephants, came to a nearby town, he loaded his wagon with eggs and vegetables and started for the market there. En route he met the circus parade led by the elephant. The farmer was enchanted but his horses were terrified. They bucked, pitched, overturned the wagon and ran away, scattering broken eggs and bruised vegetables over the countryside. "I don't give a hang," said the farmer, "I have seen the elephant."

Referring to the account of a 1841–1842 trip across Texas to Santa Fe by a New Orleans newspaper editor named Kendall, George P. Hammond, the director of the Bancroft Library at the University of California at Berkeley, explained the evolution of the phrase in a 1964 speech to the California Historical Society:

> It is clear that by Kendall's time the phrase had already taken on its secondary and more famous meaning, namely, of having seen and been disappointed, of having labored and failed. The idea of anticipation, of seeing the supernatural, of attaining the wonderful was undoubtedly the basic or original signifi- cance of the term, but as bright shiny dreams tarnish, so did the optimism of the original meaning. The hard realities of life rarely come up to man's hopes.

Nor to a woman's expectations, I should add. At the age of thirty, Louisa Clapp sailed to San Francisco in 1849 to see the elephant. She had been raised in Amherst, Massachusetts. Her parents died while she was young, but her guardian saw to it that she received a proper educa- tion for a young lady. On the verge of spinsterhood, Clapp married a medical student who was five years younger. She was described as "small, fair, and golden haired, delicately beautiful and not physically strong." Frequent headaches were among her afflictions. But Clapp loved travel and adventure.

The couple lived for a while in San Francisco and then the Sacra- mento Valley before setting off on mules for Rich Bar, where they arrived in September 1851. Clapp was one of three women in the settlement. A fourth died that same month of peritonitis. The couple stayed at the Empire, a two-story hotel. Clapp had difficulty sleeping, what with the loud swearing that went on all night in the downstairs bar. She was enchanted by the colorful language. For instance, the phrase "honest Indian" indicated doubt. "Whether this phrase is a slur or a compliment to the aborigines of this country, I do not know," she wrote.

They moved into a newly built log cabin at Indian Bar, within walking distance of Rich Bar. She described the frenetic activity:

> At every step gold diggers or their operation greet your vision. Sometimes in the form of a dam; sometimes in that of a river, turned slightly from its channel, to aid the indefatigable gold hunters in their mining projects. Now, on the side of a hill you

will see a long-tom,—a huge machine invented to facilitate the separation of the ore from its native element; or a man busily engaged in working a rocker,—a much smaller and simpler machine, used for the same object; or more primitive still, some solitary prospector, with a pan of dirt in his hands, which he is carefully washing at the water's edge, to see if he can "get the color," as it is technically phrased, which means literally the smallest particle of gold.

The scenery was strange and wondrous to this New Englander. "Not a spot of verdure is to be seen in this place; but the glorious hills rising on every side vested in foliage of living green, make ample amends for the sterility of the tiny level upon which we camp." Wildlife was abundant. A miner shot a female grizzly bear and captured her two cubs. He sold one for fifty dollars. "They are certainly the funniest-looking things that I ever saw, and the oddest possible pets," wrote Clapp.

There were "darker shades of our mountain life," she said, such as earthquakes, landslides, and a hanging. The hanging victim, a Swede, had stolen some gold dust. After a quick trial, the jury found a nearby tree and hanged the petty thief. "In truth, life was only crushed out of him, by hauling the writhing body up and down several times in succession, by the rope which was round a large bough of his green-leafed gallows." The body was left hanging from the tree, and the falling snow formed a "soft, white shroud" upon the dead man.

Indians and Mexicans were about; the Chinese did not arrive at the diggings until later in 1852. "As well as I can judge, there are upon this river as many foreigners as Americans. The former, with a few exceptions, are extremely ignorant and degraded." As for the latter, "the majority are of the better class of mechanics." There were also farmers, sailors, merchants, four doctors, and one lawyer. "Our countrymen are the most discontented of mortals. They are always longing for 'big strikes,' " Clapp observed.

When it became apparent in the spring of 1852 that there would be no such bonanzas, racism reared its ugly head. An Anglo stabbed a Mexican, who had politely asked that a monetary debt be repaid. No action was taken against the assailant. "Foreigners," meaning Mexicans as well as others, were barred from mining at Rich Bar despite the fact that many Mexicans, by virtue of the terms of the Treaty of Guadalupe Hidalgo, were

United States citizens. The Anglo miners at Rich Bar and elsewhere followed the lead of the state legislature, which had imposed a mining tax on foreigners.

Then violence exploded in Rich Bar on July 4, 1852. Miners "drunk with whiskey and patriotism" attacked a group of Mexicans, and two or three were seriously hurt. A few days later a group of armed Mexicans entered the settlement. A Mexican stabbed an Anglo and swam safely across the river amid a hail of bullets. The Anglo, a man of Irish descent, was in the company of a Mexican woman. He was laid out in a bakery and died with the distraught woman at his side. Miners swarmed down from the hills armed with rifles, pistols, clubs, swords, knives, and other weapons, while others barricaded themselves in a saloon. The rumor circulated that the Mexicans were out to kill all the Americans, who cried for vengeance and the blood of foreigners.

From the hillside, Clapp watched the action unfold. There were shots. The mob ran. One wounded man was led to a log cabin, another to a saloon frequented by Mexicans. A gun had accidentally discharged in a struggle, wounding an Anglo and an Argentinean. The latter eventually died of his wounds.

A vigilante committee was formed to find the killers of the Irishman. After a gunfight, they captured the Mexican woman. Believing she was the cause of his death, some said she should be hanged. Instead, she was banished from the community.

The vigilantes rounded up a half-dozen other Mexicans, and the mob yelled for their deaths. A compromise was struck: four were banished and two were sentenced to be whipped. One of those to be whipped implored that he be hanged instead. Both were beaten. Clapp, who flinched from little, described the scene:

> I had never thought that I should be compelled to hear such fearful sounds, and, although I immediately buried my head in a shawl, nothing can efface from memory the disgust and horror of that moment. I had heard of such things, but heretofore had not realized, that in the nineteenth century, men could be beaten like dogs, much less that other men, not only could sentence such barbarism, but could actually stand by and see their own manhood degraded in such disgraceful manner.

During this time, two other foreigners, a Frenchman and a black man, also lost their lives in violent incidents.

One by one the flume companies failed, as did the individual miners and the businesses that depended upon them in Rich Bar. There was an exodus in the early fall. Few wanted to spend another depressing winter in that place. Those who remained hurled lawsuits and insults at one another. Piles of gravel and trash lay about. The Clapps departed, and their marriage dissolved soon after. Louisa Clapp remained in San Francisco, where she taught school. She ended her days back east. Clapp, who wrote nothing more of consequence, had seen the elephant.

Throughout the Sierra Nevada others quit the diggings. "Nine-tenths of the miners are sick at heart," wrote one such person. Some left the state, departing by ship or crossing Donner Pass faster and more lightly burdened than before. Donner Pass saw less and less traffic in the 1850s until it was virtually abandoned, in favor of alternate passes.

Then began the probes for a railroad right-of-way. Although there had been some prior talk about a transcontinental railroad, it was not until 1850 and statehood that Congress took the matter seriously. The Sierra Nevada was considered to be an insurmountable barrier, so the first recommendations for routes were to the north and south of the range. Then one of those nineteenth-century engineering visionaries arrived on the scene, and once again attention was focused on the gap above the lake.

By the time Theodore D. Judah arrived in California at the age of twenty-eight in 1854, he had been employed by three railroads, built another, worked on the Erie Canal, and supervised construction of a bridge. Judah traveled to California by ship to build the first railroad on the West Coast—twenty-five miles of flat track from Sacramento to Folsom. He soon became obsessed with the idea of a transcontinental railroad. Judah's wife later wrote, "Everything he did from the time he went to California to the day of his death was for the great continental Pacific Railway. Time, money, brains, strength, body and soul were absorbed."

During the 1850s two routes to the south of Donner Pass, Johnson's Cutoff (the present-day Highway 50 over Echo Pass) and the Placer County Emigrant Road (crossing the crest of the Sierra Nevada at the head of Squaw Valley), were the preferred routes. They followed canyon systems on both sides of the Sierra, which meant a large drop between canyon and crest. A railroad needed a more gradual grade. A druggist in the mining community of Dutch Flat who was seeking to increase business invited Judah to investigate the more gradual ridge route that lay between the Yuba and Bear Rivers on the north and the American River on the south. The upper portion of this route was the old emigrant trail.

Judah crossed the Sierra Nevada twenty-three times on foot, horseback, and in a light one-horse wagon, making his calculations with a compass, odometer, and barometer. He found that the route traversing Donner Pass via the long, gently angled ridge on the west and the slopes of shorter, steeper ridges to the east was the most practical for a railroad. The engineer described the constraints of geography thusly:

> When it is considered that the average length of the western slope of the Sierra Nevada Mountains, from summit to base, is only about 70 miles and the general height of its lowest passes about 7,000 ft, the difficulty of locating a railroad line with 100-ft [per mile] grades is correspondingly increased, as it becomes absolutely necessary to find ground upon which to preserve a general uniform of grade.

Judah sought financial backing in San Francisco but failed because he appealed to the public spirit of the monied men. In Sacramento, he made no such tactical mistake and cited the potential profits to be made from such a venture to four shopkeepers: Charles Crocker, Mark Hopkins, Collis P. Huntington, and Leland Stanford. The four merchants, or the "associates," as they came to call themselves, were cautious at first, since their combined assets, meaning mostly their businesses, totaled only about $100,000. Eventually they would amass personal fortunes totaling $200 million from the venture initiated by Judah, who profited least.

To the associates, the railroad's attraction was in the money to be made from transporting supplies to the booming Nevada mines. To the federal government facing the prospect of a civil war, such a railroad was appealing because gold and silver could be transported east to finance the war, and troops shipped westward, thus keeping California and its riches within the Union. The railroad across Donner Summit became a reality because of commercial and military imperatives.

To get the legislation passed, there was massive collusion between the public and private sectors. Judah worked with the congressman who carried the bill in the House, he drafted the Senate version, and was on the staff of the key House and Senate committees. California historian Carl I. Wheat wrote:

> The importance of these appointments and their bearing upon the fate of the Pacific Railroad movement can hardly be over-estimated, and it is probable that without such action, which thus gave Judah a semi-official standing before Congress, the Bill would not have gone through to enactment.

The bill contained subsidies for the railroad that would later be substantially increased. Fifty years before Watergate, Stuart Daggett, a professor of railroad economics at the University of California at Berkeley, characterized Judah's legislative role as of "questionable propriety." Certainly in his own time (when there existed a convergence of interests) and perhaps for all time, Judah was the most effective lobbyist to walk the halls of Congress. After passage of the bill in 1862, sixty-one Northern representatives and senators wrote Judah to thank him for his "indefatigable exertions and intelligent explanations." On the matter of subsidies, which included money and land, Daggett commented in 1922, "The federal government seems in these matters to have assumed the major portion of the risk, and the associates seem to have derived the profits."

Within California the gold rush and Manifest Destiny mentalities were coupled to the railroad era, and the massive project was pushed by all means possible through the narrow gap in the Sierra Nevada. Stanford was elected governor, and at the same time retained his position as president of the Central Pacific Railroad, later to become the Southern Pacific. State subsidies to the railroad were increased accordingly. Charles Crocker's brother, Edwin, the railroad's chief legal counsel, was appointed by Governor Stanford to the State Supreme Court; and he, too, retained his job with the railroad.

First, the associates built the Dutch Flat & Donner Lake Wagon Road, using parts of the original emigrant trail in order to quickly tap into the lucrative wagon and stagecoach traffic to the Comstock mines in Nevada and to haul supplies to the end of the rails. "Shortest, best, and cheapest route," boasted Charles Crocker, president of the company that built the road, in an advertisement. The toll road, opened in June 1864, belonged to the associates, not the railroad, and the profits accrued to them. Judah, ever the engineer, objected to such shady practices. He wrote a friend:

> I had a blow-out about two weeks ago and freed my mind so much so that I looked for instant decapitation. I called things by their right name and invited war, but counsels of peace prevailed and my head is still on but my hands are tied, however. We have no meeting of the board nowadays, except the regular monthly meeting, which, however, was not had this month; but there have been any quantity of private conferences to which I have not been invited.

Judah left on a ship for the East to raise money to buy out the four associates, but he contracted yellow fever in Panama and died a few days after reaching New York City. The board passed a resolution praising his ability; employees of the Southern Pacific later erected a monument of Sierra granite in his memory, Mount Judah was named in his honor, and *Civil Engineering* magazine rated the building of the railroad "the greatest engineering exploit of the early West." Other accolades portrayed Judah as David among the money changers.

Caucasians engineered, financed, supervised, and profited from the construction of the railroad. The Chinese, as usual, did the actual building, and their hand-hewn works stand to this day. The record of their labors is slight, since the Chinese did not write about their experiences, and the railroad either destroyed many documents to avoid financial embarrassment or the records went up in flames during the San Francisco earthquake and fire of 1906.

Charlie Crocker, who had been an iron miner, gold miner, and drygoods merchant, was in charge of construction. He was a bundle of energy, a self-described "wild bull" of a man who was nearly six feet tall and weighed two hundred fifty pounds. Crocker's theory about labor relations was to rule "with an iron fist." If the white workers couldn't hack it, why, they could just quit; and he would get some Chinamen, who

A rock wall built by the Chinese

became known as "Charlie Crocker's pets." Eventually between ten and twelve thousand "Celestials" (as the Chinese were called), a number equivalent to 90 percent of the total workforce, labored on the railroad. They were not treated as pets, either by Crocker or the Chinese labor contractors.

The Chinese were docile and hardworking. They outworked the best Cornish miners. Had they gotten the same pay and benefits, labor costs would have been increased by one-third, according to one estimate. The railroad reaped profits of $130,000 a month by using Chinese laborers, according to another estimate. Additional advantages were that they were barred by federal and state laws from citizenship, meaning they could not vote, and they were not allowed to testify in court. It was, in effect, open season on the Chinese.

The Chinese struck once for higher pay, eight-hour shifts, and a halt in the practice of whipping those persons who wanted to quit. Crocker later explained how he quickly ended the short strike: "I then went up there and made them a little war speech and told them they could not control the works, that no one made laws there but me." He also cut off their food supply and threatened to fine them the cost of keeping the white foremen and all the horses idle for a week. The Chinese returned to work.

The loss of life was heavy. Nobody knows how many Chinese workers perished in the frenzied drive to push the railroad over Donner Pass. They were suspended in wicker baskets over granite cliffs. They chipped away at the exceedingly hard rock to form ledges on which to stand and holes in which to stuff blasting powder or the more volatile nitroglycerin. When ropes failed, they plunged to their deaths. Others took their places.

The mountains rang with the sounds of constant blasts. From February to April 1867, there were two thousand such explosions. Five hundred kegs of blasting powder were consumed every day. In June 1867 six Chinese laborers were killed when a blast detonated prematurely. There were rumors of Chinese being left in tunnels when explosions were set. "Many an honest John went to China feet first," said an engineer. The *Truckee Republican* gave the deaths of the Chinese laborers a patriotic twist:

> Early Truckeeites never seemed to realize that these were the people whose ancestors built the Great Wall of China and who daily risked their lives chiseling out tunnels, inch by inch, with only hammer, chisel, and black powder, who froze and died in the deep snows of the mountains, who were making American history by welding together the Union.

A severe winter had its special hazards. "Avalanches sweep over the shanties of the laborers," reported John R. Gilliss, an assistant engineer. "At tunnel 10, some fifteen or twenty Chinamen were killed by a slide." An accurate count was kept of white men trapped in avalanches, but not of the Chinese lost during the terrible winter of 1866–1867. Gilliss, who gave a report on the construction of the railroad at the 1870 meeting of the American Society of Civil Engineers, commented, "The names given to pass, peak, and lake is itself the record of a tragedy."

There were forty-four storms that winter. One storm dumped six feet of snow, another ten feet. There was a total of forty feet of snow on the ground, and massive drifts approached one hundred feet in depth. The temperatures were severe. "No one can face these storms when they are in earnest," noted Gilliss. Nor could animals. Oxen floundered, and impatient drivers twisted their tails clear off in desperate attempts to get the beasts of burden moving again through the deep snow.

To a person who was comfortable, the storms could be beautiful to behold. Gilliss rhapsodized, "These storms were grand. They always began with a fall in the barometer and a strong wind from the south-west, hurrying up the tattered rain-clouds or storm-scud in heavy masses." He gazed upon the scene below him from the vantage of the wagon road and wrote:

> From this road the scene was strangely beautiful at night. The tall firs, though drooping under their heavy burdens, pointed to the mountains that overhung them, where the fires that lit seven tunnels shone like stars on their snowy sides. The only sound that came down to break the stillness of the winter night was the sharp ring of hammer on steel, or the heavy reports of the blasts.

Below him, the Chinese crews were locked in mortal combat with the mountain pass. They were human moles: They rarely saw daylight, working in granite tunnels and living in shacks buried beneath the snow. They walked to work through tunnels dug in the snow. They became lost, and their bodies were not discovered until the spring thaw.

Tunnel No. 6 at the summit was 1,659 feet long and 124 feet below the surface at its deepest point. One crew worked from each end of the tunnel, and crews rotated around the clock. A white foreman, called a "China herder," was in charge of each crew, which consisted of thirty to forty laborers. Half the men in each crew worked on the heading, the others cleared away the debris.

The progress was excruciatingly slow. The average daily gain was 1.18 feet using powder and 1.82 feet using the more dangerous nitroglycerin, which was mixed at the site. There were at least two accidents with the new explosive. Nitroglycerin was preferable in terms of time and cost, Gilliss told his fellow engineers, who took note of the cheapness of the labor and the type of materials and applied the lessons to their construction projects.

To hasten progress a shaft was sunk from the surface at the average rate of 0.85 foot per twenty-four hours. It took eighty-five days to sink the shaft. When the shaft reached the grade of the tunnel, two additional crews fanned out and worked toward the crews inching forward from both ends of the tunnel. "The Chinamen were as steady, hard-working a set of men as could be found," said the engineer. It took two years to hole through the tunnel. Two months later a locomotive traversed it for the first time. From there it was all downhill to Promontory, Utah Territory, where the Central Pacific met the Union Pacific Railroad in May 1869.

The rock they had to contend with—the very backbone of the Sierra Nevada—came from deep within the bowels of the earth as long ago as 210 million years. It bubbled up in pulses over a period of 130 million years, no single batch being the same, which accounts for the different varieties of granitic rocks and their different shades of color—from near white to dark gray. The granite at Donner Pass is darker than the granite of Yosemite. Granite defines the Sierra, overpowering the darker volcanic rocks that are also woven into the fabric of the mountains.

At temperatures of one thousand degrees Celsius the magma deformed and metamorphosed the ancient sedimentary rock that created a cap over the granitic intrusion. Erosion wore down the metamorphosed rock in places, and some sixty-five million years ago the sparkling rock of the pure granite core of the Sierra batholith was exposed. This low area that was to become a mountain range was gradually uplifted and tilted toward the west, beginning some twenty-five million years ago.

The Appalachians and the Rockies are dark mountain systems because of the different composition of the rocks, the soil, and the resulting vegetation; they absorb rather than reflect light. Light reflects from the granite of the Sierra Nevada as it does from sequins. The word *granite* comes from the Italian *granulo,* meaning grain or particle. The different salt-and-pepper patterns of granite are readily apparent identifying characteristics. The igneous rock is composed of five minerals: quartz, two varieties of feldspar, biotite, and hornblende. Granite is hard and heavy, its hardness being a function of the amount and type of feldspar and its weight being almost two tons per cubic yard.

Early road and railroad builders knew the hardness of granite, as did more recent contractors, such as those who laid an underground fiberoptic cable across Donner Pass. Their modern tools quickly dulled or broke when they came in contact with the rock.

One of the first to write about the granite of the Sierra was William Brewer. He crossed the southern Sierra in June 1864 near where the mountains reach their climax:

The granite of Yosemite

The rocks are granite, very light colored, the soil light-gray granite sand. Here and there are granite knobs or domes, their sides covered with loose angular boulders, among which grow bushes, or here and there a tree. Sometimes there are great slopes of granite, almost destitute of soil, with only an occasional bush or tree that gets a rooting in some crevice. Behind all this rise the sharp peaks of the crest, bare and desolate, streaked with snow; and, since the storms, often great banks of clouds curl around their summits.

Rock climbers depend on the stability of granite for their lives when they dangle from great heights, using new techniques and equipment developed in California in response to the nature of the rock. Architects use granite to instill a sense of massiveness and timelessness in structures. From the quarries at Rocklin, Penryn, and Raymond came the rock for such neoclassical structures as the state capitol and various city halls, commercial banks, federal reserve banks, cathedrals, post offices, monuments, and mausoleums throughout California.

Sierra granite, the railroad, and the same cemetery are all Charles Crocker and Frank Norris have in common. Both are memorialized by granite markers in Mountain View Cemetery, near Oakland. Crocker lies in a Greek temple commissioned by his wife, Mary, who died while supervising its construction. The inscription for Norris, the author of *The Octopus,* which was his metaphor for the railroad that Crocker built, fits the state's reputation for erratic death rituals. It reads: "Beloved by his brothers in Phi Gamma Delta who cherish his memory and testify their gratitude for his devotion to the fraternity." Norris, who died young, was a golden-haired college boy who took on a railroad.

Sierra granite was used for the base of the huge monument erected on the site of the Schallenberger-Breen cabin. The heroic monument commemorated the Donner party and others who entered California in the early years. Only white people were known as "pioneers" or "native" sons or daughters. They had their own organizations and their own versions of history. For a long time their narrow viewpoints dominated the written history of California. Their influence remains, only slightly diminished to this day—its most visible form being the sanitized wording of the roadside historical markers scattered throughout the state. There were, of course, other pioneers—the Indians, the Mexicans, and the Chinese—who were not part of these organizations or their histories.

The bronze figures gazing toward Donner Pass atop the monument, dedicated in 1918 by C. F. McGlashan before six hundred native sons and daughters of the Golden West, represent an idealized version of a pioneer family. A few miles up old Highway 40, called the Donner Pass Road since Interstate 80 replaced the highway in 1964, is a vista point overlooking the lake. It was dedicated in 1986 by "all the citizens of Truckee" to the memory of Charles Fayette McGlashan. A roadside marker set in a base of Sierra granite at the vista point identified McGlashan as "Truckee's Patriarch, Historian, Author, Editor, Attorney, Legislator, Inventor, Entomologist, Astronomer."

Like other pioneers in good standing, McGlashan entered California at the age of six in a wagon that traversed the Sierra Nevada in the summer of 1854. His family, of Scotch ancestry, was poor. McGlashan worked his way through a private school in Healdsburg, a town north of San Francisco, and attended Williston Academy in Massachusetts, where he was expelled in his final year for challenging the authority of the headmaster. Returning to California with eastern polish but no degree, he taught school in Placerville in the Gold Country. In 1872, McGlashan took a job as principal and superintendent of the school in the railroad boomtown of Truckee, three miles east of Donner Lake.

Truckee was overrun by undesirables, a term used by McGlashan's adoring granddaughter and biographer, M. Nona McGlashan. She placed such people in two categories: white hoodlums and Chinese coolies. In reference to the former, a literary society was formed in Truckee; one of the topics they debated was "Should California establish the public whipping post for the suppression of hudlumism?" As for the latter, Miss McGlashan wrote:

> The coolies had been tolerated when the C.P. [Central Pacific] brought them from Canton to hew and chew the railroad's path through granite cliffs. That done, they should vanish, go home—just go. But they had returned to Truckee for the same reason white men did. They could earn a living there.

To deal with the undesirables, the good citizens of Truckee formed a vigilante committee called the "601." It was patterned after a similarly named group in Virginia City, a wide-open mining town in Nevada. McGlashan, a joiner of all groups of significance in town—including the Masons, the Knights Templar, and the Knights of Pythias—was also a member of the 601. One night they donned black masks to rid the town of "shoddy characters," as Miss McGlashan called them. A vigilante fired at a moving shadow; the shape fell. It wore a mask. When the face was unveiled, there lay the town's weekly newspaper editor, shot dead in an alley.

There was an immediate job opening. McGlashan shifted from teaching to newspaper editing, selling real estate, and, after reading law books and passing the bar examination, lawyering. He soon purchased the *Truckee Republican*. The town prospered. The railroad provided jobs. A booming ice industry supplied ice from Donner Lake to keep produce fresh in railroad cars and to grace crystal glasses in New York City's finest hotels.

Ice was a renewable resource; trees were not immediately replaceable. Fourteen sawmills supplied the mines and the railroad with support timbers and railroad ties. Each mile of railroad track needed 2,700 ties, and lumber from the eastern slope of the Sierra supplied most of the needs of the western and midwestern states at a time when almost every community sought access to a railroad. At the height of the demand, fifty carloads of railroad ties were shipped daily from Truckee during the summer months. The surrounding slopes were stripped of sixty-six million board-feet of timber. The hillsides were denuded, and the Truckee River was choked with sawdust.

McGlashan favored this type of progress. As his granddaughter wrote:

> With his generation he used terms like "illimitable" forests— and believed, in truth, they were. Not until the Big Bonanza ended and the long logging flumes decayed, would they note uneasily the black, stump-dotted hills where no tree would ever grow back.

McGlashan, an imposing figure with a large, bushy mustache, was a dynamo. He was an inventor of gadgets, an enthusiastic butterfly collector, an avid astronomer, a deputy game warden, an early promoter of winter sports, and the first chronicler in book form of the Donner party tragedy. His book, *History of the Donner Party*, was an instant success and is still in print, despite its florid style. McGlashan stressed the heroic nature of the Donner party. California historian Hubert H. Bancroft wrote of the book at the time of its first publication, "I think McGlashan has done wisely in suppressing disagreeable details and dwelling on the noble deeds of each member."

As publisher and editor, as well as being a mover and shaker within the town's hierarchy, McGlashan used his power as other newspaper publishers have through the years: advancing the town's business interests, and incidentally his own, by promoting advertisers in the news columns. His granddaughter noted, "This was not only a shrewd business stroke— it was much more than that, as he was much more than a businessman. With these stories, he strengthened Truckee's community pride and forged another link in the fraternal bond uniting his fellow townsmen."

There was a brief hiatus in McGlashan's Truckee sojourn when he left the rigors of Sierra winters to take over the newspaper in balmy Santa Barbara. The previous owner was Harrison Gray Otis, who had departed to take up the ownership of the *Los Angeles Times* and establish a newspaper dynasty to the south. Both McGlashan and Otis would assume the honorary title of general later in their careers, and both would enthusiastically advance the interests of a small portion of their respective communities.

Back in Truckee in 1883, McGlashan, a member of the local chapter of the Caucasian Leagues that had formed throughout the state to deal with the "Chinese problem," became vitriolic on the issue. (Anti-Chinese sentiment escalated in California after the Civil War. A law prohibiting blacks and Indians from testifying in court against whites was extended by the State Supreme Court to include the Chinese. Chief Justice Hugh C. Murray wrote, "The same rule that would admit them to testify, would admit them to all the equal rights of citizenship, and we might soon see them at the polls, in the jury box, upon the bench and in our legislative halls." It was no holds barred against the Chinese, providing there were no white witnesses.)

The league was not above setting fires to achieve its ends. When the shacks of Truckee's Chinese colony burned, the newspaper parodied and trumpeted the disaster in the following headline: LUCKEY TRUCKEE—CHINATOWN HOLOCAUSTED.

The league held secret meetings and sought to persuade white employers to fire their Chinese workers. One employer refused. The league burned the cabins that housed his employees and shot at them as they fled. One worker was killed. There was a trial, but no convictions were forthcoming from the all-white jury. The league, whose membership included Truckee's most prominent citizens, led other assaults on the Celestials. Truckee became known throughout the West for its aggressive stand on the Chinese issue.

A local historian noted in 1978:

> The town became famous, but this glory had nothing to do with lofty mountain tops or pure rushing water. Truckee's distinction rested on its torment of the Oriental—a persecution so effective that today you'll find little evidence that a Chinese population one thousand strong, ever existed.

McGlashan also sought a legal solution and was elected to the legislature as an assemblyman in 1884 on an anti-Chinese platform. He became one of the leaders of the exclusionary movement in the state and sought to ride the crest of the movement to greater heights. Billed as "the hero of Truckee," he spoke throughout the state. He was elected chairman of the Anti-Chinese League in San Jose. McGlashan was mentioned as a possible candidate for governor. He lost the nomination for senator and returned to

Truckee to oversee the departure of the Chinese. In a testament to his own success, McGlashan wrote in the April 5, 1893, issue of his newspaper:

> Few of the seven-year-old children of Truckee ever saw a Chinaman. Prior to eight years ago there were 2,000 Chinese in the Truckee Basin, since, not one. They were not driven out by force but simply starved out. The citizens rose up as one man and discharged every single Celestial. It took just five weeks to get rid of the entire 2,000. Not one has been in sight of Truckee for a day, an hour or a moment since March 1886. No Chinamen will ever again be employed in the region.

By being anti-Chinese, McGlashan had come to be regarded as pro–white labor, jobs being the superficial issue that divided the two races. The white railroad workers in Truckee trusted McGlashan. Unbeknownst to them, the railroad retained McGlashan's services as an attorney, and knowingly got much more than a lawyer for its money. "We hired you to work for our interests, McGlashan," railroad officials told the editor-attorney. There was a strike, and one day signs appeared around town stating:

ANYONE WHO WOULD LIKE A DAY IN THE COUNTRY
WITH FREE FOOD (ALL YOU CAN EAT), MEET TEAMS
TOMORROW AT LAST CHANCE SALOON—8:00 A.M.
SIGNED: C. F. MCGLASHAN

The striking railroad workers had a wonderful day in the country. When they returned to town, they found that their jobs had been taken by scabs. The strike was broken. "Great work, Mac," said the railroad officials, who slapped him on the back. Within a week, a lifetime railroad pass for McGlashan and his family arrived. It was not an inconsiderable reward. Nona McGlashan wrote, "I know that it paid for innumerable trips to and from college, made by nine children, including myself."

McGlashan's rewards during his lifetime were many. Following his death in 1931, there were fulsome eulogies in California newspapers. In 1968 the California Press Association elected McGlashan to the Newspaper Hall of Fame. His citation read, in part: "May men of vision and great influence continue to let their light shine from the desks of the small-town newspapers!"

At the dawn of this century the automobile came to California via Donner Pass, and the state has never been the same since then. In 1903 a doctor made the first transcontinental automobile trip in sixty-five days from San Francisco to New York. He drove a 20-horsepower, open-topped Winton and won a fifty-dollar bet.

It was a different state then. There was no smog or gridlock, no tourist camps that became cabin camps that became auto courts that became motels that assumed the name of inns, no freeways or highways, and no paved roads of any considerable length. Over Donner Pass there was just an unpaved, eroded wagon road that had gotten little use in forty years. Its sides were supported by unmortared stone walls carefully put into place years ago by Chinese laborers.

The demands of commerce and the military would change the form and name of the route numerous times during the twentieth century. The artery went from being a part of the landscape to being imposed upon it. What remained the same was the one geographical fact that could not be changed either by names or vast improvement in earthmoving equipment—the Sierra Nevada was a barrier that could be surmounted on a large scale in only one place. Lincoln Highway historian Drake Hokanson wrote of the geography of the central route across America that eventually snaked its way across Donner Pass:

> This geography—be it cultural, landform, environmental, meteorological, architectural or political—affected routes, travel patterns, and the devices and animals employed to get from one place to another. In that broad sense geography has always determined how people traveled, what they saw, and where they stopped. It determines why a road goes where it does.

Crossing the country by motorcar—and getting a highway in place that would make that crossing practicable—quickly became a commercial venture. The second transcontinental trip was sponsored by the Packard Motor Car Company, a maker of luxury cars. The public relations gimmick took fifty-one days. "I realized that we had to attract public attention with a spectacle, but it had to be a useful spectacle," said Packard's general manager. The resulting photographs were spectacular. "This performance created a very favorable public impression," he added.

The Lincoln Highway Association, a promotional group, was formed by various members of the automobile industry to designate a cross-country route and funnel donations to the various jurisdictions responsible for improving the highway. The group first envisioned a highway paved with gravel, but came to favor concrete. Henry Ford, whose cars dominated what roads there were, refused to join the association, believing that public tax monies should be the source of funding for a transcontinental highway.

Despite this setback, the "men of vision" and the "men of constructive thought," as one of the association's publications characterized board members, mounted a successful publicity campaign. Magazines donated free advertising space; "an almost unheard-of thing," crowed the association, whose publication went on to state, "No organization of altruistic purpose ever had an opportunity for greater satisfaction in the powerful widespread approval and endorsement of the correctness of its course and the soundness of its aims."

The bold red line on association maps was just that—a line—in places. But across the Sierra it followed the route of the abandoned wagon road. The Lincoln Highway came to be a combination of dirt, gravel, and pavement across the Sierra. Early motorists were amazed at how easy it was to cross Donner Pass compared to the experiences of the early emigrants.

There were difficulties, however. Teams of horses or groups of men pulled or pushed vehicles out of late-season snowbanks and mud. A dash had to be made over the railroad tracks just east of the summit lest a car be hit by a train emerging unexpectedly from a snow shed on either side of the crossing. A long-slung Winton got hung up on the tracks in the summer of 1914; the wife and child fled the car while the husband worked frantically to free it. The motorist was successful and recalled, "Needless to say, a railroad track is no place to tarry, and we were scared stiff."

Once beyond the desert and mountain barriers, the weary motorists rejoiced. Alice H. Ramsey was the first woman to cross the continent by car. She traveled with three female companions in a new 1909 Maxwell. On clearing the summit, she noted:

Majestic sugar pines, Douglas firs and redwoods lined our roads on both sides. What a land! What mountains! What blue skies and clear, sparkling water! Our hearts leapt within us. None of us had ever seen the like—and we loved it. We almost chirped as we exclaimed over the grandeur that surrounded us on all sides. We started talking over plans when the trip was completed.

World War I gave a huge boost to highway construction. The railroads could not move all the men and materials, so motortrucks came into widespread use in the East. The army seized the initiative after the war and organized a cross-country convoy. Seventy-two vehicles, mostly trucks, and two hundred ninety-five officers and enlisted men set out with great fanfare from mile zero in Washington, D.C., on July 7, 1919. The convoy arrived sixty-two days later in San Francisco after destroying and subsequently rebuilding nearly one hundred bridges designed to handle carts and wagons.

The boomers from the Lincoln Highway Association said the convoy was a great success. The association stated, "The convoy would impress upon the Nation the coming necessity for the establishment of a Federal Highway System for the military and commercial needs of the Nation." Lt. Col. Dwight D. Eisenhower, who joined the convoy just before its departure, saw it somewhat differently. He wrote:

Delays, sadly, were to be the order of the day. The convoy had been literally thrown together and there was little discernible control. All drivers had claimed lengthy experience in driving trucks; some of them, it turned out, had never handled anything more advanced than a Model T. Most colored the air with expressions concerning starting and stopping that indicated a longer association with teams of horses than with internal combustion engines.

Despite these handicaps, some of which would prove to be endemic to future military convoys, federal funds for highway construction were soon forthcoming. There was a frenzy of construction: by 1922 a total of nine coast-to-coast highways existed in name, if not in reality. Groups formed to promote a route that would benefit them. They raised money, erected

The Highway 40 bridge near Donner Summit

signs, and lobbied for federal funds. Such was the case with the Victory Highway, whose signs were placed along the Lincoln Highway across the Sierra but whose route differed in its approach through Utah and Nevada.

The jumble of names was too much for the federal government, which then devised a numbering system. So the Lincoln and Victory Highways across the Sierra became U.S. Highway 40. A two-lane, paved roadway expressly engineered for vehicles propelled by internal combustion engines was completed in 1926. In some places it superseded the wagon road, in others it carved its own path through or around the granite.

After World War II, California boomed and the traffic increased accordingly. Donner Pass became the most intensively utilized transportation and communications corridor in the nation. Concrete and asphalt, steel and wood, overhead and underground lines, cables, and pipelines were the umbilical cords that plugged California into the remainder of the nation. As motorcars and touring had captured the nation's imagination after the First World War, so the airplane was the new, glamorous kid on the block following World War II. Above Donner Pass was a major flight corridor.

George Stewart, not only a historian of the Donner party but also of Donner Pass, observed:

In general, therefore, the keynote for the whole area has been transportation. Out of the opportunity to cross the mountains at this point, the whole history of the region has developed. It has kept a rendezvous with history, and its interest to the person who passes here should be historical as much as scenic. At the summit, for instance, one can enjoy the beauty of the view, but can also see the remains of two primitive roads in addition to the present highway, can look across at the railroad, and can also know that the emigrant wagons were dragged up somewhere to reach the same gap.

In 1956, at the height of the cold war, a lieutenant colonel who had become president signed the Interstate Highway Act. Eisenhower had not forgotten his trip across the country almost forty years earlier nor how quickly the German armies had been transported on the autobahns of Europe. The highway act "promised to hurry the nation's commerce and military" across the nation, wrote historian Drake Hokanson.

To hurry the traffic over Donner Pass, a gradual grade up the side of a ridge two miles to the north of Donner Pass proper was selected by highway engineers over the more frontal assault of the old roadways. The new earthmoving equipment developed during and after World War II was at their disposal. Contours meant little. Straightness with the least amount of grade—the Judah dictum for the railroad—meant everything.

Not only terrain but history was lost in the process. Yes, speed was greatly increased, as was reliability, since it was much easier to clear snow from the interstate than from the old highway. But those who crossed Donner Pass had, through the years, lost contact with the landscape and its history as travel progressed from bare feet padding along the trail, to shod horses, to iron-rimmed wagon wheels, to the steel wheels of enclosed railroad cars, to the solid rubber and then the pneumatic tires of automobiles, which were first open to the air, then enclosed, and then equipped with heaters, air conditioners, radios, tape decks, cellular telephones, and smoked glass.

I went in search of history and picked up the old emigrant trail in Dog Valley, just inside California. I parked my camper and popped the top at a point where a map indicated the emigrant trail had been located. A dozen cattle slowly drifted by, grazing in the twilight much like the emigrants' cattle must have fed after a long day on the trail. I pretended my camper with a canvas top that was lit by a propane lantern was a wagon. The peaceful valley, midway between desert and mountains, would have been a momentary resting place before the last push. I heard the rattle of automatic weapons fire in the distance—target practice, I hoped.

The next day I managed to avoid the interstate and made my way to Truckee along the old trail, now a combination of dirt and two-lane paved roads. McGlashan, who died in 1931, did not live to see Truckee prosper. *The WPA Guide to California* described the Truckee of the 1930s thusly:

> Roundabout, pine forests cover the slopes with deep green, but Truckee itself, once a lumbering camp, now a railroad and stock-raising supply center, lacks even a sprig of green. Its ramshackle frame houses and weather-stained brick buildings sprawl over rocky slopes. On Saturday nights the cheap saloons and gambling halls overflow with lumberjacks, cowpunchers, and shepherds.

Truckee went through some rough times. I remember when half the stores along Old Highway 40, which faces the railroad tracks, were vacant during the early 1960s when I skied at nearby Squaw Valley. There was a renaissance during the back-to-the-land movement of the early 1970s, and McGlashan would be pleased with the prosperity of Truckee today. When I was last there the newspaper carried a front-page story of the town being rated a "cool" winter sports community by a national skiing magazine. The former railroad town was a good example of western chic, complete with smog from automobile exhausts and woodstoves.

Truckee's population had jumped 62 percent between the most recent decades. Shopping centers, discount malls, retirees, baby boomers, carpenters, and real estate agents were drawn to the "gateway to the splendor of the Sierra and Lake Tahoe," as a chamber of commerce publication termed it. "Today, the Range of Light is looking more like the Range of Real Estate," wrote Tom Knudsen in his Pulitzer Prize–winning series of articles on the Sierra Nevada in the *Sacramento Bee*. "Every month, Truckee looks a little less like Truckee—and a little more like everyplace else in California," he added.

I proceeded to the agricultural inspection station just west of town on Interstate 80. The threat of the alfalfa weevil invading the state prompted the inspection of vehicles on the Lincoln Highway, and at other entrances to the state, in 1921. The first Donner Pass inspection station, called a "bug station" by locals, was erected in 1924. A publication of the California Department of Agriculture noted:

> Furthermore, it became apparent that no other state is so favorably situated geographically to execute plant quarantine inspection procedures. Being bounded on the west by the Pacific Ocean, on the north and east by mountain barriers, and on the southern portion of the eastern border by deserts and rivers, it was possible to establish inspection gateways.

Inspections were not enough to keep out the dreaded hoof-and-mouth disease that threatened to topple California's agricultural economy in the mid-1920s. Millions of dollars were spent trying to eradicate the disease. More than one hundred thousand head of cattle and sheep were killed within six months. Army troops used machine guns to decimate whole herds and then burned the corpses. The disease spread to wildlife. Hunters fanned out and 22,214 deer were slaughtered.

Since then, the Colorado potato beetle, the cherry fruit fly, the thurberia weevil, the pink bollworm, and the Mediterranean fruit fly have threatened to ruin California agriculture. The day I spent at the inspection station the gypsy moth was the main target.

The inspectors concentrated on cars that had been out of the state a significant distance, meaning farther than Reno. License plates were, of course, one clue, as was the color of dirt, the type of bugs on front grilles, the amount of luggage, and whether salt had accumulated on the vehicle while passing the Great Salt Lake.

It was largely a prosperous-looking group of individuals and small groups who passed through the inspection station on that mid-July day. I detected six subspecies: there was the gambling crowd, the indoor vacationing crowd, the outdoor camping crowd, the ranching crowd, the interstate moving crowd, and the trucking crowd. They slowed, braked, crawled forward, stopped, then lurched forward into California in a dense cloud of carbon monoxide.

"Where are you coming from?"

"Bring any fresh fruit or plants with you?"

A man handed over three nectarines.

"Where are you coming from?"

Some people were primed for the second question. "No," they quickly replied.

A car was stopped and the trunk opened. Nothing but a jumper cable, shoes, and a raincoat.

"Enjoy," said a man handing over an apple before being asked. A woman stuffed as many apricots into her mouth as she could, then surrendered the rest.

"Where are you coming from?"

"Germany."

"No, I mean today."

An elderly bicyclist with EDDIE FITZGERALD—WORLD RECORD BIKE RIDE emblazoned on his red shirt passed through, leaving a newspaper clipping with the startled inspector that explained he was trying to better his entry in the *Guinness Book of World Records*. Fitzgerald, in his fifties, had ridden close to 50,000 miles. He headed up the interstate toward Donner Pass on his mountain bike.

I took a break. In the office there was a memo on the bulletin board from the station manager to all employees:

> I talked to Mr. Sandige on the 29th of June and passed on the great job I thought you were doing this summer. He was very impressed with the interceptions and pests being found. So far this month we have more pests than were intercepted in the months of June and July of last year. Bill Sandige said to tell all of you that you are doing an outstanding job and that he will mention that fact at the next staff meeting on Monday morning in Sacramento.

I went outside again. The vehicular movement was incessant on this windless, hot day. The fumes soiled the mountain air and made me nauseated. I was fortunate, I thought, to have entered California via a different route and at a different time.

III

Land of Fire

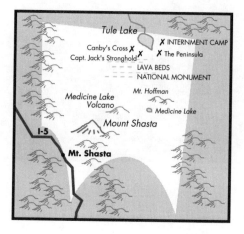

Northeastern California is a combination of myths and hard-edged realities. Mount Shasta looks down from a towering height upon the State of California like some unblinking, protuberant third eye. Its geologic past has been obscured by the myths that have ebbed and flowed around this magic mountain. If California is "America's most psychic, occult, and mystic state," as one author who surveyed California's penchant for the mystical described it, then Mount Shasta is the "new Mecca of American spiritualism—as shrouded by legend at its peak as it is crowded with free-lance mystics, seers, metaphysicians, and mediums at its base," according to a national newsmagazine.

Mount Shasta has attracted more stories and legends than any other wilderness area in the country. One bibliography lists sixty-four books or articles dealing with the metaphysical aspects alone of Mount Shasta. The articles appeared in such publications as *Fate, Amazing Stories, Saucer News, UFO Review,* and *Ancient Astronauts.* Another bibliography lists thirty-four books or articles on the scientific nature of Mount Shasta, the latter published in such journals as *Geology, California Geology, Bulletin*

Mount Shasta

Volcanologique, and *Zeitschrift Vulkanologic.* There seems to be no middle ground in the way people view Mount Shasta.

The legends go back to the Indians, who supposedly believed in a Great Spirit that made the mountain, the rivers, the fish in the rivers, the birds, and the beasts. The Great Spirit, liking what he created, then remade the mountain in the shape of a wigwam and built a fire in the midst of it, where he lived forever after with his family. This story came from the fanciful pen of the writer and poet Joaquin Miller, who lived with the Indians of the Mount Shasta area during the 1850s. Miller, however, was a free spirit who was not bound by factual matters.

His tale suits those who promote the metaphysical aspects of the mountain. The concept of an all-pervading Great Spirit, however, was actually closer to the monotheism of European-based cultures than it was to the diffused animism of the Indians. The Shasta Indians did not believe in a single creator. They had a host of stories concerning the mountain, among them a flood story involving Coyote and Mount Shasta. The Indians also thought the mountain was inhabited by mysterious powers who took the form of short people found near rocks, cliffs, lakes, rivers, and mountain summits. They were known by the Indian word for "pain."

The first white man to see Mount Shasta reported a volcanic eruption, an indication of the physical nature of this land of fire. Sailing down the coast from Alaska to California on September 7, 1786, the French explorer Jean-François de Galaup, Comte de Lapérouse, wrote in his journal, "The flame was very vivid, but a thick fog soon concealed it from our sight." Mount Shasta was one hundred miles to the east. Lapérouse had one chance to see it, and that was through a slot in the Coast Range where the Klamath River entered the Pacific. The expedition, sailing in two ships, boasted some of the best scientific minds of the Age of Enlightenment. It was fated to disappear in the mists of the Pacific Ocean.

To California, via the Pacific, came the myth that sent thousands of believers into the foothills and onto the flanks of Mount Shasta in search of the survivors of the lost continent of Lemuria. Lemuria, also known as Mu, is the Pacific Ocean's equivalent of the Atlantis legend, with some variations. While stories of Atlantis date back to Plato, who may have advanced his account either as history or allegory, the concept of Lemuria first arose as scientific speculation in the middle of the last century. Like Atlantis, it combined two lasting motifs of mythic origin—that of the Great Flood and of paradise lost.

The existence of a lost continent in the Indian Ocean was first hypothesized when certain geological and zoological similarities among Africa, India, and the Malay Peninsula were noted. The name was derived from the lemur, a monkeylike animal once widely distributed throughout the region and now endangered. The lost continent of Lemuria was thought to be the cradle of civilization. The concept of such a continent then shifted eastward to the Pacific Ocean, where esotericists and clairvoyants in the nineteenth century used its supposed existence to explain the scattered archipelagoes of the South Pacific.

The islands, it was maintained, were the mountaintop remnants of a great civilization composed of gigantic cities and huge inhabitants whose skin was tinged blue and whose third eye was located in back of their heads. Lemuria sank in a volcanic cataclysm, but an eastern outpost remained on Mount Shasta. A variant of this theme was later used to advance a theory that denigrated the achievements of people of color. The Pacific was supposedly first populated by a "white, fair-haired race which owes nothing to European admixture." As Lemuria began to sink, according to this account, the aristocracy of fair-skinned people, who were adept at engineering and had created such wonders as the figures on Easter Island, migrated eastward to Peru and eventually California.

Although a few obscure accounts had placed the Lemurians on Mount Shasta prior to the early 1930s, the Rosicrucians, a mystic order headquartered in San Jose, California, and dating back to ancient Egypt, and the *Los Angeles Times* popularized the concept and placed it squarely on Mount Shasta.

In 1931 the Rosicrucian Press published a book entitled *Lemuria, The Lost Continent of the Pacific,* by Wishar S. Cerve, a pen name for the order's First Imperator, H. Spencer Lewis. According to Lewis's account, remnants of the Lemurian race fled the final catastrophe and found sanctuary on Mount Shasta. They had huge heads and high foreheads, in the midst of which was a protrusion the size of a walnut, a sixth-sense organ. These people were about seven feet tall and well-muscled. They were self-sustaining and quite capable, through their advanced engineering talents, of deriving energy from "radioactive metals and cosmic sources." They also manufactured gold. Their size, formidable powers, and riches echoed the Spanish story of Califia's Amazons, with which Lewis was acquainted. Lewis said the Terrestrial Paradise mentioned in the Spanish myth was Lemuria.

The book sold widely in a number of printings in this country, Europe, and Asia. It probably served as the source for the fanciful newspaper account the following year, since all of the details and some of the phraseology were the same. What was missing was Lewis's account of flying boats "over the hills and valleys of California" and his boosterism of the state.

On May 22, 1932, the magazine section of the Sunday *Los Angeles Times,* under the headline "A People of Mystery" and next to a short piece of detective fiction, carried a story by one Edward Lanser. On the train to Portland, Lanser had seen the southern side of Mount Shasta "ablaze with a strange reddish-green light."

The conductor explained: "Lemurians. They hold ceremonials up there."

After taking care of business in Portland, Lanser returned to the mountain. He related stories of exotic ceremonies, rich gold mines, and the miraculous ability of the Lemurians to snuff out forest fires and make themselves invisible. The tall, barefoot men—no women were mentioned—with close-cropped hair and spotless white robes strode into town on occasion and overpaid for the goods they purchased with gold nuggets. Then in the last paragraph the author did, indeed, recount a true miracle. Lanser wrote:

> The really incredible thing is that these staunch descendants of that vanished race have succeeded in secluding themselves in the midst of our teeming State and that they have managed through some marvelous sorcery to keep highways, hot-dog establishments, filling stations and other ugly counterparts of our tourist system out of their sacred precincts.

The Forest Service was inundated by Lemurian seekers. The federal agency insisted there were no such people on the public lands it administered. Lewis thought it best to call it quits. A delegation of Lemurians had visited him in his San Jose office, he wrote, and said they were taking off for parts unknown. The traffic had gotten too heavy on the mountain.

The search for Lemurians on the mountain continues to this day. I recently hiked up the flanks of Mount Shasta, on both its south and north sides. I saw nothing out of the ordinary, although I recognized the luminescent, cone-shaped mountain as a place that could attract the extraordinary. I checked with the Forest Service in town. Ron Otrin, a resources officer, said of the Lemurian seekers, "They are on another plane. We are more grounded in the Forest Service." He said a man had approached him in Squaw Meadow and said that a spacecraft was in the meadow, and his soul mate was aboard it. Otrin could see no spacecraft.

I spent some time in Mount Shasta City at the base of the mountain. I arrived the night that a freight train had derailed, spilling thousands of gallons of pesticide into the Sacramento River. The accident tended to focus the attention of the inhabitants on more grounded matters. The next day at the Golden Bough Bookstore ("Mt. Shasta's Metaphysical Bookstore") people were complaining of headaches. I purchased some books on the occult nature of the place and remarked that a lot seemed to be happening there. The woman who waited on me explained that as many as seven ley lines bisected Mount Shasta.

The history of the town, it turned out, was quite prosaic. Justin Hinckley Sisson, a stern-faced New Englander, founded it. The economy was dependent upon logging, railroading, truck gardening, a state fish hatchery, and the mountain. Climbing expeditions were launched from the town. In 1875 on his second ascent in three days, naturalist John Muir and a local guide were trapped near the summit in a blizzard. They stayed alive by nestling close to steam fumaroles on the summit; Muir's frostbitten feet never fully recovered. To Muir, Shasta was a "fire mountain."

The early industries declined. Now the town catered to more nebulous matters: angelic voices, dolphin dreaming, child light, attuning with divine beings, and creative harmonics. An advertisement in *City of Light* magazine urged, "Discover the magic of Mt. Shasta through Mountain Gate Properties."

The sheer physical presence of Mount Shasta was enough to draw sustained attention. That intrepid traveler throughout the state, William Brewer, noted:

> And how long and earnestly we gazed on the mountain! Nothing else could be talked about. At sunset it was tinged with a lovely alpenglow; and then twilight deepened, and then the moon rose, and we sat around our cheerful campfire (for it was cold) and gazed still. And I got up from my blankets late in the night, when the moon's illumination was finer, to look at it by that light.

On closer examination, Brewer found the mountain lacking. He climbed to the summit in September 1862, eight years after the first recorded ascent, and noted, "The Alps are beautifully grand, Mount Shasta sublimely desolate. Geologically, it is nearly as barren as it is botanically." He mistook Mount Shasta as the highest mountain in the state; at a height of 14,161 feet, it ranks sixth.

When Brewer rode into the country surrounding Shasta, he noticed that the land was encrusted with lava. The mountain is but one of some seventy-six volcanoes, most of them located in the northeast corner of California, that have erupted and spewed lava over the state in the last ten thousand years. During the last 3,400 years, Shasta erupted either ten or eleven times, the geologic evidence not being conclusive on the matter. It erupted at least three times in the last seven hundred fifty years, the last time being in 1786 as Lapérouse sailed down the coast. Scientists calculate that there is a one in three or four chance that it will erupt within a given person's lifetime.

There are four other historically active volcanoes in California, all in the northeastern portion of the state: Glass Mountain on the Medicine Lake Volcano, Lassen Peak, Chaos Crags, and Cinder Cone. These volcanoes are at the southern tip of the Cascade Range, which extends northward through the states of Oregon and Washington. With the exception of Mount St. Helens, which erupted in 1980, the other Cascade volcanoes have remained fairly quiescent in recent years. But a cataclysm of tragic proportions could occur at any moment. There is this scientific scenario for Shastina, a crater on the west side of Mount Shasta:

> At exactly 6:21 on the morning of September 30, a violent explosion shatters the great dome that had almost filled Shastina's crater. An immense avalanche of seething gas and molten blocks from the dome's interior sweeps down the west slope of Shastina, spreading out over the volcano's western flank. In moments the towns of Weed and Mt. Shasta are engulfed in the pyroclastic flow and totally consumed. The entire area bursts into flames with temperatures in the center of the holocaust approaching 2,000 degrees Fahrenheit. Iron roofs, steel frame buildings, and glass windows melt into unrecognizable heaps. Most wooden structures simply vaporize. Smoke from burning forest and buildings darkens the dense haze of ash settling from the glowing cloud.

These volcanoes are part of a chain known as the Ring of Fire that encircles the fringes of the Pacific Ocean and accounts for about 75 percent of the world's active volcanoes. According to the theory of plate tectonics, the earth's crust is composed of a dozen plates that ride atop a molten mass of basalt. Where the Pacific Plate is slowly sliding, or being subducted, under the North American Plate at a speed of one to two inches a year, the magma is forced to the surface, and thus volcanoes. There is no subduction zone south of Lassen Peak or Cape Mendocino, which is approximately at the same latitude on the coast. Instead of the west-to-east movement in the zone of subduction, the dominant movement along the San Andreas Fault below the cape is from south to north. Everything in the state is subject to displacement.

There is something in this scientific theory for believers in the lost continent of Lemuria. Within the zone of subduction, which includes Mount Shasta, sea floor sediments have been forced eastward over millions of years until they have been grafted upon the western edge of the continent. Known as exotic terrains, these remnants of the ocean floor did not sink under the continental plate but, being lighter, rose above it and became parts of Northern California and the Northwest. It takes about ten million years for the sea floor to move 320 miles to the east.

Mount Shasta attracts its own distinctive weather pattern. Dark masses of towering cumulus clouds issued from its heights on a summer afternoon, much like an ash cloud from a volcanic eruption.

I watched Shasta from the summit of Little Mount Hoffman, some thirty-five miles distant, for a number of days. I am mesmerized by the latent power of volcanoes. I have traveled to Hawaii to watch the lava, blinking red in the night, creep down the side of Kilauea on the Big Island and devour everything within its reach before meeting the ocean in a steaming froth. As a young man, I climbed a perfect Mount St. Helens; in middle age, I returned to wander through the dead trees, felled like so many victims of a nuclear blast.

There was an unobstructed view of Mount Shasta from the cinder cone that sits on the lip of the Medicine Lake Caldera. Medicine Lake, a low shield volcano of greater volume than any other volcano in the Cascade Range, is attached to Shasta by a series of vents in the form of bare hillocks, pine forests, and ropy extrusions of lava that constitute a ridge. The Medicine Lake Volcano merges with the undulating Modoc Plateau to the north and east. The descent from Mount Shasta to the plateau is steep at first, then gradual.

The plateau, a tableland varying from four to five thousand feet in elevation, was formed by a series of lava flows. It is cut off from the remainder of California by the southern extension of the Cascade Range. The Medicine Lake Volcano is a long, dark presence that sits atop the tableland like an inverted saucer. It has erupted seventeen times in the last eleven thousand years.

Five neighboring tribes, including the Modocs and the Shastas, shared Medicine Lake in the summer months when the highland was free of snow. The Modocs camped near Mount Hoffman on the northeast slope of the volcano, which was closest to their territory. Here among the sugar pines and firs, away from the hot valleys, there was game to hunt. There was also a plentiful supply of obsidian on Glass Mountain—a valuable trade item that could be fashioned into sharp blades for weapons or tools. They held their religious ceremonies around the lake that lies within the caldera and whose water is always clear, although there is no outlet. Here they fasted, conducted puberty rites, and experienced their power dreams.

I sat still. A young mule deer, whose tan hide caught the late afternoon sun, passed before my eyes without warning. It seemed like a spectral presence. I thought of the Modoc Indians and the Japanese-Americans whose histories were linked to this volcanic land. War was in the air. That night, lightning flickered about Shasta. It reminded me of the night-firing of artillery in South Vietnam. Perhaps this was a power dream.

At daybreak there was not a cloud in the sky, and Shasta was reborn in the shape of a broad arrowhead unattached to any shaft. It rose without any visible means of support and floated above the horizon on a sea of gray-blue vapor. The mountain was backlit by a light pink sky. The first

The deer and a lava flow

touch of sun upon the summit was like a match to a candle, and California came alive for another day. I drove down the dirt track and into geohistory.

The geology of war goes back to the late Pleistocene, perhaps ten thousand years ago, when a six-mile chain of vents along the northeast side of the Medicine Lake Volcano erupted within a one-hundred-year time span and spewed out basaltic lava via long, sinuous tubes over two-thirds of the present-day Lava Beds National Monument. Most likely the lava that formed the stronghold where the Modoc Indians held the U.S. Army at bay for five months (December 1872 to April 1873) came from Modoc Crater, although there are some who believe that nearby Mammoth Crater was the source.

As the molten lava emerged from the vents and spilled down the volcano, the sides and tops of the tubes congealed. Fresh lava extended the tubes. Some eventually reached fifteen miles in length. Where the roofs of the tubes later collapsed, there were entrances to caves; some of the underground cavities were a few miles in length, while others were just large enough for an Indian family to find shelter.

In the last stages of the eruption, the Modoc stronghold was formed. As flow followed flow through the tubes, a low plateau emerged at the edge of Tule Lake, a shallow body of water bordering the stronghold on the north and serving as the source of drinking water for the Indians and the troops. At the edge of the flow the fingerlike tongues of crust turned down and cracks and fissures formed in the flaps of lava, the natural trenches utilized by the Modocs. These interlocking trenches were continuous along three sides of the Modoc stronghold. Outside this inner perimeter, detached portions of the plateau—also laced with fissures—served as outposts.

The result, after a veneer of vegetation was acquired, was almost total camouflage. The stronghold was a complex terrain that appeared from a short distance away to be a smooth, easily accessible rise in the landscape. Actually the plateau was a tortuous labyrinth whose congealed lava had sharp edges that could easily pierce the skin of defender and attacker alike.

Water mixed with fire to form a shallow lake that covered 1,096 square miles in the Klamath River Basin during the late Pleistocene. Lava flows burst through the lake's water, ash from eruptions at nearby Crater Lake and elsewhere blanketed its surface, and a lava flow that snaked its way down the eastern flank of Medicine Lake Volcano blocked the outlet to the south. The level of the lake rose and fell, and the early Indians adjusted to the varying heights by moving their villages accordingly. With the coming of a drier climate, the water level dropped one hundred feet and the large lake split into three parts: Upper and Lower Klamath Lakes and Tule Lake in the southeastern portion of the basin.

Two large white crosses were planted in the dark lava that spilled from the Medicine Lake Volcano into the Tule Lake Basin. The crosses commemorated two of the racial conflicts that have been such a dominant yet neglected part of California's history.

The first cross is in a flat meadow. It marks the spot where a general was shot during the Modoc Indian War. The raised black printing along the horizontal member of the white cross reads: GEN CANBY USA WAS MURDERED HERE BY THE MODOCS APRIL 11, 1873. Someone used a black felt-tip pen to raise an interesting question on the vertical member of the cross: WHY ARE INDIANS KILLED AND WHITES MURDERED? SMALLPOX, MEASLES, DIPHTHERIA, SYPHILIS WERE BROUGHT TO THE NEW WORLD BY THE WHITES. When I returned a few months later, the message had been erased.

The second cross is planted a few miles to the east on the Peninsula, an eroded tuff cone. Magma from the Medicine Lake Volcano forced its way to the surface of the ancient lake, and with a spectacular outpouring of gray steam and ash some two hundred fifty thousand years ago, formed a low mound that jutted out into the lake. The cross sits atop a sheer cliff overlooking the site of the Tule Lake War Relocation Camp, where Japanese-Americans were held during World War II.

Natural and human history are jumbled together here. Bald eagles nest along the cliffs of the Peninsula. On the face of one cliff is a dense cluster of Indian petroglyphs, one of which is thought to depict people protecting themselves in a fortified place. Aaron C. Waters of the U.S. Geological Survey studied the geology of the place and the history of the Indian war. He wrote of the large losses inflicted on the army by the small band of Native Americans:

> One part of the answer to this question is that the Indians chose a superb, but by no means unique, natural fortress in which to make their stand. This fortress, however, is only one part of the total equation. The vital point is that the Modocs

knew thoroughly and in detail the nature of the terrain south
of the shore line of Tule Lake; the army was totally ignorant of
this landscape's military advantages.

On the other side of the Peninsula are the remnants of a stone guardhouse that overlooked the relocation camp, situated in northwestern California because of its isolation. The Japanese-American prisoners
climbed to the top to gaze at Mount Shasta, its snow-topped cone a
reminder of Mount Fuji, another volcanic edifice. Incised some twenty
feet up a sheer cliff on another side of the Peninsula is an American flag
with six stripes. Like ley lines, the histories of three cultures and the fiery
landscape intersected here.

The heartland of the Modoc Indians was Tule Lake and its surrounding
sagebrush steppe, but the Modocs roamed as far as Mount Shasta to the
west and Goose Lake to the east. They may have numbered between
twelve hundred and two thousand before the coming of the whites. They
were known as "lake" to their neighbors and archenemies—the Shasta
Indians—to the west. The Modocs lived off the bountiful wildlife in and
around the lake; the tule reeds that grew in the shallow water served as
material for clothing, baskets, and boats.

They were not an easy people. The mainstay of their economy was
slavery, and they were merciless in their raids on the Pit Indians to the
south, who usually ran, and the Shastas, who resisted but eventually fell
to the Modocs. The battles were brief and bloody. The Modocs took one
scalp per battle. They killed the enemy's wounded but did not harm
women or children, who were valuable chattel. They burned villages.

Sometimes the Klamath Indians to the north joined the Modocs
against the Shastas; most of the time the relationship between the two
neighboring tribes, the only ones that understood each other's language, was uneasy and was maintained primarily for business purposes.
The Modocs traded their slaves to the Klamaths, who shipped them farther north. The word among the various Indian tribes was that the
Modocs were not to be trusted; it was impossible to conclude a peace
with them.

There was no one chief who ruled over the autonomous Modoc bands
that roamed across northeastern California. Leadership was dependent

on who could sway whom by oratory, wealth, and performance, and was divided among religious, war, and political leaders within the various bands. The Modocs adapted quickly, first to the horses and buckskin that came from the Plains Indians and then to the weapons, tools, and clothes of the white man. They paid for being "white" Indians. In 1847 an epidemic caused by a disease brought into the region by Anglos killed one-third to one-half the tribe.

The whites were not an easy people, either. Their advantage was that they were more numerous and more technically advanced; their disadvantage was that they were newcomers to a strange terrain. John C. Frémont, a headstrong military man, passed through the region in 1846 and camped on the shoreline of Tule Lake while searching for a lost member of his party. The hunter finally stumbled into camp and said that he had eluded a party of Indians. Frémont named the body of water Lake Rhett, after a boyhood friend, but the name did not stick.

The party broke camp and traveled north into the territory of the Klamath Indians, who greeted the outsiders coolly. Four of Frémont's men were killed by the Indians. Frémont's vengeful scout, Kit Carson, retaliated by burning and pillaging an innocent Indian village. Frémont then rode south to foment a revolution against the Mexicans who governed Alta California.

The pattern was set, as it had been and was to be elsewhere: Killing followed killing. Fear spread across the territory just as a huge influx of whites was poised to descend upon California. No noble red men or noble white men were involved in the struggle, although heroes were subsequently created by both sides. One hundred years before the Vietnam War the United States government conducted a morally ambiguous conflict in a strange land against a people of a different race.

How that war was conducted was determined not only by the landscape, but also to a great extent by those outside the region who were fed information by highly competitive newspaper correspondents more interested in making a name for themselves than in supplying reliable reports. It was also determined by a military caste educated at West Point, where there were plenty of classes on how to conduct a European-style war but none on the guerrilla type of Indian warfare prevalent in the West for the better part of the nineteenth century.

It began as many wars do—inadvertently. A breakaway band of Modocs, under the leadership of Captain Jack, left the Indian reservation that the Modocs had shared uneasily with the Klamath Indians. The Modocs desired a reservation in their home territory, but white settlers had preempted their lands and now wanted the Indians removed. An Indian agent who had little understanding or sympathy for the Indians issued the call to arms.

An Indian account and a government account of the causes of the war have survived. A Modoc, Bogus Charley, said:

> Government he say he give grub. Give beef once—no game, hungry—stay two moons. Captain Jack say go back Tule— go there—settler there—game gone. Settler says, "Go away." Captain Jack say, "No, both stay." No grub—hungry—kill settler's cattle—soldier come, drive us back—fight long time.

The annual report of the Commissioner of Indian Affairs for 1873 stated that the Klamath Indians had demanded tribute from the Modocs and they had paid it. Despite the tribute, the Modocs were still harassed by their neighbors. There was no government money for rations for the Modocs so they returned to their home territory "where they became a serious annoyance to the whites, who had in the meanwhile settled on their ceded lands."

Soldiers and armed civilians attacked Captain Jack and another nearby band of Modocs in late November of 1872, killing Indian men, women, and children. The other band of Indians, led by Hooker Jim, retaliated by killing male settlers. Both bands fled to the lava beds, where they were joined by a third band of Modocs in their natural redoubt, a place of refuge that over time had been used and improved.

The army marshaled its forces and marched south to the lava beds. Brig. Gen. E.R.S. Canby, who was in charge of the military department responsible for the district, foresaw a quick end to hostilities. The Modocs were to be "prosecuted as vigorously as possible until they are destroyed or captured." Thus were General Canby and Captain Jack joined in the most costly Indian war, in terms of money spent and lives lost, in the history of the West. At its conclusion Canby's superior, Maj. Gen. J. M. Schofield, wrote, "The Department of the Columbia has been the scene of a conflict more remarkable in some respects than any other before known in American history."

Over the years I have spent a fair amount of time sitting in Captain Jack's cave in the lava beds attempting to feel the climactic event of that tragic war. I say feel because what exactly propelled Jack toward his destiny has been lost to history. But its context—the cave, a collapsed lava tube, the setting for the event—is still available for meditating about place and destiny.

The cave resembled a shell hole in the grass and sagebrush-rimmed plateau. There was a small mound of pahoehoe (roped) lava at its entrance that served as the rostrum for group meetings. The hole dropped from a view of Mount Shasta, fifty-five miles distant, into the earth and disappeared for a short distance in the direction of the mountain. It was a roomy cave. Jack had an extended family that traveled with him: a young wife, an old wife, a small daughter, mothers-in-law, a grandmother, a sister and her family, and others.

There is a description of the cave by Edward Fox, the *New York Herald* correspondent, who scored a scoop by tagging along with the peace commissioners when they went to talk with Captain Jack. Fox, a Britisher

The cave with Mount Shasta and the Medicine Lake Volcano in the distance

with army experience, was the paper's yachting correspondent before being sent west to cover the war. The longhaired correspondent displayed dash, a *Herald* characteristic. At the time, Fox was compared to Henry Morton Stanley, another *Herald* correspondent who had found Livingston two years earlier in Africa. Fox's sympathetic articles describing the plight of the Modocs stirred up the peace societies on the East Coast.

Of the cave, the correspondent, who wrote of himself in the first and third persons, said:

> I followed our guide, and after clambering up the rough walls of one chasm we walked, or rather crawled, about one hundred yards over some broken rocks, when the guide suddenly disappeared down a dark hole. The *Herald* correspondent followed, but, not being acquainted with the nature of the country, went down faster than necessary, and found himself in a large cave, lit up by the blaze of a fire, which was burning in the centre and gave sufficient light to enable me to see fifty or sixty Modocs seated round in circles four or five deep.

The smoke of many fires blackened the ceiling. The fires needed to be constant, or as constant as the gathering of firewood in wartime permitted. The surface of Tule Lake froze in the winter, and frequently the temperature dipped below zero. The wind from off the distant Pacific Ocean howled across that open expanse. The rocks that now litter the floor would have been cleared away then, with sleeping platforms arranged along the sides of the cave. This cave and others like it shared by the Modocs were far more weathertight than the army's exposed canvas tents. They were also effective shelters during the periodic artillery barrages that fell indiscriminately upon the stronghold.

Dust motes rose in the early morning shaft of sunlight. Spider webs spanned the black rocks splotched with the white droppings of bats. I tried to conjure up Captain Jack. It was not a clear portrait. First, there was the matter of his name. One account said he was given the name because he admired the brass buttons on the uniform of an officer of the same name. Another said a white man gave him the name of a miner he resembled. In any case, his name was less pejorative than other names bestowed on the Modocs by the whites: Hooker, Shacknasty, Scarface, and Bogus, for example.

In the loose organizational structure of the Modocs, Curly-Headed Doctor was the shaman, the religious leader; John Schonchin represented the military faction; and Captain Jack, who rose to political leadership at the time when the treaty with the whites was the main issue, was a conciliator and an organizer. But the whites, ignoring the fractured nature of their own government, insisted on dealing with Jack as the single leader, not realizing there were other factions in the band.

Jack possessed the characteristics of a political chief. He was a talker. Canby said Jack was "most prosey." Jack's oration at his trial was a masterpiece, if the translation and rendering of the transcript into perfect English was to be trusted. He did not speak English.

Captain Jack was nearly fifty at the time of the war, although he looked younger—perhaps forty, a reporter guessed. Jack was five feet eight inches tall and had a compact body. He was dignified. A jailhouse photo showed a wistful, strong, handsome face at a time when he knew his fate. Jack had once seen an Indian lynched at Klamath Lake. It was the first time he had witnessed this undignified type of punishment that compelled a man to dance upon the air.

It came to me while I was sitting in the cool recess of the cave that Captain Jack's dilemma symbolized the quandary of all leaders who serve at the sufferance of their people—do they lead or are they led?

The Modocs hunkered down in their caves for most of December and January and came out only to forage or to fight. The army gathered its forces and prepared to attack. The commander wrote, "We leave for Capt. Jack's Gibraltar tomorrow morning and a more enthusiastic jolly set of Regulars and Volunteers I never had the pleasure to command." There were three hundred thirty army regulars and volunteers from Oregon and California. Eight years after the Civil War had ended, the quality of soldiers was poor and morale was low. A recent reduction in pay had contributed to a 20 percent desertion rate. The nearby gold mines were an additional incentive.

On January 16, 1873, the troops advanced slowly in the tule fog that hugged the ground in the winter months. Objects were ghostly. Rocks looked like Indians, who actually had melted away from their outposts and were drawing the soldiers inward toward their perimeter of interconnecting trenches. The shaman placed a painted, tule-fiber rope around the inner perimeter, stating that no soldier would cross it.

The soldiers halted for the night, made sagebrush fires, then advanced in a skirmish line early the next morning. They were pinned down by the Indians' rifle fire and refused orders to move. Those Indians who spoke English taunted them, then picked them off one by one. The soldiers retreated in disorder. The commander exaggerated the size of the Indian forces and asked that his troop strength be doubled. On the Indian side, the shaman took credit for the victory. No soldier had crossed his rope, and there were no Indian casualties.

There is nothing like a military defeat and bad press to turn governmental policy toward a peace initiative, and that is exactly what happened after this first battle. Indian policy during the administration of President U. S. Grant was badly fragmented. The Department of the Interior favored the Indian agent approach, which meant government aid and reservations. The War Department favored the sword and the gun. Lastly, there was the Quaker policy favored by those who wanted a peaceful solution that supposedly lacked the corruption inherent in the first approach. The Indians would be turned over to missionaries, who would then remake them into Christians and docile citizens. At the end of January, the policy of the Grant administration tilted toward the Quaker solution, and a peace commission was appointed. Grant ordered Canby to take a defensive posture. In the words of one local observer, "Jawbone is cheaper than ammunition."

General Canby arrived in the camp under the bluff and near the windblown waters of Tule Lake to personally direct the troops and peace commissioners. He was a handsome, imposing man in his fifty-sixth year. Canby was married, but had no children. He was a West Point graduate who had served in all the requisite wars of his time. Canby had sought the command on the West Coast as a rest stop near the end of a long, active military career marked by the absence of enemies within the service, by competency, and by a proclivity toward administrative positions. He suffered from rheumatism and neuralgia, which were not helped by the biting winter cold.

When Canby first arrived in Oregon to command the Department of the Columbia, he had sought a peaceful solution with the Modocs, then turned toward force when the conflict escalated and the Indians remained intransigent. The general was convinced that the Modocs had to be removed from their territory. Concerning the Modocs, Canby wrote his wife:

> They are the strangest mixture of insolence and arrogance, ignorance and superstition that I have ever seen, even among Indians, and from this cause results great difficulty in dealing with them in any way but by force. They have no faith in themselves and have no confidence in anyone else. Treacherous themselves, they suspect treachery in everything.

There was delay on both sides. It took three weeks to assemble the three members of the peace commission. The Modocs were divided and argued among themselves. Jack was ill. He vacillated. The immediate commander of the troops, Col. Alvan C. Gillam, was ill. His officers were in disarray. There were now some seven hundred soldiers plus Indian scouts from a tribe that lived north of the Klamaths. The Modocs had between fifty and seventy warriors.

The newspaper correspondents were restive. They were enraged when the peace commission held secret meetings and refused to give them information. They retaliated in print by calling one commissioner a "Micawber politician," accusing the second of corrupt dealings with the Indians, and stating that the third could "talk the legs off a cast-iron pot in just ten minutes." The commission as a whole was referred to as the "High Old Joint Commission," or simply "High Old Joints." The stories were read in Washington, D.C., where the order went out to Canby to control the unruly commissioners.

The Modocs were being squeezed by a policy of containment. Their leaders had been indicted by an Oregon grand jury for the murder of the settlers. There were parlays. The Indians feared being lynched. The army, it was offered, would protect them on a reservation in Arizona. The Modocs had no concept of Arizona. They knew only Tule Lake. They refused to leave. They delayed. Using a tactic called "gradual compression," the troops ringed the stronghold and slowly moved closer to the Modocs' inner sanctum. Their horses were captured. The Indians became desperate.

The Modocs held their fateful council at Jack's cave. They sat against the walls or stood along its lip. The early spring wind buffeted the fire, and the light flickered unevenly across impassive faces that hid deep emotions. The shaman had scripted the scene. John Schonchin, the war leader, was the front man. I have participated in or observed such governmental councils that were held in less exotic settings, such as the state capitol in Sacramento, but were equally well-choreographed.

They waited quietly until thoughts appropriate to the moment had been gathered. Then Schonchin seized the initiative and jumped atop the low mound of lava at the entrance to the cave. He said:

> My people, I am old. I have been trapped and fooled by the white people many times. I do not intend to be fooled again. You all see the aim of these so-called peace commissioners. They are just leading us Indians on to make time to get more soldiers here. When they think there is enough men here, they will jump on us and kill the last soul of us.

As planned, Black Jim, so named for his complexion, followed. He said:

> Schonchin, you see things right. I for one am not going to be decoyed and shot like a dog by the soldiers. I am going to kill my man before they get me. I say we kill them peace-makers the next time we meet them in council. We just as well die in a few days from now, as to die a few weeks from now.

There was a stir. Black Jim called for all those in favor of killing the peace commissioners to step forward. A majority of the warriors did so. Captain Jack remained outside the firelight. His advocate stepped forward to talk. William Faithful defended Jack's leadership and said he should be heard. Black Jim readily agreed. Jack had fallen into the trap.

Jack emerged slowly from the darkness. He searched every face and felt the hostility. He talked long, and he talked peace. But the crowd tasted blood. It was restive.

Hooker Jim stepped forward with an ultimatum: "You will kill or be killed by your own men."

"Why do you want to force me to do a coward's act?"

"It is not a coward's act we ask of you. It will be brave to kill Canby in the presence of all these soldiers. You show them you dare to do anything when the time comes."

Jack vacillated. The plotters were prepared for this moment. They tripped him and threw him to the ground. A squaw's hat and shawl were thrust on Captain Jack's head and shoulders. There was laughter at his ridiculous figure and shouts of "Woman!" "Coward!" "White-faced squaw!"

Jack agreed to kill Canby.

Captain Jack sulked in his cave for the next few days, planning how to get out of his promise. But the others would not release him from his pledge. They kept taunting him. Months later at his trial, he testified:

> Hooker Jim said, "You are like an old squaw. You have never done any fighting yet. We have done the fighting, and you are our chief. You are not fit to be a chief." I told him that I was not ashamed of it; that I knew I had not killed anybody, and I did not want to kill anybody, and I would have felt sorry if I had killed any white people. They told me that I was laying around in camp and did not do anything, but lay there like a log, and they were traveling around and killing people and stealing things. They said, "What do you want with a gun? You don't shoot anything with it. You don't go anyplace to do anything. You are sitting around on the rocks."

The army employed a white man by the name of Frank Riddle and his Modoc wife, Toby, as translators. Riddle, derided by the military for being a squaw man, was a miner and a farmer who had been unlucky at most of his undertakings. His wife was a stalwart woman. On visiting the stronghold after the council, she had been warned by William Faithful of the plan. (I have often wondered if this wasn't an attempt by Captain Jack to negate his unwanted mission.)

The Riddles pleaded with the commissioners not to go to the meeting. General Canby refused to listen to the couple, whom he did not trust, and put his faith instead in a display of troops at a distance. He told a commissioner, "I think there is no danger, although I have no more confidence in these Indians than you have. I think them capable of it, but they dare not do it."

One commissioner, a clergyman, put his faith in God. The other two commissioners armed themselves with derringers for the meeting with the Indians in the meadow about halfway between the lava beds and the army camp. One of the conditions of the meeting was that the parties be unarmed. On reaching an army tent that had previously been pitched in the field for shelter, it soon became apparent to the party of four whites and the two interpreters that the eight Indians—three more than the number agreed upon—were armed with revolvers. There were bulges in their clothing. The two armed commissioners did not think the Indians were aware of their smaller weapons.

Eleven days into April 1873 and it was a beautiful Good Friday morning. A small sagebrush fire burned near the tent. Canby handed out cigars, and all except the clergyman smoked. The air was pungent with the smell of tobacco smoke and sagebrush.

Hooker Jim took the overcoat of one of the peace commissioners, put it on, and buttoned it all the way up. He turned to another Indian and asked if he didn't look like that old man, or that he would be that old man. The exact meaning wasn't quite clear, but the intent was. The commissioner glanced at Canby to see if he had caught the meaning. He said later, "I am very confident, although no words were passed, that General Canby understood the act and knew what it meant."

The general made a speech, saying that when he was younger he had helped resettle some Indians in the South, and that they had resented him at first but had grown to like him. He had never deceived the Indians and had dealt fairly with them. He would not remove the troops until this matter was settled. Jack and Schonchin said they wanted the troops removed and a reservation in their home territory.

Jack left the circle, ostensibly to relieve himself. Two Indians immediately rose from the surrounding brush and advanced toward the group, their arms laden with weapons. Jack quickly returned to the circle, drew his revolver, said "all ready," and walked straight up to the transfixed Canby and pulled the trigger. The gun misfired. Jack cocked the revolver again. This time it fired. The bullet struck Canby in the left eye and traveled through his brain, killing him instantly. Shots erupted on all sides. Besides Canby, the clergyman was killed, a second peace commissioner was badly wounded, and the third ran like a rabbit and was unharmed. The Riddles, who were not targets of the Indians, escaped uninjured.

All the clothes were stripped from the bodies of Canby and the clergyman. The wounded commissioner lay half-scalped and unconscious in his underwear on the ground. The troops were alerted by the signal officer, who had the parlay under his glass. There was indecision and delay, but they eventually arrived at the scene on the double quick. Two newspaper correspondents accompanied them to the gruesome scene.

The reaction of the press was predictable. There was a call to arms. The story in the *San Francisco Chronicle* appeared under a headline that screamed THE RED JUDAS. The headlines in descending one-column order over Fox's story in the *New York Herald* read MASSACRE/ BLOODY TREACHERY OF THE LAVA BED INDIANS/ GENERAL CANBY AND THE REV. DR. THOMAS BUTCHERED. In his story, Fox referred to Canby as "the noble old gentleman" and "the hero of many a fight." Abandoning his former sympathy for the Modocs, Fox's story began:

> Peace policy and the Indian Bureau have accomplished the bitter end, and offered as martyrs to the cause the lives of General E.R.S. Canby, commanding the District of the Columbia, and the Rev. Mr. Thomas, of Petaluma, Cal., Presiding Elder of the Presbyterian Church. As my courier leaves instantly, having eighty miles to ride, I can only give brief details of one of the most treacherous massacres ever perpetrated by the Indians.

The reaction elsewhere was as immediate as the telegraph lines permitted. President Grant was beseeched by Quakers and other humanitarian groups in the East "not to allow the clamor of an ungodly press or the passionate reasoning of those about him" to sway him from following his peace policy. In the West, the dominant attitude toward the Indians differed. The Yreka newspaper, which served the California town closest to the lava beds, said the fate of the Modocs "should be such as to strike a salutary terror among other Indians."

General William T. Sherman, the commander of the army, immediately met with Grant, who was Sherman's and Canby's commander during the Civil War. The president, through Sherman, ordered the "utter extermination" of the Modocs. General Sherman notified his commanders in the field, "The President now sanctions the most severe punishment of the Modocs and I hope to hear that they have met the doom they so richly have earned by their violence and their perfidy."

Sherman believed that treachery was inherent in the character of Indians. He told a reporter from *The New York Times* that he did not think that the Modocs would be able to escape "although the topography of the lava beds is something of a puzzle." A lesson was to be made of the Modocs that would not escape the notice of other rebellious Indian tribes. Three years later the Sioux at Little Bighorn were familiar with the Modoc experience.

General Canby was treated as a national hero. The day after his death Canby's troops gave him final military honors within sight of the lava beds. His body was then taken to Yreka, where it lay in state in the Masonic Temple, and from there to Portland, where the community's respects were equal to those given President Lincoln. The body was then shipped to San Francisco, where it lay in state for two days. The governor of California and the mayor of San Francisco paid their respects. All the city's flags flew at half-mast. The casket was loaded onto a special Central Pacific Railroad car and transported east to Indianapolis.

There were numerous stops; and homages were paid along the way to the general's final resting place, where services were conducted by Baptist, Episcopal, Methodist, and Presbyterian ministers. Four of the army's highest ranking generals, including Sherman, were at the funeral. Canby was buried under a large granite marker. A fort at the mouth of the Columbia River was renamed Fort Canby; and some minor landmarks in what was to become known as Modoc County, California, were also to bear his name.

The general died penniless. In addition to her small widow's pension, his wife was granted an additional pension by Congress; and she drew upon the interest generated by five thousand dollars raised for her by the citizens of Portland. Canby's brother, upon hearing the news, went into shock and was committed to a state lunatic asylum, where he died.

The treatment of Captain Jack, his family, and his tribe was to be quite different.

There was never a question that some Indians would hang for this crime; it was merely a question of who, how many, and how quickly. It would take the army fifty more days to capture Captain Jack. In the process the army suffered far more casualties than did the Indians even though it was better provisioned, not hampered by women and children, and had far more troops and horses. After the capture of Captain Jack and his followers on June 1, the Modocs were brought to a camp on the Peninsula where carpenters immediately set about constructing a large gallows. The commanding officer planned to hang between seven and ten Modocs immediately, but his superiors opted for a trial.

The Modocs were taken to Fort Klamath at the northern end of the basin in Oregon. It was an idyllic location. Compared to the harsh lava beds, the land was soft. The fort was located in a flat grassland that bordered the marshes of Klamath Lake. The white clapboard structures were arranged in neat precision. The prisoners were shackled in the stuffy guardhouse. They could barely move about in their minute cells. They were slowly dying. The curious came and gaped. Admission was free.

The four-day trial was a charade. There was a great rush to get the proceedings under way and completed with dispatch. The competency and fairness of the Riddles' translations were questioned. While they were paid ten dollars a day to translate by the government, they were also witnesses for the prosecution. The Indians understood very little of the legal proceedings: They did not question any government witnesses. Their lawyer was not informed when they were to be tried. He arrived at Fort Klamath the night the trial ended. The judge advocate spent an inordinate amount of time excusing the army of any wrongdoing.

The verdict was guilty. The punishment was to be hanging, a particularly abhorrent death for the six Modoc defendants, two of whom were notified of their pardons in the early morning hours of the day of execution, although the pardons had been issued earlier by the president. Those to be executed were Captain Jack, John Schonchin, Black Jim, and Boston Charley—all of whom had taken part in the assassination of Canby. Hooker Jim, who had helped the army capture Jack, and the other Modocs, who had sold their services to the army, strode about the fort as free men.

October 3, 1873, was a clear fall day—a good day for a hanging. The *Herald* correspondent wrote that "a horse race and a hanging are two of the most popular amusements" in the West. On this day there was to be a quadruple hanging, and two thousand spectators were on hand at Fort Klamath. A correspondent for a Vermont paper wrote:

The mountain roads from Jacksonville and Ashland to this place are reported as filled with Oregonians, hurrying, in all descriptions of vehicles, to be in at the death. Many of the settlers in the Rogue River Valley are on their way here. The "reliable gentleman" from Ashland tells me that one school in that town has been given a week's holiday, so as to enable the preceptor and pupils to come here and gloat over the ghastly scene. But few from Yreka will be on hand, as the Siskiyou County Fair begins this week.

There was a festive atmosphere. Certain people were out to make money. The bodies of the Modocs were much in demand for display, and an embalmer said he had made a deal with Captain Jack for his body. He showed a receipt. An order was issued not to mutilate the bodies of the Modocs after the execution, but apparently it did not apply to the army. About one hour before the execution, an officer visited Jack in his cell and, guiding Jack's hand, obtained for sale a dozen pieces of paper autographed with his mark—two triangles joined at their apexes.

The insistent beat of muffled drums marked the procession to the gallows in an open field to the south of the fort. On this warm day, Jack kept a blanket drawn up to his ears and tightly wrapped about him. He had to be helped to the gallows by a corporal, as he was weakened by four months of close confinement.

The garrison dogs darted back and forth among the Indians, white civilians, and the troops who stood at attention. An officer read the orders. A chaplain read the Episcopal service. The four Modocs looked around occasionally. They were offered water. A slight breeze stirred the dry grass.

The nooses were placed over their heads and adjusted. Next came the black caps made from old army haversacks and dyed for the occasion. They removed Jack's blindfold and a corporal clipped his hair, ostensibly so that the noose would fit better. Locks of Jack's hair, pieces of the hanging rope, and Modoc artifacts were hawked by officers and others after the execution.

At a handkerchief signal, the bodies dropped, convulsed, twitched, and finally swung back and forth. The command rang out, "In parade, rest!" Breaking the decorum of this solemn rite was the wild dash on horses from the scene of three couriers who carried the dispatches of the newspaper correspondents to the nearest telegraph, some ninety miles distant. Carrier pigeons and tricks were used to get a jump on the competition. RETRIBUTION proclaimed the *Herald* in bold-faced type the next day.

The gallows were constructed so they could be seen from the stockade where the other captured Modocs were being held prisoner. The wailing of women and children could be heard from that quarter. As a sign of mourning, the Indians blackened their faces, which had paled during the long confinement.

After a half hour, the bodies were cut down, placed in coffins, and carried to a closed tent where there was a long table, similar to a dissecting table used in medical schools, covered with India rubber. Nearby were surgical instruments. The heads were secretly separated from the bodies and shipped in a barrel filled with alcohol to Washington, D.C. The *San Francisco Chronicle* reporter who stumbled upon the tent and its grisly contents wrote, "Language cannot be found severe enough to denounce this act." He asked, "Where is our boasted civilization, when, after dealing out strict justice to these captives, we shamefully mutilate their remains?"

The four bodies were decapitated at the request of Dr. George Alexander Otis, a native of Boston and a graduate of Princeton University and the University of Pennsylvania Medical School. Otis was the curator of the Army Medical Museum; his hobby was the study of the cranium, especially that of the North American Indian. He had amassed a large collection of skulls at the museum. Hearing of the imminent demise of the four infamous Modocs, the Washington doctor arranged with the post surgeon at Fort Klamath to have the heads shipped east.

The four crania, with numbers stenciled on their foreheads, became the only heads among some twenty-one thousand North and South American Indian skulls held at the Smithsonian Institution of Natural History whose identities were known. The Modocs believed, with good cause, that the remainder of the four bodies were spirited off by other whites before or shortly after burial. The skulls were returned to the Modocs a few years ago.

Fifteen days after the executions a sad procession headed south. One hundred fifty-five Modocs, many chained together in wagons and with their faces blackened, passed before assembled knots of curious citizens as troops escorted them to the railroad at Redding, California. An equal number of Modocs who had not participated in the war remained on the

Klamath reservation. The captives, including Jack's family, eventually wound up in Oklahoma, where diseases killed one-quarter of them in the first four years. On hearing of the death of a former Modoc warrior, a California newspaper commented, "We have some respect for an Indian who is dead." A corrupt Quaker Indian agent contributed to the hard times of the Modocs. By the turn of the century, only fifty Modocs were alive in Oklahoma.

Ninety years later their descendants returned to the stronghold for the first time and held a reunion. Indians and whites held hands in a circle at the welcoming ceremony. Gerald Jackson, a Modoc from nearby Klamath Falls, smiled, injected wry humor into the proceedings, and made white visitors feel at home. Jackson said, "I am not better than you are. You are no better than me."

Jackson said he thought we were there because of our ancestors. The Modocs wanted to acknowledge the spirit of their ancestors, not with a memorial but with a celebration of life and a search for meaning. Perhaps the presence of whites had more to do with racial guilt, and in that way was ancestral.

Perched on the craggy volcanic formations that surround Tule Lake is the largest winter concentration of bald eagles in the contiguous United States.

The bald eagle is a ferocious-looking bird with its crinkled talons, curved beak, and angry, unblinking eyes. The feet, beak, and eyes are yellow; its head and tail are white; its body is dark brown. The bald eagle is a fitting symbol for a professional football team, an elite military unit, or a nation of warriors. The Amazons of California had their gryphons—half eagle, half lion—and we have our bald eagle, whose lineage dates back twenty-five million years to the ancient sea eagles of Southeast Asia.

Down through the centuries, the eagle has represented a dual nature: The eagle has been equated with pride, strength, victory, and the type of spiritual qualities that can ascend unblinkingly toward the sun. The eagle was the solar bird on which Vishnu rode; and it was the bearer of lightning for Jupiter, the storm god. But the eagle was also one of the four beasts of the apocalypse and has been associated with the raven, another eater of carrion. John James Audubon found the feeding habits of the bald eagle repulsive and cowardly—not worthy of a national symbol. As for its proclivity to steal the food of other birds, Benjamin Franklin declared, "He is a bird of bad moral character."

In his multi-volume work *The Birds of California*, William Leon Dawson wrote in 1923, "We will grant without debate that the Bald Eagle is a bad actor. He eats fish—a most reprehensible practice—and he occasionally captures game birds, which we would prefer to do to death by our own peculiar artistry." Sometimes bald eagles fed on lambs and fawns, in addition to which "he has been known to carry off babies—say in two or three really authenticated instances in our national history," wrote Dawson. Other authorities say there are no proven baby snatches, although bald eagles have been known to make off with domestic pets and small pigs.

As it begins to get cold in the breeding grounds of interior Canada and along the coasts of British Columbia and southeastern Alaska, the eagles migrate southward, starting to arrive at Tule Lake in November. Their population builds to a peak of some five hundred birds in January and then declines from mid-February through March. They roost in large trees along the sheltered sides of such volcanic features as Mount Dome, Three Sisters, Caldwell Butte, and Cougar Butte.

Bald eagles are substantial birds with wingspans of up to eight feet. They need old-growth trees to accommodate their weight. Little foliage and sturdy branches make for a clear flight path and a secure place on which to roost. Great heights are needed from which to search for prey. From mature ponderosa pine trees on the slopes of the Medicine Lake Volcano they have an unobstructed view of Tule Lake.

The eagles prey upon the plentiful waterfowl that use the lake as a rest stop. There are peak migrations of one million birds through the Klamath Basin National Wildlife Refuges, of which Tule Lake is one. The basin, located along the Pacific Flyway, hosts one of the largest populations of waterfowl in the country. I have seen sheets of snow geese rise from the waters of the lake, with the implosion of a single wing beat, and dissolve into the sky. Backlit against a sun beginning to set behind the sere volcanic cones and long ridges, grebes trail an iridescent wake in the darkening water. In winter the lake freezes over, and a few pintail ducks and other waterfowl are caught in its icy grasp.

Bald eagles have keen eyesight and stealth in flight. With a swift plunge, the eagle grasps its prey with large talons that can crush bones and pierce flesh. It tears with its sharp beak. The bald eagles are on the ice before dawn, feeding on sick, trapped, or even healthy birds. Feathers are scattered about the dead bird, whose broken neck flops at an unnatural angle. A single talon pins down the carcass as the eagle casts its yellow eye about, then dips his sharp beak into the warm breast. When the eagle has taken his fill, he flies to a nearby tree and preens. There he remains on his perch until sunset, when he flies off to the southwest in the direction of his roost on Mount Dome. There can be hundreds of birds on a single roost. This is their social time, and standing is determined by age and aggressiveness.

Tule Lake and the other refuges are managed for the benefit of farmers as well as wildlife. They are small ponds in the midst of large farms. In *Silent Spring*, Rachel Carson described the deaths of hundreds of fish-eating birds due to pesticides being dumped into Tule Lake in 1960. Because of the decreased use of DDT, increases in some breeding populations of bald eagles have been noticed. There is some talk of taking them off the endangered list. Now the greatest single cause of premature death is shooting.

The conflict between the needs of wildlife and agriculture dominated the history of northeastern California between the two events that attracted the attention of the outside world. It was a nature photographer who was responsible for drawing President Theodore Roosevelt's attention to the Klamath Basin. Roosevelt created the first refuge in 1908 on the basis of the images produced by William L. Finley, of whom the bird authority Roger Tory Peterson wrote, "Finley's contribution to environmental awareness can be equated to that of only one other naturalist who was on the scene in the West during those early years—John Muir." In later life, Finley went on to document the condor in Southern California.

Finley was attracted to Tule and nearby Lower Klamath Lakes by the shooting of plume birds. Like other ornithologists of the time, Finley's hands were not clean. He had gotten his start collecting and selling bird eggs, feathers, and skins. When a national tide of revulsion arose over killing birds and collecting their eggs and plumes, Finley renounced his past activities. He said, "The taking of eggs for the purpose of selling them for a few paltry dollars is an outrage often perpetuated under the guise of collecting for scientific purposes."

Wildlife photography was just beginning to evolve. Englishman Eadweard J. Muybridge, who had documented the more stationary aspects of the Modoc Indian War, had made his studies of animals in motion. Another English photographer had managed to photograph birds in flight in 1888. The wide-angle and telephoto lenses were just coming into use. Camera speeds were faster, but film was still slow; and the glass negative plates and large view cameras were cumbersome to lug about in the field.

Two years after graduating from the University of California, Finley and his boyhood collecting friend and photographic partner, Herman Bohlman, were dispatched by the National Audubon Society to the Klamath Basin in 1905 to investigate the shooting of waterfowl for hat plumes and for meat. They found no plume hunters but discovered that one hundred twenty tons of ducks and geese had been shipped south the previous fall to the markets and restaurants of the San Francisco Bay Area.

The two young men clowned about, worked hard, and eventually got their pictures by waiting patiently in a blind made out of an old wagon umbrella with a green cloth attached to the ribs; a large camera stuck out of a flap. The ducks floated in and out of the viewfinder. The photographer pressed the shutter release, changed plates, refocused, and released the shutter again—a much more cumbersome process than the simple aiming of the small motor-driven automatic cameras of today.

Despite being a wildlife refuge, the U.S. Reclamation Service closed off the flow of water to the shallow lake in order to create more farmland for a land development company in 1917. Finley was bitterly disappointed. He wrote:

> Man has a peculiar habit of building something with his hands and, at the same time, kicking it to pieces with his feet. In no direction is this more graphically illustrated than in conservation and use of natural resources. With grandiloquent inconsistency we move to conserve or develop one resource while, at the same time, we are destroying another. This is, no doubt, all a part of man's imperfection, and is comparable with his inability to avert war or solve, with his vaunted intelligence, the problem of changing social and economic order. No more graphic example of this can be found than the destruction in the name of reclamation of important areas in southern Oregon and northern California.

By the early 1940s, when the first Japanese-Americans arrived at the relocation camp, Tule Lake was less than one-fifth its original size. The descendants of the original white settlers were established farmers, and others had moved into the basin to homestead land the Modocs would not have recognized. What the Bureau of Reclamation called "sunbaked prairie and worthless swamps" had become "the present broad expanse of lush, green fields." Tule Lake was used as a sump for the other lakes in the irrigation project and received their residue of pesticides and other wastes.

War returned to the volcanic terrain in 1942. Shortly after the bombing of Pearl Harbor, ninety-five thousand Japanese-Americans were rounded up in California and, along with others from elsewhere on the West Coast, shipped off to ten internment camps located in the less desirable portions of the interior West. There were two camps in California. One was located at Manzanar in the Owens Valley and the other in the small community of Newell, bordering Tule Lake. Those judged to be the most intransigent and dangerous were eventually sent to the Tule Lake War Relocation Camp, the largest of all the camps. Tule Lake, surrounded by the somber relics of volcanism, was far from any urban center.

The settlement of Newell, just to the east of the Peninsula, was named after the first reclamation director. The Bureau of Reclamation made the land available for the Tule Lake War Relocation Camp—a total of 1,100 acres just to the east of the Peninsula for the camp itself and an additional 3,575 acres that internees would farm. Local residents bitterly opposed the camp. Representatives of the chambers of commerce of the small farming settlements, the local grange, and the American Legion met in the town of Tulelake on April 11, 1942, to draft a resolution. (Tulelake is the town; Tule Lake is the lake and basin.) The local newspaper reported:

> The resolution as ordered by the group Monday noon will point out that the Tule Lake basin was settled by whites and has no orientals or negroes among its residents. It will voice the desire of the Tulelake chamber to maintain the present character of the population. Southend residents and officials who attended the meeting pointed out in general discussion

that the Tule Lake area was settled and developed by Caucasians and that it would be obviously unfair to deprive them
of the rich land in favor of the Japanese. They also pointed out
that American-born Japs would be more apt to remain in the
basin after the war.

These sentiments were not limited to the Tule Lake Basin. A representative of agricultural interests in the Central Valley was quoted at
about the same time in the *Saturday Evening Post* as stating, "We're
charged with wanting to get rid of the Japs for selfish reasons. We might
as well be honest. We do. It's a question of whether the white man lives
on the Pacific Coast or the brown man. They came into this valley to
work, and they stayed to take over."

The local congressman agreed with his Tule Lake constituents. Secretary of War Henry L. Stimson brushed aside these arguments. He wrote
the congressman, "Our experience has shown that so rarely has any local
group approved of the establishment of the centers that it is impractica-
ble and unwise to seek approval." He then added a note of reassurance.
The camp would be in a prohibited zone and "an external guard of military police will assure full protection to nearby communities and to the
Japanese themselves." The congressman was assured by the Bureau of
Reclamation that the Japanese would acquire no rights to the land they
would farm.

The government quickly got the message across that the region would
benefit economically from the camp. In the first public acknowledgment
that there would be such an entity, it sent purchasing agents into Klamath
Falls to buy one hundred thousand board-feet of lumber. There was a
one-month completion date for the center. The local paper noted,
" 'Gone to the Jap camp' has become a by-word with farmers in the south
end of the basin."

Farm laborers and store clerks deserted their employers to work at
higher-paying jobs in the camp. The relocation center, consisting of some
one thousand buildings, was completed in thirty days; and the advance
party of nearly five hundred internees arrived on May 27, 1942. In early
June, larger groups began to arrive by railroad. The camp would eventually hold more than 18,000 of the 110,000 Japanese-Americans—two-
thirds of whom were American citizens—who were relocated during
World War II.

To a people who worshiped orderliness and had a highly developed sense of aesthetics, the camp was a messy jail. The flat one-and-a-half-square-mile site, rated as "cheap pasture" by the Bureau of Reclamation, ran east from the Peninsula to Horse Mountain. Tule Lake had once covered the site of the camp. The sandy soil contained the shells of freshwater snails, and the resourceful internees made jewelry from them.

The camp was bounded on the west by the Canby-Hatfield Highway and the Central Pacific Railroad, from which a spur track ran into the center. With military rigidity, ten gigantic wards were laid out into nine blocks apiece. Each block consisted of sixteen tar-papered army barracks, a mess hall, recreation hall, toilets and showers for men and women, and laundry facilities. All were ringed by a tall barbed-wire fence and guard towers. For a culture that depended on the family unit for stability, the communal aspects of military architecture set amid the alien landscape would be nearly ruinous.

The first shock came upon arrival. The windows in the ancient Pullman cars were blacked out during the all-night and all-day train rides from Seattle, Portland, Sacramento, Oakland, Los Angeles, or wherever. In the harsh morning light the well-dressed evacuees descended directly from the cars onto a flat sea of sagebrush. One evacuee wrote, "Another processing—family number, contents of luggage, etc., etc. Tired, sleepy, and uninspired by that process, I got off the train. Lo and behold, I was greeted by a hill overshadowing the camp, and beyond that hill was the snow-clad peak of Mount Shasta."

Of the mountains that were in sight of the Manzanar and the Tule Lake centers, another evacuee later wrote, "They also represented those forces in nature, those powerful and inevitable forces that cannot be resisted, reminding a man that sometimes he must simply endure that which cannot be changed." The Peninsula was called Castle Rock for its crenellated rock formation that is now topped by the cross. When the security loosened up a bit, the prisoners were allowed to climb the hill. The adults searched for Indian artifacts while the young boys played war.

To these people who had lived on the coast or in the irrigated interior valleys of California, this land was true desert. It was dry, at least when it wasn't raining; when it rained, it was a quagmire. There was no middle ground. Bleak, barren, windy, isolated, frightening, disorienting, depressing, dusty—these were the words that came to mind. There was the biting cold of winter and the merciless sun of summer—a sun that baked the skin a bronze color, a color that earned the epithet "California nigger" for some. But mostly it was the ever-present dust that sapped the spirit. One Tulean recalled:

> Dust. Dust. The weather of Tule Lake, as unpredictable as a woman in a millinery shop. Snow in May, Indian Summer in November—but all year round, wind, wind, and more wind. Wind gentle as a baby's breath; strong enough to rattle the windows; wild enough to shriek between the telephone wires—whirling dust and papers like a miniature tornado—sending fine dust particles seeping through the windows; blanketing furniture and floor with a coating of white. Dust. Dust. Dust.

All lines moved slowly: lines to be fingerprinted, lines to be photographed, lines to receive bedding, and lines to eat the strange food in this strange place. A woman recalled years later, "I just remember when we got there how bare it was, just a few cots. What I remember about the first day: when we went to dinner, my younger sister got lost. We had to cross the firebreak to another block. That was the fear of getting lost; you'd go to another block and it looked like the same thing."

Tule Lake was a prison that would later be called a concentration camp. "We didn't dare go near the fence for fear of being shot at, and there were instances of that. But in addition to the physical confinement, there was the fence around our spirit, and this imprisonment of the spirit was the most ravaging part of the evacuation experience," said one man.

From outside the fence, there was a different view. The local farmers gave the Japanese-Americans good marks for their farming techniques. But they were bitter because the evacuees were farming the land they had leased from the government. The locals thought the prisoners got all the sugar, coffee, and meat they wanted, and thus, that these rationed items were wasted. They also thought that all of the locally bottled soda pop was going to the camp, and they resented this favoritism. "Why on God's green earth should the yellow man get first choice and the American dealers and the public get second best?" asked a local resident. Those who lived near the camp were jittery. They kept guns by their beds in case there was a mass breakout.

To the Caucasian workers inside the camp, the tension and bitterness were palpable. There were advantages. It was an opportunity to have cheap servants. "Well, you know, each morning we had two or maybe three women that came to our house and cleaned. Oh, it was the life of Riley. I've never had it so good," said the wife of a civilian worker.

Other prisoners were about. For a brief time, Italian prisoners of war were brought in to help the farmers harvest their crops. In town there was a German prisoner-of-war camp. The four hundred prisoners lived in tents in the town park or were housed in the old Civilian Conservation Corps camp to the west of town. The Germans helped the farmers harvest their crops. It was a "friendly situation," recalled the son of a farmer who employed the prisoners. The army guards, if they liked, could work alongside the prisoners. Both were paid $1.25 a day, far more than the sixteen dollars a month paid the Japanese-Americans for similar work in the camp's fields.

A farmer took his German workers a case of beer, soda, or some coffee; they reciprocated with handmade gifts, such as paintings. The Germans rode their bicycles into town (from which the Japanese-Americans were barred) to make purchases in the stores. They picnicked in the fields wearing blue jeans and shirts on which P.O.W. was stenciled.

Inside the camp, tension kept building. There were strikes, incidents of vandalism, and beatings. Factions arose, coalesced, and evaporated to re-form into different cliques. Fear was a constant, especially, according to one inmate who was a trained sociologist, "the fear of *inu* [dogs] who were thought to be spying on the evacuees and passing on damaging information to the authorities. There were said to be one-hundred-dollar *inu*, one-hundred-fifty-dollar *inu*, and so on, the different figures representing the alleged bounties paid by the FBI to the informers. When someone was beaten, it was often justified on the grounds that the person was an *inu*."

Then there were rumors. Rumors spread like molten lava throughout the camp and solidified into granite that could not be dislodged: Japan was winning the war. Tule Lake would be a hospital for wounded American troops. Freight cars full of crutches (actually wooden laundry racks for the evacuees) had arrived. Food supplies in the central warehouses would soon be depleted. (In anticipation of food shortages, leftovers from the mess halls were saved and dried, and there was a run on canned food that could be purchased from the canteen.) Tule Lake was within the forbidden coastal zone; and transfer, it was thought, was imminent to Arkansas, which was farther away from home.

A movement—or rather a number of them—sprang up within the camp to fill the vacuum of leadership that white authorities could not fill because of their lack of understanding of a different culture. The War Relocation Authority (WRA) officials saw it as a guerrilla movement headed by outcasts—*rowdies, fanatics,* and *cranks* were some of the words used in official reports—that was designed to wrest control of the center from them. The movements were never that cohesive or well articulated. Rather, most were aimed at the issues of the moment, which had more to do with improving conditions. Misunderstandings became antagonisms that escalated to violence.

The more militaristic groups marched early in the morning. Goose-stepping buglers led the way; the ranks were filled with white-robed young men with a white band, called a *hachimaki,* tied about their heads. The sentries and military police, some of whom had served in the South Pacific, watched coldly from the perimeter of the camp.

An inmate, who was a minister, commented on the dissensions that were rife throughout the camp:

> The relocation center community was in an explosive condition. Nothing, no matter how good in its own right, was good enough to please everybody in the camp. We had one strike after another: kitchen crews, coal workers, construction crews, farm workers. No one knew what was going to happen from day to day.

The entry in the diary of a Tulean for September 8, 1943, noted:

> Brought in 6 tanks for use against colonists in Tule just in case. Damn those guys anyway. What the hell can we do against two battalions armed with machine guns. As it goes we're prisoners of war and no doubt about it. Double fence, guard towers, soldiers, tanks, armored cars, jeeps, electrically charged fence, etc. Damn their hide!

Within a year of establishing the centers it became clear to the Roosevelt administration that internment was not the best solution for all Japanese-Americans. Some had skills that were badly needed in the civilian sector, and the army was beginning to recruit Japanese-Americans for active duty in a separate unit. The problem was determining who was loyal and who was disloyal. The solution was a questionnaire. The result was a long document with some forty questions that were poorly worded, complex, and filled with ambiguities and Catch-22s. After all that had been done to them, who could trust the white man's sheet of paper? The possible consequences of the wrong answer seemed terrifying: no homeland, separation of families, a son killed in the war, serving a country that hated them, the label of *inu*, evacuation to Arkansas, and so on. Some preferred life in Japan.

A thirty-two-year-old woman who refused to answer the questionnaire, which automatically conferred upon her the status of "disloyal," was asked for her reasons by a WRA worker at Tule Lake. She answered:

A. I have American citizenship. It's no good, so what's the use?
Q. Has the evacuation caused you to lose faith?
A. I feel that we're not wanted in this country any longer. Before the evacuation I had thought that we were Americans, but our features are against us.

Did she want her child to be brought up in Japan?

A. Yes. I found out about being an American. It's too late for me, but at least I can bring up my children so that they won't have to face the same kind of trouble I've experienced.
Q. You realize that you will have difficulty in adjusting to life in Japan?
A. I know that, but I'm willing to try it anyway. It's too late for me. The important thing is that my children will not have to go through the same experiences as I have.

When the results were in, Tule Lake had the highest number of "disloyals." The solution was to change its status from a relocation center to a segregation center. All those in the other camps who had not sworn absolute allegiance to the United States would be sent to Tule Lake, while the "loyals" at Tule Lake would be dispersed to other camps. This massive reshuffling took place in September and October 1943. After some thirty-three train trips, nine thousand internees were shipped into and out of Tule Lake. Of the original occupants, some six thousand who wound up in the "disloyal" category remained there. In the view of Washington, all the bad apples were now in the same basket. The problem was that the basket was very crowded and badly divided within itself.

George Kuratomi was an example of a "disloyal." He arrived at Tule Lake with a group from the Jerome camp in Arkansas. On his second day in the new camp the twenty-eight-year-old *Kibei* (a person of Japanese descent who was born in the United States but educated in Japan) went to see the director, Raymond R. Best, in his office in the administration building. Kuratomi had been involved in organizing activities at Jerome, and Best had been apprised of this fact. Best was a former marine who had served in World War I. He had a son in the marine corps. Best had been in the Soil Conservation Service before switching to the WRA, where a colleague had characterized him as "prone to jump to conclusions" and "the enthusiastic backslapping Rotarian type."

Best began the conversation. "I don't recognize any group activity," he said. "I don't care what you have done in the past, but as far as this center is concerned, you shall represent no group or groups of people. I am not interested in your demands." That ended the meeting.

Kuratomi was astounded. He had been given an introduction to Best from the assistant director of the Jerome camp, who had accompanied the internees to Tule Lake. He had simply wanted to meet Best, not make any demands. He felt that Best's behavior was rude and that he did not understand Japanese-Americans. This encounter did not bode well for the newly reconstituted camp, Kuratomi thought.

He wrote to a friend at the Jerome camp, but the letter never arrived at the Arkansas center. First-class mail was censored. The letter was placed in his dossier at Tule Lake. "I don't know the reason for his hot-tempered behavior, but if he thinks he can employ such high-handed methods in administering us, I'm afraid he is making a big mistake," wrote Kuratomi.

Kuratomi, who was born in San Diego, was taken to Japan at the age of seven. He was educated there until he was fifteen years old and then returned to complete high school in San Diego, where he graduated with honors. After another brief trip to Japan, he returned to San Diego and, by dint of his quick intelligence and hard work, had progressed from being a clerk in a fruit-and-vegetable store to managing a retail store to ownership of a wholesale produce market.

Kuratomi was a devout Buddhist who taught Sunday school and was a group leader in his church. Slender, of medium height, and high-strung on occasion, Kuratomi had a dignified manner. Although he displayed flashes of arrogance that did not sit well with his captors, he was an effective orator in both English and Japanese. Kuratomi had been a foreman of a logging crew at Jerome, where he had quickly risen to a position of leadership on the questionnaire issue. At Tule Lake he had a fiancée and a baby.

To Kuratomi the initial evacuation in April 1942 to the Santa Anita Assembly Center, a racetrack in Los Angeles County, was proof of his second-class citizenship. From time to time he had flirted with the idea of returning to Japan but had rejected it. He would not renounce his citizenship, no matter what its status. Kuratomi wanted to work within the system, to move it in a direction that would benefit his people.

The more conservative elements in Tule Lake did not want to make any waves at all, while the hotheads wanted to create the sovereign state of Japan within the barbed wire. Kuratomi fell somewhere in between. Jerome had been a hellhole, Kuratomi wrote his friend; but Tule Lake was worse. Toilet and shower facilities were unsanitary, and it was badly overcrowded. "The most pressing item appears to be dust control," he wrote. These were typical of his concerns.

The camp administrators equated all organizers with the most militant factions. The WRA semi-annual report that covered the activities of the last six months of 1943 stated that during October at Tule Lake "an undercover movement to get control of the center was in progress." The report described Kuratomi and other leaders thusly:

> The leaders were not old residents of Tule Lake, but men who had gained some prominence as minority leaders in the centers from which they were transferred to Tule Lake; chiefly they were from Jerome, Poston, and Heart Mountain. They were men who in pre-evacuation days had failed to achieve leadership in their communities, some of them having been repudiated as fanatics and cranks. Under the tension and stresses of the evacuation, they had managed to win minority

leaderships within the relocation centers, and there they had gained expert knowledge of center politics and evacuee psychology.

On October 15, 1943, an incident occurred that galvanized the camp. A truck carrying twenty-nine farmworkers overturned and five were seriously injured. One died shortly thereafter. The truck driver was a teenager. Eight hundred farmworkers refused to go to work, protesting the lack of safety measures. The harvesting of crops that fed Tule Lake and other relocation centers was jeopardized.

The farmers wanted to hold a public funeral service for their coworker. Director Best refused permission. A public funeral was held anyway, and when a member of the center's staff showed up to take photographs, he was roughed up. The administration turned off the electrical power. The public address system went dead. The WRA decided that it would pay the widow and children of the dead farmworker a monthly compensation of two-thirds of his monthly pay of sixteen dollars.

On the day after the truck accident, and without the permission of the administration, a center-wide election to the *Daihyo Sha Kai* (representative group) was held; and Kuratomi was elected chairman by a wide margin. Kuratomi's immediate concern was to halt future truck accidents. The group's negotiating committee, with Kuratomi heading it, met with Best on October 26. Kuratomi wanted the director to issue a statement of regret about the accident. Best refused, but he agreed to look into the food situation and to improve sanitary conditions.

The committee was heartened, but it was not until two days after the meeting that they learned of the importation of strikebreakers. Best enticed Japanese-Americans from other camps to harvest the crops at one dollar an hour—a pay scale that would enable them to earn in two days what a Tulean made in one month. All the Tulean farmworkers were fired. Warehouse workers discovered that certain luxury food items denied to Tuleans were being taken by truck to the separate quarters of the strikebreakers. The Tuleans were angered.

Dillon Myer, the national director of WRA, visited the camp on November 1. After lunch a crowd estimated to number between five thousand and ten thousand gathered about the administration building. The presence of the elderly, the young, and mothers holding their babies in their arms convinced Myer and Best that violence was not intended. The negotiating committee entered the building. The crowd outside was kept orderly by members of the *Daihyo Sha Kai*. No Caucasian employees were allowed out of the building. A WRA employee kept in touch by phone with the military, which was in readiness nearby with tank engines running.

Shortly after the meeting began there was a telephone call from the hospital. The chief medical officer had been beaten by a group of young men. Kuratomi was surprised and said, "We will stop it." Someone was dispatched.

The doctor was treated for minor bruises and cuts. Later it was determined he had struck an internee first. The doctor was then thrown to the floor and kicked a couple of times. The medical staff was suspect in the eyes of the internees. They questioned the competency of the doctors and thought that medication for use in the center was being sold on the black market. They believed that a pregnant mother had been injected with morphine; her baby was stillborn. Just that morning another child had died of burns. Two appendicitis patients were kept waiting by the chief medical officer until it was almost too late. They wondered if all the Caucasian doctors were licensed to practice.

After the doctor's safety was assured, the meeting in Best's office proceeded. Kuratomi was aggressive in his demands. He wanted all the Caucasian doctors and nurses dismissed.

Myer said, "That is impossible because I have been on the project only six or seven hours and haven't even had a chance to look around."

Kuratomi warned, "I don't want to see any violence take place, but I cannot guarantee what the people will do if we have to give them this answer."

"I have never taken any action under threat or duress," said Myer.

"It is not a threat. It is a fact. I am explaining the actual tension," said Kuratomi.

The two-and-a-half-hour meeting ended with no agreement. Both Kuratomi and Myer spoke in conciliatory tones to the crowd that had been patiently waiting outside the administration building. The crowd bowed in unison to Myer and their leader and then departed. The only violence that day had been the scuffle with the doctor.

The newspapers saw it otherwise. They proclaimed the incident a "Jap Riot!" The doctor was badly beaten by the "Judo boys" and a nurse had her hair pulled and was slapped around. Oil-soaked straw was placed around the administration building. The Japs brandished long, curved knives and swords. "An unruly mob of Japs" held the national director of the WRA and his staff prisoner. The problem was that the newspapers had no reporters at the scene and had relied on badly frightened employees, whose reports were distorted. Also, the center was without a public information officer at the time.

The news from Tule Lake displaced the war news on the front pages of West Coast newspapers for one week. Tokyo was blamed for fomenting the riot. The WRA was accused of coddling the internees. Governor Earl Warren requested that the military take over the camp, which he said was a threat to the state and the war effort. Various congressional and California legislative committees geared up for hearings at Tule Lake.

The camp employees demanded more troops, machine guns, tanks, and a barbed-wire fence separating their living and work quarters from the internees. Willard E. Schmidt, the national acting director of internal security for the WRA, took charge of internal security at Tule Lake. "Huck" Schmidt was an eighteen-year veteran of the Berkeley, California, police force. He was a large, beefy man—"not the type of individual the Japanese would be eager to impede," commented an FBI agent. Patrols were increased and large gatherings were prohibited. The Spanish consul, who represented the interests of the Japanese government, visited the camp and spoke with Kuratomi.

On November 3, some ten thousand internees took part in an elaborate ceremony commemorating the birthday of the late emperor, the grandfather of Emperor Hirohito. They did not have the permission of the camp administration, and the army and internal security officers looked on with rising anger. The next day there was a rumor that trucks driven by Anglos were taking more luxury food items to the strikebreakers.

Events escalated rapidly. Younger internees armed themselves with baseball bats, pick handles, and pieces of lumber. Trucks driven by inmates chased cars and trucks driven by Caucasians. An unruly crowd headed toward the director's house. Schmidt drove to Best's house. The WRA's semiannual report stated:

> He parked his car across the road from the house and walked toward the house, suddenly perceiving 30 or 40 men with clubs in the shadows. Six of these attacked him, but he used

the judo hold on two of these, wrenching an arm from each socket. In the lull following this feat, he got back to his car, hearing the men yell in English: "Get Best! Take Best!" He started in his car for the military area and outmaneuvered the driver of the pick-up who tried to cut him off, reaching the military area to call in the Army a few minutes after the Project Director had made the same request by telephone.

At ten P.M. the army—behind tanks and tear gas, with most of the center's occupants asleep and not aware of what was happening—took control of the camp. Six shots were fired by the soldiers, but no one was hit. By coincidence, a meeting of the *Daihyo Sha Kai* was being held at the same time. The minutes of the group noted that "the thundering roar of the tanks, armored trucks, and jeeps" could be heard outside the mess hall. Kuratomi and his group had done nothing. "We had everything to lose and nothing to gain by having such a fight take place at this crucial time," he said.

Schmidt and his men moved quickly. Eighteen young men were rounded up. FBI internal reports noted that Schmidt took an active part in questioning the men that night in a supply room that had no windows. One report stated, "Before going into the room, Schmidt apparently asked [name blacked out] a question, and when he received what he considered an unsatisfactory answer, he twisted [x's] arm behind his back and as he increased the pressure he kept saying, 'Don't give me that s——. Don't give me that s——.' " The next day the internee was observed with his arm in a sling.

A camp doctor heard screams and cries from the room and wanted to leave, but he was kept there by an army officer who thought that his services might be needed. The doctor watched while an internal security officer repeatedly hit a young man, who had called him a "son of a bitch," with a baseball bat. The officer kept asking, "Who is a son of a bitch now?" Turning to another internal security officer, the first officer said it was "open season" on Japanese, and would he like to try his hand with the bat? The prisoners were kicked and hit repeatedly with bats and fists. Said one officer, "I had been itching to take a sock at the Jap, so I shoved [y] aside and hit [z] a hard-right blow to the jaw with my fist."

It was a long night. As dawn began to break, Schmidt, the other internal security officers, and the military officers who had been on hand during the beatings walked over to the mess hall for coffee. All eighteen internees were injured, some required many months of hospitalization, and one suffered permanent brain damage. In Washington, Senator Albert B. ("Happy") Chandler of Kentucky, who would become commissioner of baseball, said those prisoners who had rioted at Tule Lake should be removed to the Aleutian Islands in Alaska. His wife's solution, voiced when she and her husband visited the Manzanar camp after a disturbance, was to put all Japanese-Americans on a ship and dump them in the ocean.

The army, the WRA, and the negotiating committee called a mass meeting for November 14. Without notifying the first two parties, Kuratomi and his group pulled out of the meeting and word circulated throughout the camp that no one should attend. An army colonel and a WRA official showed up at the appointed place and walked onto a stage surrounded by soldiers with fixed bayonets. Armored cars, Jeeps, and more soldiers were poised nearby.

There was not a single evacuee in attendance. The colonel gave his speech to the air and then departed, greatly angered. Someone would pay for this. A massive search was launched for Kuratomi and others. A few were caught, but Kuratomi went into hiding in the camp. He successfully eluded numerous manhunts, but eventually surrendered on December 1. Kuratomi was placed in an unheated tent in subzero weather for eleven days, after which he was transferred to the barracks in the stockade.

The Japanese government protested the use of troops and the treatment of the prisoners at Tule Lake. The most tragic outcome of the disturbances was that the Japanese permanently canceled all further prisoner-of-war exchanges, and American lives were lost as a consequence.

From November 1943 to late August 1944, some three hundred fifty men were detained in the stockade for varying lengths of time without having any charges made against them, any formal hearings, or any trials. During this time the stockade grew from a single tent to a separate village within a city. Eventually there were five barracks, a mess hall, and a building with toilets and showers. Other than the showers, there was no place to wash. Four blankets were issued per inmate, but there were no sheets or pillows. Carrots alone or carrots and rice comprised the sole diet for weeks at a time. Sugar was almost nonexistent. Inmates ate in shifts because there were not enough utensils.

Prisoners were cut off from their families. No visitors were allowed and all mail was heavily censored. Although Kuratomi and the other members of *Daihyo Sha Kai* became martyrs while in the stockade, their influence in camp affairs atrophied; a more conciliatory group replaced them. In time, more militant, nationalistic groups would be ascendant.

Carl Mydans, a photographer for *Life* magazine who had spent sixteen months in Japanese internment camps, visited Tule Lake. In the stockade he encountered a young prisoner with a guitar who was singing "Home on the Range." Mydans wrote:

> He sang it like an American. There was no Japanese accent. He looked at me the same way I guess I looked at a Japanese official when he came to check on me at Camp Santo Tomás in Manila. At the back of my mind was the thought, "Come on, get it over and get out. Leave me alone." This boy felt the same way. He was just waiting, killing time.

What was hardest on most prisoners was that many had no idea why they were there or when they would be released. They conducted hunger strikes that were short-lived and failed to budge the administration. Finally, the American Civil Liberties Union was contacted, and Ernest Besig, director of the Northern California chapter, arrived in Tule Lake in July 1944. He was not allowed to interview his clients without a WRA official present. Before he completed his interviews, Best told him to leave. Besig barely made it back to San Francisco. Someone at Tule Lake had dumped the contents of two bags of salt as well as the bags themselves into the gas tank of his car.

One of the matters pending during Besig's short visit was a murder charge against Kuratomi. Camp authorities had begun the process of having Kuratomi indicted for the murder of "Public *Inu* Number One." While Kuratomi was in the stockade and thereby denied any communication with internees outside the facility, the Japanese-American manager of the camp's cooperative enterprises had been stabbed to death. He was widely disliked and his death evoked little sympathy. The camp administration tried to pin the murder on Kuratomi, who said, "When I was questioned, I saw the statement accusing me of murder and conspiracy of murder and asking the Modoc County Grand Jury to indict me. One thing I am on the lookout for is a frame-up." There was no indictment, and the matter was eventually dropped.

Back in San Francisco, Besig notified the press of the fact that the prisoners were being held without charges or hearings. He was quoted as stating, "These people are entitled to a hearing and a trial. If they are guilty, put them in prison or anywhere. But they have the basic right of every citizen for a fair trial." Meanwhile, the WRA was building a new concrete stockade "to keep you there for the duration of the war," Kuratomi was told by an internal security officer.

The sixteen remaining prisoners in the old stockade embarked on a hunger strike. The press was informed of it. On the eleventh day the prisoners were taken to the hospital for physical checkups and forced to eat. Besig threatened to file suit. The pressure on the WRA mounted. On August 24, Kuratomi, whose health had suffered, and the few remaining prisoners were released from the stockade after nine months of incarceration.

Kuratomi never regained his position of prominence. He was released from Tule Lake in 1946 and resettled in an eastern city, where he died a short time later. Of those who spent considerable time in the stockade, most chose to repatriate to Japan. More than half of the eight thousand internees from all the camps who chose repatriation came from Tule Lake. Seven out of every ten citizens at Tule Lake gave up their citizenship.

The perimeter lights went out for the last time on March 20, 1946. It was the last center to close.

Only one Japanese-American family from Tule Lake made their home in Modoc County. Best, the center's director, went on to head the American refugee camps in Germany, Austria, and Italy. His boss, Myer, became commissioner of the Bureau of Indian Affairs, where he dealt with the Modoc Indians.

For a brief time in 1952, at the height of the McCarthy era and the cold war, the federal government considered reopening a portion of the decaying camp at Tule Lake as a prison camp for Communist subversives. To be run by the U.S. Bureau of Prisons, the camp would have been located on eleven acres leased to the Tulelake Growers Association for a farm labor camp. Rumors swept the region that it was to be used for members of the Communist party, but that never came about.

The stockade and Castle Rock on the Peninsula

A bronze plaque marking the entrance to the camp was dedicated in 1979. The phrase "concentration camp" was revised to "American concentration camp" when Jews objected to the original wording. The local historical society refused to cosponsor any memorial with such a designation, preferring the term "prison camp." The local newspaper noted, "Many senior citizens of the county will undoubtedly disapprove of the small, bronze plaque and attendant ceremonies scheduled to take place in June."

Busloads of former occupants and their families visit the site on occasion. They also take in Captain Jack's stronghold on such reunion weekends.

Fences are the principal physical legacy of the segregation center. They come in all sizes and shapes around the small farming community of Newell. There are wooden picket fences, chain-link fences, and chain-link fences topped with barbed wire. A fence surrounds the migrant labor camp run by the county department of public works and populated mainly by Latinos. WELCOME. THIS CENTER OPENS ON MAY 15 reads the sign by the barrier gate. There are fences around the state highway maintenance yard, the elementary school, the old jail, and between housing developments and neighbors, some of whom live in converted barracks.

The concrete jail still stands, and written on its walls is the following *haiku*:

> When the golden sun has sunk beyond the desert horizon, and darkness followed, under a dim light casting my lonesome heart.

A new wave of white settlers arrived in the Tule Lake Basin following World War II. They were homesteaders, and this was some of the last land the federal government would give away.

Between 1917 and 1949, eleven sections of the Klamath Basin were offered for homesteading. The sections were opened as irrigation water became available from the Klamath Reclamation Project, one of the first such federal projects undertaken in the West. Through the years seven dams and a number of diversion channels, tunnels, and drains have been constructed. In 1957, on the fiftieth anniversary of the project, the Bureau of Reclamation boasted, "Two hundred thousand acres of fertile land have been reclaimed from swamp and arid prairie." (About one hundred thousand acres had been lost to wildlife.) Forty-four thousand acres were opened to entry by homesteaders and sixteen hundred farm families and those who served them "derive an excellent livelihood from this reclamation project." Potatoes were the principal crop in the Tulelake portion of the basin, where frost could occur any day of the year.

Tulelake is an anomaly in this state of incessant population growth. The town and the immediate surroundings have been in economic decline for the last forty years, despite Bureau of Reclamation assurances to the contrary, and it has lost population.

But in 1946, on the eve of the ninth homestead offering, Tulelake was booming. Hope was rampant. WELCOME HOMESTEADERS read the large banner spread across Main Street. The cars were parked diagonally on the wide, unpaved street; there was barely an empty space. The store-fronts ran down both sides of the street and merged in the distance. In Otis Roper Park, where the German prisoners had lived, there was an honor roll of five residents killed in World War II. Eighty-six farm units comprising 7,528 acres were sought by a record 2,028 applicants, all of whom were veterans. Other necessary qualifications were two thousand dollars in assets, two years of farming experience, letters of character reference, and good health.

Male and female applicants were to be given equal consideration. A Caucasian veteran said in his application, "I believe this will be the chance of a lifetime to become independent." A Japanese-American veteran, who had been wounded in Italy, wrote the selection board, which consisted of white males, "The fact that I am of Japanese descent should not be of issue, but in the past other agencies have passed up qualified people of my race simply because of the difference in skin color." He did not make the final drawing.

Into a three-gallon pickle jar went the names of 1,305 persons who had been winnowed from the original applicants. The drawing was held on December 18, 1946, in the Klamath Falls National Guard Armory. Local residents beat the drums heavily for the event which drew some national attention. There were high school bands and speeches and Governor Warren said, "This nation, however, is not merely giving some veterans a chance to become owners of valuable public land. You are being given the opportunity to become self-reliant, independent American farmers."

When the name of a local resident was drawn early, there was stomping, whistling, and yelling. By the time the drawing got to the fifty-eighth name, interest had lagged a bit. It was a woman's name from out of state, and that occasioned some buzzing from the capacity crowd. Eleanor Jane Bolestra was an aviation machinist's mate third-class in the WAVES during the war. She had been raised on a farm.

The Bureau of Reclamation was quick to capitalize on her publicity value. Photographs were taken and distributed to the press. The tall, handsome woman, dressed in a blouse and skirt, stood at a map and pointed to her unit. Another picture of Bolestra was taken in shirtsleeves, slacks, and sensible shoes. She wore a determined, farsighted look at the steering wheel of a tractor that she never drove. The caption for the tractor picture read:

Mrs. Eleanor Jane Bolestra of Whidby Island, Washington, only successful homesteader among 10 women applicants. Her husband was a Marine, badly shot up in the Pacific, who is regaining his health and turning into one of the best farmers in the Project, according to reports.

The couple had moved south to Salinas, California, in early December, and Bolestra was just starting a job in the post office when she got a phone call from her mother notifying her that her name had been drawn in the lottery. Bolestra and her husband immediately drove north, almost ruining their car in the subzero weather of the Klamath Basin and having to dip into the two thousand dollars of qualifying assets for repairs. "We were babes in the woods," Bolestra said. They lived in a cabin behind the Tule Lake Hotel and then moved to the Jap camp, as it was locally called. The homesteaders were given the barracks, and the Bolestras moved half a barracks and half a mess hall to their farm. They constructed a home, a barn, and a well.

The Bolestras worked hard and the bank gave them credit because everyone knew the value of the land. They borrowed enough money for a small Ford tractor and barley and wheat seed. The couple watched the other farmers and copied their moves. Charles Bolestra, who was a good mechanic, built a small combine. They raised two pigs, had a garden, exchanged eggs from their chickens for milk from their neighbors' cows, and paid back the loan. With what was left over, they bought a washing machine and a stove. Along the way they had five children.

The couple had five years to "prove up" their land, meaning they had to make certain improvements during this time; and then the land would be theirs. No taxes were levied during this period, so there were no public funds for schools. The Bolestras and their neighbors built a grade school in Newell. "We helped each other," Bolestra said. "At first we were all together. We had potlucks and overcame our feelings of inadequacy by sharing our experiences. Then it became more divided. It divided into our own areas, our own religions, our own families."

The Bolestras farmed until 1968 when Charles took a job selling real estate and Eleanor went back to school to get a teaching certificate. They leased their land to another farmer. "We could support a growing family at first. But when the equipment got more expensive and we needed more land to make a go of it, that was the end of family farms," said Bolestra.

It was the end for others too. The 634 units that were homesteaded in the Tule Lake region over the years had, through leasing and sales, shrunk to one hundred sixty farming units by 1977. The average size of a unit had grown from seventy to three hundred sixty cultivated acres. The administration of President Jimmy Carter briefly tried to make western farmers adhere to the one-hundred-sixty-acre limitation of reclamation law. The uproar in California was heard from the Imperial Valley to Tulelake, where the manager of the local irrigation district testified at a hearing: "In conclusion I would like to remind you that many of us in this room proudly wore the American man's fighting uniform to protect the very ideals and free-enterprise standards which this same Government is now attempting to strip from us." The Carter administration backed down.

The marriage also ended at this time. Bolestra got her degree, taught school, and then went to the University of California at Davis to obtain a master's degree in English. For ten years she ran a learning resource center for a community college in Alaska. Bolestra returned to Tulelake in 1984. The local paper noted, "She reports that seeing old friends is a pleasure. Eleanor, Tulelake is happy to have you back."

When we talked in her gracious home that no longer resembled an army barracks, Bolestra was preparing to go to her fiftieth high school reunion. She belonged to a great-books club in Klamath Falls and had taken Elderhostel trips to Germany, Israel, Greece, Egypt, and France. She had just come back from the Shakespeare festival at Ashland, Oregon, and in the fall would take a tramp steamer to Europe. "I feel I am just beginning to learn," she said. It was the peacefulness that kept drawing her back to Tulelake.

There were two more homestead drawings in the late 1940s, then the long decline began. There was always the struggle with obscurity. When Paramount Pictures filmed the wildlife refuge for a newsreel, it identified the town as being in Oregon. The town sits astride county as well as state lines, and sometimes tends to fall between the cracks. When it feels neglected, it makes noises about seceding to Oregon.

Then there has been just plain bad luck. Hugh Wilson, Jr., an Eagle Scout who won an appointment to the United States Military Academy, was the pride of Tulelake. He was killed in a plane crash, and the high school parent-teacher association published as a memorial a paper he had written at West Point on the Modoc Indian War. The twenty-eight-page pamphlet sold for one dollar, and the proceeds went to the scholarship committee.

As the farm population decreased, one business after another failed. Like the gaps left by extracted teeth, holes began to appear among the storefronts on Main Street, which had by now been paved. The town's main gathering place, the Sportsman's Hotel, burned to the ground. Where once there had been four auto agencies, there were now none. The last gas station closed not long ago, and although a convenience store sells gas, there is no place to get air for a tire. The movie theater closed its doors, and the foyer was reopened as a video store. In 1986 the weekly newspaper that had greeted Bolestra's return two years earlier went out of business.

But the Latino population increased. Instead of a baseball game on the Fourth of July, there now was a soccer match. In the coffee shop, which serves Mexican food, I heard some white farmers talking about falling grain prices and the "tacos," their term for the latest wave of foreigners.

On the morning of September 29, 1988, the Medicine Lake Volcano began to shake; swarms of earthquakes, peaking at sixty an hour, contin-

Tulelake

ued throughout the day and then subsided. Windows rattled in Tulelake and some plaster fell. There were smaller clusters of quakes on October 4, then the seismic activity tapered off to a few per week for the remainder of the year.

Julie M. Donnelly-Nolan, a geologist with the U.S. Geological Survey who has done the most work on the Medicine Lake Volcano, hurried north from the San Francisco Bay Area when the earth began to shake. Donnelly-Nolan later wrote in a technical journal, "The eruptive activity is probably driven by intrusions of basalt that occur as a result of crustal extension." Being the careful scientist that she is, Donnelly-Nolan shied away from any prediction of future volcanic activity.

I came upon the scientist at the Visitor Center of Lava Beds National Monument early one morning when she was giving a talk on the volcanism of the region to a group of youths who were working in the monument that summer. "I'm intrigued by volcanoes and volcanic processes," she said. The thin, bespeckled geologist has been working in the region for the last thirteen years. Each time Donnelly-Nolan drives into northeastern California she feels that she has entered a different world.

Like Donnelly-Nolan, I have been drawn to the Modoc Plateau for a number of years. The mix of landscape and history is quite noticeable here. It was from one of Donnelly-Nolan's many scientific publications that I came across the term *geohistory* in reference to the Modoc Indian War.

IV

Land of Water

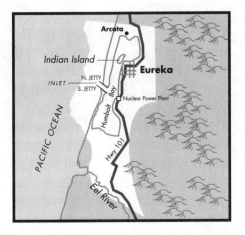

The northwest wind was relentless. It swept down the steep headland leaving the golden grass rippling in its wake. It bent around Sugar Loaf Rock off Cape Mendocino, the most seaward point on the West Coast, and buffeted the driftwood shelter behind which I crouched. The sand blew into my eyes; the wind-driven spume dampened my clothes.

It was the type of constant barrage, along with the sea pounding the headland, that wore a person, a piece of wood, a rock, and even plastic down to a nub, a grain, and then nothing. The constant wind reminded me of the regions around Cape Horn, where I had once visited. That was another extension into the sea around which sailors attempted to make their way in the face of implacable forces.

The never-ending ranks of waves that had journeyed unobstructed across the wide North Pacific broke upon the black-sand beach and rolled to within a few feet of my shelter. The black rocks in the surf line were ground and polished to a gleaming, satiny finish. Soon they would be black sand, warm underfoot during the few days when the sun shone on this foggy coast. This was one of those days.

Cape Mendocino

A seal drifted slowly by on the inshore current, its head emerging every few moments to glance about. A colony of seagulls gathered where the salt water of the ocean met the fresh water of a creek. They rose like so much confetti blown by the wind.

I felt a slight earthquake at 6:29 P.M. on that waning August day. It traveled up through the earth and entered me through the driftwood on which I sat. Aftershocks followed. Such tremors were not unusual occurrences here, where three tectonic plates and their respective fault systems, called the Mendocino Triple Junction by geologists, met and radiated from the fractured ocean floor off the cape. The seagulls, forewarned by some mechanism that I lacked, fluttered back and resumed their beachcombing.

Capes, like mountains, collect unusual geology, weather, and history. The waters off Cape Mendocino are notoriously stormy. The current edition of the *United States Coast Pilot,* the navigators' bible, states:

> The cape is the turning point for nearly all vessels bound N or S. In view of the dangers in the vicinity, it should be approached with considerable caution in thick weather; the bottom and the currents are very irregular. It is in the latitude of great climactic change.

A bronze plaque near where I camped that night declared that the "perilous waters" off Cape Mendocino had claimed nine ships. I suspected the number was higher, especially if you counted the mythical ships that had attempted to round this point. There is a local legend about a Spanish galleon being wrecked near Cape Mendocino. The Indians carried the treasure to a cave, but no one has found it.

The Spaniards, with their foreign navigators and impressed crews, were the first European sailors to sight California. Their first landfall, after the long journey from Manila in galleons laden with bullion, silks, damask, spices, jewelry, china, and bric-a-brac, was usually at or near Cape Mendocino. One bulky galleon made one round-trip a year from Acapulco, beginning in 1565. The journey could be tremendously profitable, extremely horrible, or perhaps a mixture of both. Thirst, scurvy, or beriberi reduced crews to half their number. One ship was found drifting, the entire crew dead.

By the time they neared the California coastline, they were on the homeward leg of the journey. The signs of land were known: Some one hundred leagues offshore there were the *perillos*, sea creatures "with the head and ears like a dog and a tail like that they paint the mermaids." Farther in was *porras*, giant kelp, and closer yet to shore *balsas*, large bunches of sea grass. Once they spotted a known landmark, such as Cape Mendocino, they stayed just far enough offshore to pick up other landmarks, the next being Point Reyes. Usually in thick fog they passed to the west of the Farallon Islands, thus missing the entrance to San Francisco Bay, just as they had missed the entrance to Humboldt Bay, twenty-one miles north of Cape Mendocino. A navigator of the time described the journey down the coast:

> Without losing sight of land, the ship coasts along it with the NW, NNW, and N winds, which gradually prevail on the coast, blowing by day toward the land and by night toward the sea again.

Because the coast was dangerous and there was a need to resupply the galleons after their long sea voyages, King Philip II of Spain dispatched Sebastían Rodríguez Cermeño, a Portuguese navigator ("because no Castilians were suited for the work"), in the Manila galleon of 1595 to explore the California coast for a suitable harbor of refuge. This was the first attempt by Europeans to settle California.

Making a landfall somewhere north of Cape Mendocino on November 4, 1595, Cermeño proceeded down what the chronicler of the voyage described as "a bold and dangerous coast." Heavy surf pounded the small islets and reefs just offshore. There were signs of humans on the land; smoke and fires rose from the heavily forested shoreline. They rounded Blunts Reef off Cape Mendocino and headed for the bight of land behind it—where I now sat—thinking it might be a suitable anchorage. Just in time the lookouts spied rocks, and the *San Agustín* quickly put about and headed out to sea, where it encountered the worst storm of the voyage.

The crew suffered greatly. The ship leaked badly. Both survived, but just barely. The men were terrorized by their near brushes with death along this coast. They petitioned Cermeño to remain far offshore for the duration of the voyage. "Notwithstanding this," wrote the scrivener, Cermeño kept the ship close to land on its southward journey.

The lookouts spied a high point of land behind which was a broad bay. They anchored and called it *La Baya de San Francisco*. It was not San Francisco Bay, which was a few miles to the south. It was Drakes Bay, behind the hook of Point Reyes. The Spanish, who claimed the land for their king, did not know that Englishman Francis Drake had landed there sixteen years earlier in June 1579 and claimed the land for his queen; the Indians were oblivious to both these facts. The bay was not the safe port for which the Spanish had searched, and their ship was wrecked there. Cermeño and his crew barely made it back to Mexico in a small launch. The Spanish never had much luck in California.

Sebastían Vizcaíno sailed north on the same mission seven years later. He had additional instructions from King Philip III, who had discovered among his father's secret papers a sworn declaration that some foreigners, on passing through the mythical Strait of Amin (also known as the Northwest Passage), had encountered a storm. They sought shelter and entered a "copious river, on which they came in sight of a populated and rich city named Quivera, well filled with civilized, courteous, and very literate people wearing clothes, and well fortified and surrounded by a wall." Quivera, a companion to the Califia myth, was supposedly in the same latitude as Cape Mendocino.

By now the land that lay between Cabo San Lucas at the southern tip of Baja California and Cape Mendocino had come to be known to the

Spanish as the Kingdom of California. It was reputed to be "a long country well inhabited by Indians and other nations, and contained great riches in pearls, silver, gold, and amber," according to the friar who accompanied Vizcaíno and wrote the narrative of the voyage.

From a distance they spied a blunt cape rising from the sea. Snow lay upon the coastal mountains on a stormy day in January 1603. "The sea became so wild that it seemed that with each wave the ship would be lost or submerged," he wrote. It was cold and dark. There were only six crewmen to sail the ship, the remainder dead, dying, or ill.

They were driven north of the cape and entered the mouth of the Eel River, where they were greeted by "uncultured" Indians. Had they gone a few miles farther they would have found the entrance to Humboldt Bay, the only sizable, protected body of water between San Francisco Bay and Coos Bay, Oregon—fulfilling at least part of their mission. However, there still would have been no Strait of Amin, no City of Quivera.

Vizcaíno abandoned the effort and returned to Mexico, sailing past San Francisco Bay. The Spanish would not discover the bay until 1769. By 1776, when the thirteen original colonies set about forcibly disengaging themselves from England, there were rude Spanish settlements at San Francisco, Monterey, and San Diego. Although ordered to put in to Monterey to aid the struggling colony, most galleon captains preferred to keep sailing south.

The afterglow from the sunset that I watched at Cape Mendocino was an exercise in extended luminosity. Twenty minutes after the sun dropped beneath the thin fog bank that lay offshore, the saffron sky began to radiate as if the molten orb was about to rise again and repeat the day in reverse. The sky turned brilliant orange. The colors were the orange of the sky, the gray of the fog, the dark blue of the ocean, and the startling white of the surf as it broke over the bow of Steamboat Rock, a quarter-mile offshore.

The rock resembled the loaded hull of a freighter doomed to plow north against the prevailing wind—the Sisyphus of ships. Seabirds perched on the higher portions of the rock and stained its superstructure white with their droppings. Forward, there was a long, dark foredeck.

Another ten minutes and the sky was almost drained of color; details faded, became abstract. The huge bullet shape of Sugar Loaf Rock seemed to rise like an apocalyptic missile from the thrashing surf.

For one summer night I was the westernmost person in the forty-eight states. I checked the newspaper the next day and saw that California had recorded two other national records at places that I had previously visited on this counterclockwise journey around the state. It had been 120 degrees in Death Valley and 36 degrees at Lake Tahoe the day I was sandblasted at Cape Mendocino.

The first recorded exploration of Humboldt Bay occurred in 1806 when Jonathan Winship of Boston, who was hunting sea otter for the Russians, dispatched a number of Aleut hunters in kayaks into the bay. His ship remained outside the treacherous-looking entrance. The similarity with San Francisco Bay was noted, "except that the entrance to it for vessels of large class is not convenient, and with strong southwest winds is even impassable for any kind of vessel. The depth of the entrance is two fathoms, and therefore the ocean waves break into surf."

Little has changed in nearly two hundred years. The current edition of the *Coast Pilot* states, "In the past Humboldt Bar was considered treacherous and dangerous, and many disasters have occurred there. Even with present improvements, mariners are still advised to use extreme caution on the bar."

The narrow inlet between the ocean and the bay was the eye of the needle through which the abundant natural resources of the northwest were shipped to other ports in the state and throughout the world. Manufactured goods and supplies arrived by water, as did the thick redwood logs cut the summer before and flushed down the coastal streams by the heavy winter rains.

The control of water was the determining factor here: quiet water for safe navigation, running water for transporting logs, and the correct seasonal flow of water for salmon. People needed to be protected from floodwaters and the giant tsunamis that were generated by distant earthquakes.

The rivers carved the interior, and the ocean shaped the coast. The result was a coarse, tumbled landscape that insulated the northwest from the remainder of California for many years. The railroad finally reached Eureka in 1915 and the state highway was completed only in 1920. The region is still isolated, and its inhabitants are relatively poor. It is California's Appalachia.

There are deep river gorges within the fastness of the Klamath Range, and the swift waters have sliced the tableland into many different sub-ranges: the Siskiyou, Salmon, Marble, Scott, South Fork, and Trinity Mountains. Most of the sparse population collects along the river bottoms. A few cling tenaciously to hillsides. Heavy rains transport the most erosive soils in the nation toward the sea. Where the rivers have deposited the mountain soils at their mouths, people now live on the wide interstices. The rocky headlands of the coast present a fortresslike presence to would-be intruders from the sea.

William Brewer passed through the eastern edge of the province in the fall of 1863 and noted from a mountaintop:

> The day was very smoky, and the landscape spread out around us rough in the extreme—the whole region a mountainous one—the peaks five thousand to seven thousand feet high, some indeed much higher—and all furrowed into deep canyons and sharp ridges, many of the former over two thousand feet deep. The hills are covered with scattered timber, not dense enough to be called forests, or in places with shrubby chaparral. With the exception of the ranch below us there is no tillable land; there is nothing to make the region ever a desireable home for any considerable population.

He plunged on, heading westward through an almost impenetrable tangle. Deep inside the fastness of the Klamath River Basin, where paved roads do not penetrate and where there are rumors of large humanoid creatures, the few remaining stands of virgin trees resemble the thick woodlands that once covered western North America. They are living fossil forests.

In the Douglas-fir forest, the annual rainfall reaches one hundred forty inches, as compared to less than three inches of rainfall in portions of the California desert. When there is a drought and the dry forests burn, the dense pall of smoke can lower temperatures by as much as 30 degrees, a condition not unlike the scenario for a nuclear winter.

The grasses on the bald hills of the Coast Range sway with the ebb and flow of wind. Here in the dry summer and early fall months, the color of gold is added to the shades of green that prevail during the wet winter and early spring. The gray fog sifts through the dense ferns and mosses of the temperate rain forest along the coast. The sun breaks through the fog, and the gray sea turns a brilliant blue.

Brewer continued on to the ocean, where he encountered the coastal redwood forest containing the tallest trees in the world—the highest of these exceeding the length of a football field. He wrote:

> These forests have almost an oppressive effect upon the mind. A deep silence reigns; almost the only sound is that of some torrent coming down from the mountains, or the distant roar of the surf breaking upon the shore of the Pacific.

As time passed and the forests lessened, one of the great drainings of natural resources took place through the narrow confines of the inlet that connects Humboldt Bay to the Pacific Ocean. It was through this inlet that lumber was shipped southward and westward. The economy of a region and wood for the world were dependent on controlling a few hundred feet of shifting sand.

From the interior came the sand, and from offshore came the waves and currents that shaped the sand into a barrier reef. The reef was punctured by storms and kept open by diurnal tides.

The Eel River, which empties into the ocean just south of Humboldt Bay, carries more silt for its size than any other river in North America. The soils of the 3,113-square-mile watershed of the Eel are sedimentary sandstones that have been highly fractured. In addition to naturally occurring landslides, such as those triggered by an abundance of rain on steep slopes, the topsoil was shaken loose in the last one hundred fifty years by extensive logging, grazing, and road building. The sediment is transported to the mouth of the river by winter floods. The prevailing southerly storm winds generate countercurrents to the north, and the sand is deposited along the twelve-mile barrier beach behind which lies Humboldt Bay.

The north coast is an outsized land. Everything seems bigger: the trees, the rivers, the wind, the rain, the fog, and the waves that pound the beaches and breakwaters. In 1913 the keeper of the Trinity Head lighthouse, just north of Humboldt Bay, recorded a 195-foot wave in his log thusly:

At 4:40 P.M. I observed a wave of unusual height. When it struck the bluff, the jar was heavy. The lens immediately stopped revolving. The sea shot up the face of the bluff and over it, until solid sea seemed to me to be on a level with where I stood by the lantern.

In 1923 near Crescent City, just south of the Oregon border, breaking waves swept over the lighthouse tower on St. George Reef, some seventy feet above sea level, and tore a donkey engine from its foundation. A different type of wave—a tsunami generated by an earthquake in Alaska—bore down upon the California coast at five hundred miles per hour in March 1964. It destroyed twenty-nine blocks of downtown Crescent City and drowned eleven people. The Coast Guard station was swept to sea. The material losses were estimated at $16 million, a staggering blow to the small logging and fishing community.

More commonly, the maximum height of waves generated by distant storms is likely to be about twenty-four feet, while local storms can produce nineteen-foot waves at the entrance of Humboldt Bay. These are still terrifying waves, especially when viewed from the violently pitching deck of a ship about to cross the bar.

The shifting sand and pounding surf have been a lethal combination at the entrance to the bay. Some forty vessels have sunk at or near the Humboldt Bar with a resulting high loss of life. Northern California's maritime tradition, which flowered between 1850 and 1920, was an extravagant exercise in waste. Men and ships were thrown at the sea in order to transport the resource. They were forced to make do with an inhospitable coastline. They crept around points and capes, past reefs, over sandbars, and into tiny dog-hole ports to deliver manufactured goods and depart with the lumber that would build a state but deplete a region.

The ships, many of them built on that same coast, sank or were left to rot when their usefulness ended. Sailing vessels were replaced with wooden steam-powered schooners that were, in turn, usurped by steel-hulled vessels. Coal replaced wind, and was in turn displaced by oil as the principal locomotive force. In what came to be called the Scandinavian Navy because of the predominance of sailors of those nationalities, men with names like Hurry-up Bostrom, Race Horse John, and Hog Aleck were expendable.

When the paved roads and railroads reached Humboldt Bay, the lumber vessels became garbage scows in their home state, gunboats in foreign lands, or hulks on Rotton Row in the Oakland estuary; that is, if they had managed to survive the rigors of the coastal trade. Few did. A history of ships on the redwood coast noted:

> The entrance to Humboldt Bay is strewn with the wreckage of coastwise ships that failed to cross the bar safely, and even San Francisco's own entrance is not without ship loss from the same reason. Heavy seas picked up the small ships and slammed them against rocks or sandbars, breaking their backs and smashing them as if they were but toys.

The discovery, extraction, and use of natural resources in California has been a story of initial abundance, the need to transport them elsewhere, rapacious depletion, and hand-wringing; but little has been done to reverse the trend. California has gone from a resource-rich to a resource-dependent economy. In the process, the state has lost control over its destiny and, in turn, drained the natural coffers of more distant states. From gold to raw land, known as "real estate," and from trees to water to fossil fuels to such commercially exploitable wildlife as salmon, the story has been much the same.

The first person to perpetuate the myth of everlasting abundance in a forceful and widespread manner was James S. Hittel, who cast his eyes about the state during the early years of its existence and saw nothing but a cornucopia of natural resources. As a recent college graduate from Ohio, Hittel, who spoke four languages, walked most of the distance from his home state and arrived in California in the gold rush year of 1849.

After a brief fling in the gold fields, he traveled to San Francisco and began his lengthy writing career with the *Alta California,* the most influential newspaper at the time. Of his contemporaries he wrote, "We were the flower of the West"; all were "full of hope and enthusiasm." Hittel became a roving correspondent for the newspaper and traveled widely throughout the state.

From his dispatches, Hittel put together a book titled *The Resources of California,* first published in 1862 and then reprinted in seven editions in the next seventeen years. The state awarded him a five-hundred-dollar prize for the best book written, printed, and published in California. Hittel wrote in the preface, "I write of a land of wonders"; a land, he said, that for its size contained more strongly delineated topographical divisions than any other place in the world.

There were chapters on climate, geology, botany, zoology, scenery, society, laws, agriculture, and mining; under "Other Branches of Industry," there was a section on logging. There was no place else in the world that cut such thick slabs of wood as the sawmills around Humboldt Bay, Hittel observed. Why, the redwood was so straight and tall that "it was a pleasure to strike an axe into such a tree." The trees will last for years, he predicted, there being no end in sight of their availability. Hittel, blinded by the optimism of the times, wrote:

> Our houses are built of lumber, our streets are planked with lumber, our fields are fenced with lumber, and our flumes and sluices are made of lumber. And some parts of the state are very rich in timber and can readily supply the whole demand.

The journalist and author, who has been virtually forgotten, was one of the most prolific, widely read, and influential writers of his time. In

A pulp mill

addition to his weekly columns for the newspaper on the commercial aspects of the state, he wrote books or pamphlets on phrenology, Christianity, the papacy, and immigration, plus regional guides and histories. Hittel also wrote and edited for Hubert H. Bancroft's history factory, and contributed to the work of other authors, such as J. Ross Browne's book *Resources of the Pacific Slope.* Some of his writings were translated into French, German, Dutch, Norwegian, Swedish, and Polish, thus inducing immigration from those sectors of the world.

One of Hittel's themes was the need for cheap, transient labor from the Orient. "We want muscle, not citizens," Hittel wrote, and his words were quoted in congressional testimony and enacted into law. Of the native Indians, Hittel had little good to say, writing that they lived "in a state of complete savageness." Hittel, noticing the high incidence of violence in the state, asked: Why so many murders in California? He answered: "The first and great cause is the high temper of the people. The Americans are an arrogant race."

Humboldt Bay was finally settled when Hans H. Buhne piloted the *Laura Virginia,* formerly engaged in the slave trade, across the bar in 1850. Buhne, the second officer, first sounded the entrance in a small boat that was twice swamped. The ship's passengers were greatly intimidated by the "terrific breakers" that obscured the entrance to the bay. Buhne guided two small boats loaded with supplies and passengers through the entrance on April 9. They made camp among the high dunes on the North Spit, where they claimed the land and quickly laid out a townsite.

A few days later Buhne guided the schooner across the bar and into the quiet waters of the bay. Over the years Buhne was to pilot many vessels into the bay, and lose a few in the process. He became the charter member of that small group who sought to improve the wealth of the region by ensuring the safe transit of the bar.

Buhne's fortunes improved with the rapid commercial growth of the region. His first pilot boat was a whaleboat pulled by Indians. By 1852 he was aboard a steam tug, the *Mary Ann,* which did double duty as a whaler in local waters. On the Fourth of July of that year Buhne attempted to take the bark *Home* out through the entrance with a load of lumber, but the ship was driven onto the South Spit. By 1853 more lumber was shipped from Humboldt Bay than any other port in the state, but there was a price to pay. Ten vessels were wrecked while crossing the bar between 1850 and 1854, and Buhne's wife was drowned when a wave swept her from the deck of a passenger ship.

During this period an obscure army captain crossed the bar twice within a five-month time span. Apparently U. S. Grant's crossings were uneventful because he did not mention them in his memoirs, nor did he mention why he left Fort Humboldt so quickly after being posted there.

The windswept fort overlooked the entrance to the bay and had a broad view of the surrounding landscape. Grant noted the redwood trees along the banks of the bay, trees that have long since disappeared. During the dreary winter of 1853–1854, the slouching Grant was a frequent visitor at a local store, whose owner kept a barrel of whiskey handy for army officers, the better to ensure the store's protection should Indians attack. There is this account of his behavior:

> At times Grant had problems getting home from the taverns after too many drinks. On one occasion he fell into the slough. On another his drunkenness caused him to fall off his mule, and he failed to return to the fort, whereupon a party went in search of him. They found him asleep in a thicket, and his trusty mule was browsing close at hand.

Captain Grant was observed standing unsteadily at the pay table one day and given the following ultimatum by his commanding officer: either be court-martialed for drunkenness or resign from the service. He chose the latter. In his memoirs, Grant said he resigned because he could not afford to bring his wife and two children to the West Coast on a captain's pay. Before he left Humboldt Bay he said, "Whoever hears of me in ten years will hear of a well-to-do old Missouri farmer." A more prescient version of his departure speech had him saying, "My day will come, they will hear from me yet." Grant left for San Francisco, where he noted there was less drinking and gambling than when he had last been there.

A lighthouse to aid such coastal traffic as bore Grant in and out of Humboldt Bay was completed in 1856 on the North Spit at a cost of fifteen thousand dollars. It was of little help. The light failed at times and was too low to be seen at other times. Although the lighthouse was constructed a half-mile from water, the surf broke about its base during high tides and winter storms. A breakwater was constructed around the base of the lighthouse. These were the first federal funds expended on what would become the most costly ongoing harbor project for any port of comparable size in the United States. The pattern was established; the sea continually wore down what the local citizens ceaselessly lobbied the federal government to rebuild.

Human as well as natural resources were depleted. There had been scattered killings of Indians and whites in northwestern California for ten years before February 25, 1860, when the Eureka newspaper, the *Humboldt Times*, issued a call to action:

> The Indians are still killing stock of the settlers in the back country and will continue to do so until they are driven from that section, or exterminated. Wednesday they killed two head of stock belonging to the brand of Larrabee and Hagans and drove off twelve others which were, however, recovered.

A vigilante committee had already been formed to deal with the Indian problem. Its members included "some of the prominent men of the county," according to newspaper reports. They were sworn, under pain of death, not to reveal their membership. It was said that they were intelligent family men who came from New England and Canada's Maritime Provinces.

A low island lay in the shape of an elongated tear just off Eureka in Humboldt Bay. At the northern end of Indian Island there was a brush-covered mound built up over the centuries by the Wiyot Indians, who had discarded their clamshells on that spot. Robert Gunther, a native of Germany who had become a prosperous logging mill engineer, bought the island on Washington's Birthday in 1860. Three days after Gunther had acquired the property and one night after the newspaper's call to action, some two hundred fifty Indians from up and down the coast were having a religious celebration on the island.

Gunther, who slept at a mill on the mainland across from the island, was awakened that night by what he thought were screams. The sounds died away, and he went back to sleep. Gunther was roused the next morning by the justice of the peace, who said he had heard that some Indians were killed on the island the previous night. They hurried across by boat. Gunther later recalled the scene that greeted them:

> But what a sight presented itself to our eyes. Corpses lying all around, and all women and children, but two. Most of them had their skulls split. One old Indian, who looked to be a hundred years old, had his skull split, and still he sat there shivering.

Three squaws survived. Gunther recounted the story of one:

> She said she saw them coming. She ran to the shanties and hallooed that white men were coming and the Indians got out as fast as they could. She stood on a low place as they came up the bank and she saw them distinctly against the sky, and she thought there were six or seven. When she saw they came to kill them, she ran for the brush which was close by, and everybody else ran. All the men got into the brush but two, but the women and children were killed. The white men did not dare to go into the brush, but left after they killed all they could find.

Gunther and the justice of the peace rowed back to the mainland, vowing to bring the guilty parties to justice. They then had second thoughts. "We soon found that we had better keep our mouths shut," Gunther said.

There were three other attacks by whites on Indian encampments in Humboldt County that night. The attacks were carefully planned, coordinated, and executed. The weapons of choice were knives and axes. Such weapons killed silently and were more personal than guns. The killers were bathed in the blood of their victims. Perhaps two hundred Indians were killed, the majority of them being elderly men, women, and children.

"It was never publicly known who did the killing," Gunther later recalled, "yet secretly the parties were pointed out." The grand jury summoned witnesses but could find no one to indict. The jurors condemned the massacre and publicly regretted that their investigation had come to naught.

A minor casualty of the massacre was young Bret Harte, who dared to write an account of the incident for the Arcata newspaper while his boss was out of town. He questioned the policy of extermination that seemingly now extended to women and children. Harte thought it best to leave town in a hurry. Also on board the schooner that passed through the inlet on its way to San Francisco was the undersheriff of Humboldt County, who had absconded with seven hundred dollars in county funds. He sent the key to the county jail back with the harbor pilot.

The out-of-town newspapers excoriated Eureka. A San Francisco paper referred to it as "Murderville." A New York paper stated, "Even the record of Spanish butchers in Mexico and Peru has nothing so diabolical." The local press was unrepentant. After a silence of three weeks, the editor of the *Humboldt Times* wrote:

> For the past four years we have advocated two—and only two—alternatives for ridding our country of Indians: either remove them to some reservation or kill them. The loss of life and destruction of property by the Indians for ten years past has not failed to convince every sensitive man that the two races cannot live together, and the recent desperate and bloody demonstrations on Indian Island and elsewhere is proof that the time has arrived when either the pale face or the savage must yield the ground.

William Brewer was in the region three years later. He wrote in a very emphatic manner: *"Murdered the whole of these people!"* and *"Over a hundred were slain!"*

To this day there is no public acknowledgment of the Indian massacres in Humboldt County. There is a stone monument at the small boat harbor that points out the location of what came to be known as Gunther Island, but it says nothing of the bloody events that took place on that swampy ground.

For the Chinese, who were brought north to work in the mines and to construct the wagon roads and logging railroads but wound up doing the jobs that no one else wanted—such as cooking, serving, and washing—the end of their stay in the county was also abrupt. A local historian, Lynwood Carranco, has suggested that the region's geographical isolation, the clannishness of its white residents, and the fact that power was held by a few—namely lumbermen and the merchants who depended on the timber trade—accounted for the blatant racism. This homogeneous group feared outsiders. They acted on the basis of their economic interests. Carranco quoted a former mayor, who was no friend of this group, as stating, "The sacred cows had to have the best of everything."

The perceived conflict with the Indians was over land; with the Chinese it was jobs in the last quarter of the nineteenth century. As the Chinese inched their way into the outer fringes of the California economy after being exiled to the most difficult tasks in the remotest areas of the state in the early years, an intense wave of xenophobia swept the state in the late 1870s and early 1880s. It reached its climax at the national level when Congress passed the Chinese Exclusion Act of 1882, barring further immigration and denying naturalized citizenship to those already in the United States. Bret Harte's single "Heathen Chinee," as his poem was titled, came to be viewed as an "industrial army of Asiatic laborers."

Whites from Truckee, Eureka, Turlock in the San Joaquin Valley, and Los Angeles, as well as many other communities throughout the state, turned to violence to rid their communities of the hated Celestials. In Humboldt County, the citizens waited for a pretext to rid themselves of the "hordes of Mongolian paupers, criminals, and prostitutes" that had inundated the county, in the words of a petition they sent Congress. The county newspapers took up the cry, "Too many laborers—starvation wages."

That pretext came at 6:05 P.M. on Friday, February 6, 1885, when two Chinese men exchanged shots. A Eureka city councilman walking to his downtown office was hit by a stray bullet and killed. Within minutes an unruly crowd was yelling for the mass execution of all Chinese and the burning of the Chinatown section of Eureka. A meeting, chaired by the mayor, was held within twenty minutes of the shooting. The crowd continued to call for the massacre of all Chinamen and the destruction of their property. Some recalled the precedent for such a massacre that had been established a quarter-century earlier on the island.

Cooler heads prevailed, and a citizens' committee was dispatched to tell the more than two hundred Chinese residents of Eureka, and perhaps an equal number in the unincorporated areas of the county, to pack up and be ready to depart within twenty-four hours. The next morning the streets of Chinatown were piled high with hastily assembled baggage. By noon the holds of two steamships, which fortunately were in port and available as transports, were loaded with the possessions of the Chinese, some of whom had been in Eureka for almost as long as the thirty-five years it had been in existence. The Chinese in the hinterlands were rounded up, but a few managed to escape the dragnet, only to be hunted down later.

A small Chinese boy, on his way to where he had been directed to go by the whites, was grabbed off the street and a noose was fitted around his neck. It led to a rope that was attached to a scaffold. A sign on the gallows read ANY CHINAMAN SEEN ON THE STREET AFTER THREE O'CLOCK WILL BE HUNG TO THIS GALLOWS. The police did nothing. The badly frightened boy was finally released on the pleadings of a clergyman.

Because of rough seas breaking on the bar at the entrance of the bay, the ships did not depart until Sunday. After a two-day voyage down the coast, the vessels arrived in San Francisco. The fearful passengers melted into that city's Chinatown. A suit seeking damages for those expelled from Eureka was filed in federal court, but it never went anywhere. Back in Eureka a mass meeting was held and resolutions—what came to be referred to as the "unwritten laws"—were passed barring the Chinese from the city. The "laws" were then extended to county territory and elsewhere on the north coast as other communities kicked out their Chinese residents.

The unwritten laws were on no official books, but they were etched into memories and passed on from generation to generation. In 1906 a salmon canning company imported twenty-three Chinese and four Japanese laborers. Upon their arrival, the *Humboldt Times* declared in a blazing headline THE CHINESE MUST GO! They went; there is a photograph of them standing on a barge being pushed down the bay by a small tug, with Eureka's stately yacht club in the background.

Curiously, the four Japanese were allowed to remain. But the region was always slightly behind the times. It had not yet caught up to the anti-Japanese feeling that was then sweeping the remainder of the state. Three years later two Japanese brothers opened a store in Eureka. A few days after the opening, a stick of dynamite left in the doorway destroyed the store. The brothers left town quickly. No arrests were made.

In 1930 the Japanese ambassador to the United States arrived by ship in San Francisco on his way to Washington, D.C. He planned to motor through Humboldt County, view the redwoods, and then depart for the capital by train from Portland. The ambassador was met at the southern border of Humboldt County by a delegation from Eureka, escorted without a stop to the northern border, and told not to return. The incident rankled the Japanese government.

On the eve of World War II, Humboldt was one of the few California counties without any Oriental residents. The crews of Chinese ships did not dare go ashore when they docked in Eureka to take on a load of lumber. Up to 1959, the city forbade the purchase of Chinese goods and the hiring of Chinese for any work done for the city "except in punishment for a crime." A few persons of Japanese and Chinese descent trickled into the county in the 1950s and 1960s, but they were mostly confined to the state college in Arcata and the restaurant business in Eureka.

Meanwhile, the bar took its toll. Between 1853 and 1880, eighty-one persons died attempting to cross the shifting sand. Sailors were swept from decks, a captain was plucked from the rigging, and one seaman drowned the day after dreaming he would die while crossing the Humboldt Bar. "Crossing the bar" became a euphemism for the passage from life to death. The United States Life-Saving Service reported that there were "more terrible shipwrecks at Humboldt Bay than at the great port [San Francisco] to the South."

The *Pomona*, a comfortably fitted-out passenger steamship that was always on a tight schedule, was one of the largest vessels to cross the bar. A passenger wrote:

> On, on we went, deeper and deeper into the mists. Higher and higher rose that terrible swell until it seemed as though a mountain of water carried its dark slope to a hidden summit ahead. In an instant we had surmounted it, to plunge down the awful declivity of the other side. With bated breath we take the plunge. A sickening sense of danger seizes every heart. That mountain has changed to a leviathan with a curling, hissing crest. On, on he comes, down the slippery steep. "Full speed ahead!" rings the Captain, but too late; the monster has fallen upon the ship's high stern. The after rails and cabin staterooms are crushed like eggshells. The good ship broaches to and trembles in every joint. If another strikes we are gone. But the vessel has recovered speed and soon rides in the calm waters of Humboldt Bay, across the bar.

Buhne and the community prospered. By the 1880s Buhne had acquired a thousand-acre ranch, thoroughbred racing stock, four thousand acres of redwood timberland, two lumber mills, gold mines, a store, a shipping line, a new tug named the *H. H. Buhne,* and one of the first telephones in the county. Captain Buhne was described as "the mainspring" of Humboldt County business enterprises.

Some six to eight hundred vessels were entering and departing from the port each year. Their departures were frequently delayed by storms, a condition referred to as being "bar bound." That was bad for business, but the chamber of commerce put the best possible face on the port in a publication that noted, "It has always been considered one of the safest harbors on the Pacific, when gained."

A breakwater would make the entrance safe and increase trade, but Buhne, who was still piloting vessels, had his doubts. "He does not speak very encouragingly of the possibility of building a breakwater," noted a local history of the time. Buhne knew firsthand the savageness of the seas that broke over the bar. He needed to be won over, and was so by the economic arguments of William Ayres, the editor of one of Eureka's newspapers.

Ayres mounted a campaign for construction of the breakwaters that would be to Eureka and the surrounding region what such gigantic public works projects as dams, aqueducts, freeways, bridges, flood-control channels, and other breakwaters would be elsewhere in the state—and that was the means toward prosperity for some.

The editor worked up a set of figures that showed that by deepening the entrance to the bay by only six feet, such costs as tug fees and seamen's wages could be reduced to save a shipper $45,000 a year. In addition, timberlands would double in value and foreign markets would be more accessible. "Thinking minds who are interested in the growth and development of this section should consider this proposition well," he wrote. Ayres suggested that jetties, similar to those used at the mouth of the Mississippi River, would do the job.

The locals quickly came to the conclusion that they did not want to pay for such an expensive and risky proposition. But they were willing for the federal government, in the form of the U.S. Army Corps of Engineers, to pay for the project from tax monies collected nationally.

Recognizing the need for a safe harbor between San Francisco Bay and the mouth of the Columbia River, the corps had already investigated

Humboldt Bay in 1871 and found it lacking. The investigating officer, who was "bar bound" for a fortnight, reported that the bar was constantly changing. He wrote, "Every severe storm changes the channel; sometimes there are two channels, and sometimes there is but one."

In 1877, a three-member examining board was unable to enter the bay in moderate weather because the pilot thought the conditions too dangerous. The board also gave a negative opinion, but this did not daunt Ayres. He finally procured a positive assessment in 1880 from Lt. Col. George Mendell of the San Francisco District. Mendell foresaw an increase in shipping and was impressed by the public displays of wealth, such as Buhne's mansion.

Ayres was replaced by another editor who continued the crusade. It was by "dint of continued agitation and effort, in the face of much opposition, that we have secured government recognition," wrote the new editor. The agitation was sustained. A contemporary account noted, "Commerce struggled over the bar for years, suffering all of the adversities of detention and shipwreck in the unequal fight, until some public-spirited citizens of Eureka were sent to Washington to importune national assistance."

The lobbying produced results. The first experimental jetty was built on the North Spit in 1881. It consisted of willow brush sandwiched between planks cut in twenty-foot lengths and driven into the sand by a pile driver. The jetty seemed to halt the erosion at the end of the North Spit. The channel was more constant. Then, in a foretaste of what was to come, storm waves and currents quickly destroyed this first frail extension into the Pacific Ocean.

The destruction of the jetty couldn't have happened at a worse time. The region was ready to surge forward. Southern California was beginning to boom. More ships were needed to carry lumber to the south. Between 1881 and 1893, fifty-six ships were built on Humboldt Bay. Meanwhile, the logs that were now arriving by railroad were piling up on the wharves where the schooners were docked bowsprit to stern.

These were bold times. There is an 1885 photograph of a long train hauling sixteen gigantic sections of a redwood tree. The engine and flatcars are decked out in numerous American flags. Within the engine, atop the coal car, standing on top and beside the massive sections of redwood (each taller than the person beside it), and sitting atop the fence and standing in the field alongside the engine and its caterpillarlike segments are the citizens of the region, all attired in their best white dresses and dark suits. In the background are the charred trunks and smoothly cut stumps of a depleted redwood forest.

The first warning of what the results of this patriotic exercise in extensive logging would be were raised during this decade in a local history book. Wallace W. Elliott wrote:

> The calamity which will befall the people of Humboldt County by the exhaustion of the forests of redwoods could be in a great measure averted if the growth of the young redwoods were fostered. But no care is taken; and, in fact, it seems that an effort is made to thoroughly eradicate all traces of the forests. The stumps are fired just to see them burn, and fire runs over the land every fall, which serves to completely destroy the young shoots.

Throughout the 1880s, the crescent-shaped sandbar at the mouth shifted back and forth, sometimes changing positions several times during a year. The redwood industry suffered accordingly. There were virtually no foreign shipments in 1885 because such heavily laden ships as were needed to carry such a cargo could not cross the bar. Other vessels had to depart with a partial load. Pilots were needed for every vessel. Hans Buhne felt confirmed in his initial opinion that the bar, which shifted like a writhing snake, could not be tamed.

The corps thought otherwise, but yearly congressional appropriations were not sufficient to build a permanent solution in a timely manner. A large sum was needed to complete the project in two years. By 1888 there was enough money for a low south jetty. Such a jetty would shelter the entrance from winter-storm waves that came from the south. A contract for $171,000 was let. The contractor defaulted, the channel continued to shift, and the partially constructed south jetty caused the North Spit to erode and actually accelerated the buildup of the bar. Nothing was predictable, except the expenditure of more money.

The solution to this predicament was to complete the south jetty and also to build a north jetty at a cost of $1.7 million. The work began in earnest, but when appropriations lagged, the work ceased. The Humboldt Chamber of Commerce resumed its lobbying efforts. The chamber pointed out that a safe entrance would improve coastal defenses and ensure reliable mail delivery. The local congressman delivered, Congress made a large appropriation, and work resumed.

Compliant men conquering nature was a dominant theme in California at this time. W. E. Dennison, the superintendent of jetty construction, wrote of the quarry where rock was obtained:

> In excavations at the base of a bold rocky bluff, 420 feet high and extending up the creek for more than three thousand feet, two hundred brawny men are working. They look like pigmies. Stupendous devastation! The bluff has been undermined again and again with steam drills until its face is bare to the roots of the giant redwoods on the brow. The work of removing the mountain has just begun.

Dennison described the quarrying and transportation of rock to the jetties: ten steam derricks, a locomotive crane, a dynamo for twenty-five arc lights to illuminate the scene, the three-mile diversion of a creek, piles of waste rock, trains of twenty-one cars from the quarry to the bay, six trains a day, seven days a week, fifteen minutes to load, barges towed to the breakwaters, cars hoisted by cable to the trestles, loads dumped by hand or hydraulic power, and, as of 1896, more than half a million tons of stone placed on sixty thousand cubic yards of brush mattresses by three hundred men risking their lives for the sake of progress.

For the uninitiated, Dennison explained in a popular magazine: "It may not be clear as to the object of all this vast expenditure of money. It may be stated in a few words: the opening and keeping open of a fine harbor." His article ended with the firm prediction, "In two years this channel can probably be declared permanent and the work completed."

The work was completed in three years at a cost of three lives and $2 million. The south jetty was 7,408 feet long; the north jetty extended 8,068 feet. There appeared to be a stable, deep channel. Everyone agreed that the project was a success. The jetties were viewed as a permanent solution, and no maintenance was provided for them. The value of goods shipped out of the bay increased by $1 million a year through the first years of the new century.

The quarry

Six years after their completion in 1899, the jetties were disintegrating and were virtually useless. They were covered with sand and the surf washed across them at normal tide levels.

It occurred to me, while camped beside the mile-long entrance channel, that I had always been drawn to powerful places where great forces were in constant flux, be they tidal, volcanic, or seismic. California had all these elemental forces, plus more.

As a young man, I had repeatedly piloted a sportfishing boat through an inlet that pierced a barrier beach on the south shore of Long Island, New York. The departure and return were always tense moments. Wait for the right wave, then gun the engine. When there was a storm, I drove to the inlet to watch the conflict between land and sea. I checked to see if the channel had shifted and listened to the fishermen reminisce about greater storms.

I moved to California and lived amid the houseboats in Sausalito, just inside the Golden Gate. In a fourteen-foot catboat—with no outboard motor, no oars, no wind, and a strong ebb current—I was pulled outside the Golden Gate. To my embarrassment, I had to be rescued by a motorboat. Later I traded up to a twenty-two-foot topsail cutter with a diesel engine and navigated the entrance to San Francisco Bay with impunity.

When I changed jobs, my wife and I sailed the cutter to Los Angeles. Point Conception is called the Cape Horn of the Pacific. Like Cape Mendocino, it collects fearsome weather and its share of shipwrecks. I passed the point on a night so calm that I had to use the engine. The lights from the few ranches on shore twinkled benevolently, as did the phosphorescent wake left by the passage of the boat. I entered Angel's Gate, the opening in the Los Angeles breakwater, and later came to live on Point Fermin, which overlooks the entrance. I navigated this other gate frequently in the cutter and, after she was sold, in an oceangoing kayak.

During storms, I walked along the Los Angeles breakwater. I felt the pressure—a sharp *whomp*—pass underneath me as the waves broke upon the huge granite blocks that had been dislodged in places. The wave energy passed through the thick breakwater and set up a reciprocal effect in the harbor. Every once in a while someone was plucked from the high stone edifice by an aberrant wave and drowned or crushed against the rocks. On calm days, I watched people fish or collect edible plant and sea life along the breakwater.

The South Spit, where I was camped, was the province of the homeless of Humboldt County. On land owned by the federal government, there were semi-permanent house trailers and plastic Portatoilets; wooden signs designated who lived down which sandy track. There were abandoned vehicles of all sizes and shapes, and garbage was strewn across the dunes. A man, his T-shirt proclaiming ECOLOGY, and his wife and two sons pulled up next to me in their van. They proceeded to strip the breakwater of orange starfish, then flung them Frisbee-style into the water. Three fishermen made their way out the jetty in three-wheeled, balloon-tired, all-terrain vehicles.

I moved around to the breakwater on the North Spit. It was littered with cautionary signs. Erected "courtesy of" the Army Corps of Engineers, the signs warned: DANGER, EXTREMELY HAZARDOUS WAVES. The concrete facing of the walkway was flayed in places, revealing a thick underpinning of bleached wooden beams. More signs warned, successively: DO NOT WALK ON STRUCTURE, HAZARDOUS WAVE CONDITIONS EXIST EVEN ON CALM DAYS, WAVES CAN WASH OVER STRUCTURE AND SWEEP PEOPLE INTO THE OCEAN. I walked out to the end and passed a lone fisherman who said, "Wish me luck." I did so, but I wasn't sure whether he was referring to the fishing or what the signs portended. I looked back and saw him scamper from a wave.

I did not tarry long at the end of the breakwater. With a southwest swell remaining from a storm front that had passed the day before, the end of the South Spit was being swept by waves. It seemed safer on the North Spit, but I did not count on that. No one was in sight. If I were swept away by a rogue wave, who would know that I had been there?

The shipwrecks continued as the two newly constructed breakwaters were battered into submission. The steamer *Corona,* sister ship to the *Pomona,* was known as a lucky ship. Built in 1888, she had almost sunk during the Alaska gold rush of 1898 when she struck a rock. A huge fish referred to as a blackfish became wedged in the hole and saved her from sinking. The *Corona's* good luck ended on the morning of March 1, 1907, as she swung to enter the inlet. The captain gave the wrong order, and the ship struck the inside of the north jetty; because the jetty was so low in the water, the *Corona* was lifted by a large swell and deposited on the outside of the breakwater, in about the same place where the steam schooner *Sequoia* had been wrecked a few months earlier.

The town turned out to gape and help. Those who owned boats made a handsome profit transporting the five thousand sightseers. Some came with picnic baskets. It took most of the day to take the passengers and crew off by lifesaving boats and breeches buoys. Two crewmen and later a salvager lost their lives. A cofferdam was built around the hull, and some machinery was salvaged. When the tide and sand conditions are right, the timbers of the cofferdam are still visible.

There are other visible reminders of past mishaps. The submarine *H-3* was looking for the entrance to Humboldt Bay in a dense fog when it ran aground some four miles to the north of the inlet on December 14, 1916. The twenty-seven crewmen were knocked about as the round hull rolled in the surf, but they were eventually rescued without undue injury. Three small naval vessels failed to pull the *H-3* free. The navy brass decided to send the U.S.S. *Milwaukee,* a cruiser, to the rescue. It had a crew of four hundred fifty men and 24,000-horsepower engines.

Two navy tugs stood offshore. They were tethered to the cruiser, which was attached to the inshore submarine. The tugs and cruiser pulled; the lines to the tugs gave way. The cruiser was dragged into shallow water by the line attached to the submarine. The crew made it to shore via boats and breeches buoys. The crowds again came to help and gape.

The cruiser was almost a total loss on the eve of World War I. It was partly cut up for scrap metal during World War II, but portions remain and are visible at extremely low tides. The submarine was eventually salvaged and put back into service. The newspapers blamed the mishap on the treacherous coast and the lack of competent commanders and sufficient crews.

The navy's public embarrassments in California had not yet run their respective courses. Seven years later on another foggy night, seven destroyers, in single file, crashed one after another into Point Pedernales, a few miles north of Point Conception. They were steaming south at twenty knots from San Francisco and executed a left turn at the wrong point to enter Santa Barbara Channel. In mindless, follow-the-leader fashion each piled upon the rocks or its predecessors.

The wreckage was visible from the main San Francisco–Los Angeles railroad line, and passengers stopped to gawk. To avoid further embarrassment, the navy dynamited the remains of the ships. But if you look long enough, as I once did, you can still see the rust-flaked shards of steel hulls.

The chamber, corps, and Congress—a different triumvirate of comparable groupings in water development circles becoming known as the iron triangle—went through their dance again. In 1910 money was appropriated to restore the jetties to their original condition. Work began on the south jetty, which had been beaten into a compact mass. Fierce winter storms and the slow appropriation of money by Congress dragged out the job. Over a three-year period 519,000 tons of rock were dumped into the ocean.

The most important and most vulnerable part of the jetty was its head; if the head eroded, then the whole structure would peel back. It needed to be invulnerable. The solution that engineers hit upon was to construct a monolithic, reinforced-concrete block weighing one thousand tons and dump twenty-ton boulders around it as additional protection. The concrete monolith was reinforced with steel rails and cables. When completed, the head was more than fifty feet wide and one hundred feet long—surely vast enough to rebuff all waves for all time, or so the engineers thought.

The south jetty was completed in 1915; but because of World War I, the north jetty was not completely repaired until 1925, when a similar concrete monolith was placed at its head. The corps of engineers referred to the job as a battle—a battle that had successfully been won. But all military victories are transitory, as are all such "conquests" of the sea. No sooner had the north jetty been completed than plans had to be made to repair the ten-year-old concrete cap on the south jetty. Portions of the cap had washed away, and the sides of the jetty were beginning to erode. Repair work would now be ongoing.

With at least some relative stability being established at the entrance to the bay, the 1920s proved to be a prosperous decade. There was plenty of lumber for the mills to cut and some forty wharves along the shoreline of the bay from which to ship it. The chamber of commerce repeatedly pointed out to the corps of engineers that it was the latter's responsibility to maintain suitable conditions for shipping, as the region had come to depend upon it. In letter after letter, public hearing after public hearing, and private conversation after private conversation, the chamber kept hammering this theme home. As one historian of the jetties, Susan J. Pritchard, pointed out, "It was the bay after all, and the commerce from it that had led to the settlement and development of the region."

Those timber interests who used the Eureka Chamber of Commerce as their point of contact with the federal government fit the profile of others throughout the American West who sought to make a living or profit from the extraction of natural resources. They personified the cult of individualism and saw themselves as torchbearers of the American way jousting with hostile market forces and distant bureaucracies. Yet they were dependent on those same markets and institutions for their largesse. It was a colonial mentality.

I sat only once, and very briefly, in their inner circle.

As assistant secretary of the California Resources Agency in the mid-1970s, I had accompanied the secretary, Claire Dedrick, on a fence-mending tour of the North Coast. The agency had overall responsibility for the state boards, commissions, and departments that dealt with environmental matters. We were invited by the powers that be in Humboldt County to dinner that night in a tastefully restored Victorian house, reserved for such occasions, in Samoa on the North Spit.

It had been an exhausting day. There had been an early morning press conference, we had visited a California Division of Forestry fire station, and then been conducted by timber industry representatives on a vigorous walking tour of the watershed above Redwood Creek. The tour of

Redwood Creek had been so vigorous that we suspected our hosts of seeking physical vengeance.

Additional lands were being considered for inclusion in Redwood National Park above the creek. That would mean a loss of revenue for the industry. We favored the addition. We had also suggested tougher logging rules and a change in membership of the State Forestry Board to reflect a conservation as well as an extraction viewpoint. In Sacramento, logging trucks, dispatched with the blessings of the same timber industry executives who sat around the dining table, had noisily and repeatedly made the rounds of the streets surrounding the state capitol in a protest heard through the faceless glass exterior of the fourteenth floor of the Resources Building.

The atmosphere at dinner was strained, at best. Who were these men? Hank Trobitz, the silver-haired and silver-tongued manager of Simpson Timber Company, was the informal leader of the group; he sat opposite Claire. They were members of the chamber, accustomed to setting the agendas for the isolated region. And they were comfortable in this setting, Samoa being a neatly laid-out company town next to a pulp mill whose effluent smelled like rotten eggs. They were equally at home in corporate boardrooms or the legislative hearing rooms of Sacramento and Washington, D.C.

We were newcomers to government. We were interlopers who were fascinated by the levers of power. Pull this one and watch the bad guys jump. We had yanked a lever, and they had jumped all over us. We represented the hostile, environmentally-oriented administration of Jerry Brown that had replaced the sympathetic viewpoint of former governor Ronald Reagan, who once had said if you have seen one redwood tree you have seen them all. Claire, a chain-smoking former vice president of the Sierra Club who held a doctorate in biology, liked to drink and had a violent temper.

At the other end of the table, I heard Claire's voice increase in volume. She had taken exception to something one of the men had said. I got up, walked over to the back of her chair, and whispered in her ear to count to ten. She calmed down and later thanked me. We left and were the eventual losers of that symbolic encounter. What we initially proposed was diluted by the time the legislation was signed into law by Governor Brown. A year or two later Claire and I were both working at different jobs.

Since my brief time in state government, the spotted owl and the U.S. Forest Service have become the centers of the latest controversies surrounding the cutting of northwestern forests. Smaller lumber companies were purchased by more distant corporate entities, who brought in their own managers. They were under pressure to cut more timber in order to satisfy the needs of their home office. Along with environmental concerns and greediness, the recession also contributed to the closing of mills, high unemployment, and a greatly depressed economy. The region turned to tourists and the selling of its refurbished image.

On this most recent journey, I ate at the nearby Samoa Cookhouse, where there is now a logging museum. In their time, loggers, mill workers, and the survivors of the wreck of the U.S.S. *Milwaukee* were fed inside the cavernous structure. Breakfast consisted of four pieces of French toast, four sausages, four scrambled eggs, orange juice, and coffee—and as many additional helpings as desired. The company at the long table, consisting of a family from New York and another from Orange County, was congenial.

With the entrance to Humboldt Bay relatively safe, the fishing fleet multiplied during the 1920s. It set sail or motored from the bay in vigorous pursuit of the ocean resource that was desirable beyond all others—the salmon, the king of edible fish. From a fleet of a dozen vessels in the 1890s, the fishing fleet had expanded to more than one hundred boats, with additional boats making the port their home during the salmon season. Half the fish consumed in California were landed at Eureka during the 1970s, when the local fleet numbered more than two hundred boats and another five hundred used the harbor on occasion. Since then, there has been a drastic decline in both the number of fish and boats.

The object of this frenetic seasonal activity was no ordinary fish. More than any other fish, the salmon has historically appealed to both the palate and the imagination. To ancient cultures, the salmon was a phallic symbol and represented fecundity. The Celts associated the salmon with wisdom, the knowledge of other worlds, and supernatural powers. Northern California Indians associated salmon with Coyote, the trickster, in a number of myths that explained natural phenomena. For the Indians, the supply of salmon seemed endless. They consumed or bartered nearly five million pounds of salmon a year.

Then the Anglos arrived and applied their laws to this valuable resource. Not surprisingly, the whites forbade the Chinese to fish for salmon, while the Indians were simply pushed aside.

Not only gold but salmon drew some emigrants to California at a time when the Atlantic Coast commercial salmon catch had all but become nonexistent. In the West there was a legacy of free and open access to natural resources, a codicil to the doctrine of Manifest Destiny—or, perhaps, the driving force behind it. There were new rivers to conquer. It was open season on salmon. "These canny fishermen were able to obtain or thwart passage of regulations as their interest dictated. They openly defied laws they were unable to change," wrote Alan Lufkin in the book *California's Salmon and Steelhead.*

The Hume family from Maine, and Richard D. Hume in particular, were a case in point. The grandfathers, father, and sons fished for Atlantic salmon until the resource was depleted on that coast; the oldest son then set off for California in 1852 with a homemade gill net. Two other brothers followed shortly thereafter; the youngest, Richard, arrived in Sacramento in 1863 at the age of eighteen. The next year the brothers, in partnership with another Maine fisherman, established the first fish cannery on the West Coast. Their efforts on the Sacramento River were unsuccessful. The salmon runs had dropped because of the damage to upstream spawning beds caused by hydraulic mining.

Two years later the brothers moved to Oregon, where they established a number of firsts in the salmon industry: They were the first to exploit the salmon runs on the Columbia River, and were quite successful. (Later they expanded their operations to Alaska.) They were the first to employ Chinese workers, who were less prone than whites to join labor unions. And they were the first to discharge the Chinese when the "Iron Chink," the first effective salmon-butchering machine, became available.

Richard Hume split off from his brothers in 1877 to establish a salmon cannery at the mouth of the Rogue River, just north of the California border. Up until his death thirty-one years later, Hume ruled the Rogue River by dint of force, litigation, and the propaganda generated by the two newspapers he owned. He was the cantankerous king of an empire of nearly one hundred thousand acres that was dependent upon the extraction of natural resources, whether it was the timber, farm products, minerals, or fish that he sought from the soil and waters of southern Oregon and northern California.

Approximately 630,000 cases of tinned salmon, with forty-eight cans to the case, were shipped from Hume's canneries in 1883. The catch would drop after that banner year. Hume wrote:

> We find all trades and professions plunging to get a whack at this new El Dorado, all seeking a fortune to be made from the capture of the scaly beauties. What a mine of wealth, that even all who might plunge might be enriched. But all good things which nature has furnished have a capacity beyond which they cannot be strained.

The solution that Hume envisioned was the creation of salmon hatcheries, seasonal regulation of the catch, and the expenditure of state funds to keep people like himself in business. Hume lobbied the Oregon legislature incessantly, and then was elected to it. He and his wife built an estate at the mouth of the Rogue, took up horse racing (the sport of kings), owned the first car in the county, and traveled to Europe. They were known behind their backs as "King Richard and Queen Mary."

Hearing that there were plentiful runs of salmon at the mouth of the Klamath River in California, Hume decided to expand his empire in the mid-1880s. He first filed a claim for land on the Klamath Indian Reservation, but it was denied by the federal government. In his autobiography, written and printed first in one of his newspapers to dispute the charges of being a monopolist (titled *The Pygmy Monopolist*), Hume wrote of the trip in his steamer, the *Thistle,* to the mouth of the Klamath River: "So taking some of the boys from the Rogue River, went down, crossed the bar, steamed up to where the soldiers and their sergeant were stationed, dipped the colors and let go the anchor."

The sergeant came on board. Hume gave him a cigar and the two men sat down to talk. Hume outlined his plans. The sergeant said he had orders to halt such operations on the reservation and mentioned something about having a fair number of guns and men who could shoot them. Hume said:

> I happened to have on board a Henry express rifle that I had bought in Edinburgh, Scotland, which carried an explosive bullet and 120 grains of powder, which had been made for a Rajah in India for tiger shooting, so showed that, and told him if he meddled with us we would treat him as a highwayman.

Chief Spot, as Hume called the Yurok Indian, then came on board and told Hume that the captain of the military detachment wanted him to leave. One-Eyed Billy, whom Hume identified as another chief, was more amenable to Hume's blandishments, so Hume dealt with him. The next day Hume and his men went fishing. "I had never seen so many [salmon] in such a small stream in my life," he wrote. Hume brought a barge down from the Rogue River and erected a two-story bunkhouse for his non-Indian employees and a salmon salting facility atop the scow. The captain avoided Hume all that summer. When the fishing season was over and Hume had departed, the military confiscated the scow.

The federal government filed suit, but the United States Attorney failed to appear at the trial. The government, acting on behalf of the Indians, lost the case. Hume complained that the suit cost him $8,000 "and I have not received a dollar's benefit." But he operated a small cannery at the mouth of the river until his death. Hume also purchased timberlands and a mill along the Smith River. In later years Hume spent the winters in San Francisco, returning to the Rogue River and what his biographer called "part feudal serfdom, part company town" in the summer months.

Canned salmon from the Klamath River, Humboldt Bay, and the Eel River was shipped to San Francisco or around the Horn to Great Britain and Western Europe, where during the Industrial Revolution it was much in demand as a cheap, nutritious food. Some found its way to New York City, where it was favored over salmon caught in the Sacramento and San Joaquin Rivers in the Central Valley. Old photographs displayed the glistening carcasses of one-hundred-pound chinook salmon strewn about the gravel banks of California rivers or dangling from heavy lines or poles. The faces of the fishermen appeared disinterested, while the set of their bodies suggested awkward pride.

The outcome of such harvests was not difficult to predict. In an 1892 report to the California Board of Fish Commissioners, David Starr Jordan, the president of Stanford University, wrote, "Just as the forests are wantonly and thoroughly destroyed by early settlers and by lumbermen, just so the fisheries of this coast will go under the hands of the canner."

Fishing, logging, road building, mining, grazing, toxic-waste dumping, and dam building contributed to the steady decline in the California salmon catch: eleven million pounds in 1880, seven million in 1923, less than three million in 1939, slightly more than two million in 1983. A report on the state of the salmon fishery by an advisory commission to the California Department of Fish and Game in 1971 was titled "An Environmental Tragedy," and a follow-up report in 1986 was called "The Tragedy Continues."

The salmon were a totem animal to the Indians; for those who came later, the salmon served as an indicator species for the health of the natural environment. Their trait of recognizing a given river or stream as their place of birth and following it upriver—providing there was enough water and no new obstacles—to fertilize an egg and then die was more in keeping with a Greek tragedy than the California way of life.

I have watched thick masses of bruised and rotting salmon fight their way upriver on the Kenai Peninsula of Alaska. Although I have kayaked down portions of such traditional salmon streams as the Eel and Sacramento Rivers and live not far from such a stream, I have never witnessed this phenomenon in California. Nor have I ever been present at the moment of creation, as described so movingly by Paul McHugh, who observed the following on a tributary of a North Coast river:

> As I came around a bend in the creek, there they were, metallic bodies flashing in the sunlight as they sparred and mated in the crystal shallows. The silvery torpedoes of the males, darting against each other, shoving competitors into jets of current where they would be washed downstream. The long large female, weighing twenty pounds or more, her body flashing rainbow markings and the bright red blotches that steelhead acquire after their return to fresh water. I watched her turn on her side to dig her redd in a shoal of gravel with powerful thrusts of her tail. The victorious male joined her then, and their bodies shivered in muscular spasm, side by side, as the eggs drifted down into the nest through a cloud of milt.

In the present decade, the federal government, with jurisdiction over the waters from three miles to two hundred miles offshore, and the state government, which oversees fisheries within the three-mile limit, closed the commercial salmon fishing season off the northern California coast and put severe limitations on recreational catches, citing the greatly decreased numbers of fish returning to spawn. Drought, habitat degradation, and overfishing had reduced the salmon to a shadow of their former presence.

By the 1930s the monoliths at the ends of both breakwaters were badly eroded; and labor problems had unsettled the region, which was essentially a number of interlocking company towns. Fifty-ton concrete blocks were dumped at the end of the jetties, only to fracture and disappear into the depths after the first winter storms. Smaller twenty-ton blocks seemed to do the job better, and they were stuffed into the widening gaps. Then the engineers, in desperation, repaired the concrete monoliths. But nothing seemed to work for long in that storm-wracked environment.

The movement of ships in and out of the harbor was also menaced by labor unrest. In 1905 Humboldt County lumber workers organized. Within one year, half the workforce belonged to a union, making it "the strongest bastion of lumber unionism in America at the time," according to Daniel A. Cornford of San Francisco State University, the author of *Workers and Dissent in the Redwood Empire.* Labor's dominance, however, was short-lived. Cornford wrote:

> The lumber companies, wielding their formidable power in the community and an array of more subtle strategies, virtually eliminated lumber trade unionism in the aftermath of a major strike in 1907.

Ships were loaded from existing stocks of lumber during the unsuccessful four-week strike. The companies refused to bargain with the unions, one of whom was the radical Industrial Workers of the World, better known as the Wobblies. Workers were thrown out of company housing. The local newspapers sided with the lumber companies. A criminal syndicalism ordinance was vigorously enforced by Humboldt County courts, and eight Wobblies were sent to prison during the 1920s. An article in the Wobblies' publication about California was titled "The Land of Sunshine and Serfdom." There was a residue of bitterness.

In June 1935, the mill workers again went on strike. They demanded more pay and a reduction in their sixty-hour, six-day workweek. Out of sympathy, union longshoremen refused to load lumber on the ships. Commerce came to a standstill. Mill owners, city officials, local newspapers, and the police retaliated quickly. The Eureka Problems Committee of the chamber of commerce, chaired by the mayor, formed a vigilante group. "If outsiders come in to take over the strike, we must be prepared," declared the mayor. The newspapers excoriated the strikers. Goons and thugs were imported by both sides. There were fights. The salesman for a leading national distributor of tear gas reported to his home office in Pennsylvania:

> I have just returned from Eureka where the general lumber strike is in progress and wish to report that the trip was most successful. I sat in on a meeting of the lumbermen's association and discussed with them at length the need of adequate equipment and suggested that they allow me to make a survey of the various mills and camps. They were most agreeable and I proceeded to follow out this plan, making my recommendations the following day.

Not only the lumber companies but also the Eureka Police, Humboldt County Sheriff's Department, and California Highway Patrol stocked up on tear gas.

On June 21, the strikers concentrated all their efforts on the Holmes-Eureka mill, situated on the bay in Eureka. The pickets taunted strikebreakers, and rocks were thrown. The police arrived in a Packard sedan, which was blocked by the pickets. Chief George Littlefield climbed out of the vehicle and fired his pistol into the ground, shouting: "Who's going to stop me?" Someone inside the car fired a tear gas gun out the window, and the canister wounded a woman. The pickets thought it was a shotgun blast. They yelled, "Let's get the bastards," and charged the car, hurling rocks.

The police officers and mill security guards, the latter crouching behind stacks of lumber, fired into the crowd. A tear gas grenade went off inside the car, and policemen blindly stumbled out. The chief and others were beaten with their own weapons. A patrolman got off a few rounds from his submachine gun before it jammed. When the tear gas cleared, the dead, dying, and wounded lay on the ground. Three pickets were killed, scores were wounded, and five policemen, including the chief, received minor injuries.

The police, firemen, and vigilantes scoured the small city, the largest in California north of Sacramento at the time. A shantytown near the railroad yard was burned to the ground. "With the demolition of 'Jungletown' it is believed that the city was purged of much of the undesirable element that has contributed to the unrest leading to the strike," reported the *Humboldt Times* of the incident that was reminiscent of the Indian massacre and the ouster of the Chinese. "No longer will the odor of Mulligan stew mix with the fetid odors from the nearby swamp."

The local unit of the National Guard was placed on alert by the governor, liquor sales were prohibited, and vigilantes turned away from Eureka those cars whose occupants looked like "radical sympathizers." There were more than one hundred arrests, including the president of the local union who hadn't been at the mill. His bail was set exceptionally high because a packed suitcase, subsequently determined to belong to a friend, was found in his car. The union had to shift gears and defend its members. The strike was broken, and the ships sailed once again.

The repair work on both jetties that had accumulated over thirty years was completed in 1963, just in time for the gigantic storms and floods of December 1964. The floods were the second water-related disaster to have occurred that year, the first being the tsunamis that devastated Crescent City in March. It rained hard in November and early December; the ground was saturated and the snow lay deep upon the higher elevations of the Coast and Klamath Ranges. Then came warm rain, lots and lots of it.

"The December 19–23 storm was of unprecedented intensity in the region," stated a U.S. Geological Survey report. Fifty inches of rain fell in one river basin, fifteen inches within twenty-four hours. Rainfall during this period averaged 260 percent of normal and reached a high of 300 percent in the Klamath River Basin. Rivers rose to heights beyond the wildest guesses of what might be possible, way beyond the peak flows of the 1955 flood, labeled "The Big Flood" in a report by the California Disaster Office.

"Flood damage in the Eel River Basin was catastrophic," stated the geological survey. There were nineteen deaths. Towns rebuilt after 1955 were obliterated in 1964; others were severely damaged. Bridges, the crucial links in this land of water, were washed away like so many Erector-set facsimiles. Small communities—indeed, a whole region whose prime characteristic was isolation, even in the best of times—were separated from larger ones or hung on by the tenuous thread of a single telephone line or a muddy back road. Farmhouses on flat ground, together with cattle, were washed away, as was the soil itself. The bloated carcasses and rotting vegetation produced a miasmic stench.

The naturally erosive landscape had been loosened over the years by the tearing down of trees and associated road building. An estimated one hundred forty million tons of sediment, which translated into eighty-four tons an acre within the Eel River watershed, passed one point on the river. Millions of board-feet of lumber and logs stored at streamside mills and the loose debris of the forests were washed downstream to the Pacific, where they floated on the ocean currents as far north as the mouth of the Columbia River and were a menace to navigation for years to come.

The floods of 1955 and 1964, however, were not the first high waters the region had experienced. People forgot or were new to the state, and, in any case, Californians did not dwell upon or learn from their natural disasters. The aftermath of the 1861–1862 floods was recorded by William Brewer thusly:

> The floods of two years ago brought down an immense amount of driftwood from all the rivers along the coast, and it was cast up along this part of the coast in quantities that stagger belief. It looked to me as if I saw enough in ten miles along the shore to make a million cords of wood. It is flung up in great piles, often half a mile long, and the size of some of these logs is tremendous.

Brewer measured logs that were one hundred fifty to two hundred ten feet long and three and a half to four feet in diameter at the tapered end. He didn't give the measurements for the large end, perhaps because he had no measuring device that long and, being a scientist, refused to guess.

Similar-size logs from the 1964 storm battered the ends of the two jetties that were supposedly protected by one-hundred-ton concrete blocks. They were soon reduced to rubble. By 1970 a new rescue effort had to be mounted. The Army Corps of Engineers was determined to find, once again, a lasting solution. A scale model of the bay the size of a football field was constructed in Mississippi, and various tests were conducted. The tests showed that an interlocking, reinforced-concrete object called a dolos, after the anklebone of a goat which it resembled, worked best. Dolosse had functioned effectively in South Africa, where they were invented.

For Humboldt Bay, the engineers settled on a forty-three-ton dolos, twice the size of those used to protect a South African harbor. Into the concrete for each dolos went 1,400 pounds of steel reinforcement. By 1973 the last of 4,796 dolosse were placed on the ends of the two jetties. A single dolos was deposited alongside the Eureka Chamber of Commerce building, where it was dedicated to the seafaring men of Humboldt County—"past, present, and future."

There were problems in the early 1980s. The dolosse held firm, but they were beginning to settle into the sand, and waves broke across the jetties. One thousand more dolosse were cast from concrete made from rock obtained in the same quarry that supplied the rock for the original jetties nearly one hundred years earlier. The repairs and maintenance would be ongoing, apparently forever. In addition, the offshore bar, entrance channel, and harbor needed to be dredged every year.

Inside the breakwaters, the shape of Humboldt Bay has changed over the years. From the two-hundred-thirty-three-square-mile watershed of the bay came massive amounts of sedimentation that choked the tidelands and constricted harbor channels. At high tide there were twenty-eight square miles of water; at low tide 70 percent of the bay was exposed mudflats. Over a period of one hundred years the twin jetties concentrated

The dolosse

the erosive power of the waves on the shoreline opposite the entrance and washed away two hundred acres of Buhne Point. The shoreline was graded and lined with rock to protect a subdivision, a railroad, a road, and a nuclear power plant.

The latter was a study in yet another of the many failures that plagued the region. The Pacific Gas & Electric Company (PGE), the giant utility that serves most of northern and central California, purchased 137 acres of land at Buhne Point in 1952. The plant went into operation in 1964 at a time of great promise for nuclear power. A company pamphlet predicted that after 1970 many of PGE's new electrical generating facilities would be nuclear.

The plant was one of the first of its type to go into service and was the first to close down. While in operation, the gray edifice with the candy-striped stack was labeled the "dirtiest" such plant in the nation by *Science* magazine. Twelve years after start-up, the plant ceased operating because of seismic concerns—it lay at the junction of two earthquake faults.

Since there was no national repository for high-level nuclear wastes, the spent fuel rods sat in a twenty-six-foot-deep concrete container lined with stainless steel. The lip of the tank was fourteen inches above the level of the water. At 2:28 A.M. on November 8, 1980, an earthquake that registered 7.0 on the Richter scale jolted the region.

A nearby freeway overpass collapsed, and the rock revetment that protected Buhne Drive slid into the bay. The radioactive water within the tank sloshed about and came within three inches of spilling. Small cracks appeared in the concrete and masonry of the plant. Fissures and slumps tore apart the earth around it. PGE said there was no danger to the public.

There is no overt indication that a nuclear dinosaur sits on the shore of Humboldt Bay. The sign on the highway states only PGE HUMBOLDT BAY FACILITY. The company no longer gives tours of the plant.

Willard Bascom foresaw the eventual failure of the jetties. Bascom arrived at Humboldt Bay during World War II as part of a University of California team studying waves. He and others blithely drove into and out of the surf in amphibious vehicles called "ducks." With radios and a team on shore, they took measurements.

They were tossed about like rubber toys in the roiled water of a bath-tub. The glass windshields were shattered and the canvas awnings were ripped to shreds. Fortunately, no one was injured. "We were young and lucky, and in return for the risks we had the fun of challenging the break-ers," said Bascom.

But Bascom learned firsthand the power of waves and currents. He went on to a distinguished career at the Scripps Institution of Oceanog-raphy in San Diego, where he wrote of the Humboldt jetties in his clas-sic work on West Coast shoreline processes, *Waves and Beaches*:

> The ocean is huge, powerful, and eternal. Puny humans can scarcely expect to win by overpowering it, and anyone who counters its attack with brute-force solutions is doomed to expensive disappointment. Rather, the engineer must try to understand how the sea acts and learn to take advantage of the geographic and oceanographic conditions so that every-thing possible is in his favor. Then, on a battleground of his choosing for the short span of human interest, he may be able to hold his own. The first and most valuable lesson one can learn about the sea is to respect it.

That the harbor would bring prosperity was a myth that never became a reality. After the logging boom of the 1950s and 1960s went bust, the number of vessels using Humboldt Bay dropped precipitously. The hope for bringing prosperity to this minor West Coast port was containeriza-tion facilities. Hope had sprung eternal over the years. The carriers of the torch this time were two men who were the commercial and navigational descendants of Hans Buhne.

In Les Westfall's office at Westfall Stevedore Company, the pho-tographs on the wall celebrated the past glory of the port. There were pic-tures of ships being built, tied up at wharves, and under way. Piles of lumber were stacked high at bayside mills. There was even a color-tinted picture of the old passenger steamship the *Pomona*. The *Pomona* was no more, fewer vessels loaded at the handful of wharves, and the high of three hundred fifty sawmills in the county in the 1950s was now down to nine.

Still, Westfall, who started in the stevedoring business in 1950, made his annual pilgrimages to Washington, D.C., to seek money for dredging and repairs to the jetties. The secret of his success, he said, was never to ask for money but just state the need for harbor improvements to the appropriate subcommittee. Local congressmen and California senators had always been supportive, as had the corps of engineers. A tight-knit group within the chamber of commerce made up of Westfall, the manager of the local Caterpillar dealership, the publisher of the newspaper, and the owner of the local television station were responsible for pursuing the annual appropriations. "We work at it. And we get a lot more money per ton than any other port on the West Coast," said Westfall.

While Westfall was in charge of handling the arrangements for ships on land, Burt Bessellieu guided them through the entrance to the docks. Bessellieu first went to sea on a cargo ship in World War II. By 1968 he had become chief pilot at Humboldt Bay. Bessellieu knew the power of water. He had seen the end of the jetty destroyed overnight, monster waves break over the bar, white water (termed "popcorn") boil up on the ebb current, mirages of ships not due to arrive for hours loom on the horizon, and the set of the powerful current from the south.

The ships have gotten bigger, deeper, and less powerful over the years, and that made the job more difficult. "Some skippers go on vaca-

The port

tion when they hear they are scheduled for Eureka," said Bessellieu. The approach to the entrance was a dogleg. Ships approached from the west; the jetties were lined up in a northwest-southeast direction. The jog had to be timed perfectly and with hard left rudder—the opposite from what one might suppose—and as much speed as could be mustered in order to counter the grip of the current. Otherwise, the current swung the stern toward the north jetty and the bow headed toward the south jetty and potential disaster.

Bessellieu, who had some close calls over the years, handled ships' captains in the following manner:

> We scare them in the beginning. We say, "Okay, Cap, how fast do you go?"
>
> "Well, my maneuvering speed is about 10½ knots and I can probably give you 11."
>
> And we say, "Well, that's not enough. You have to give us 12 or 13 or you will have to stay out here and wait until the next tide."
>
> Well, they wonder why. There's too much current. Most of these newer ships have computers in the engine room; and when they put the ship on full ahead and they want more, they can't override the computer unless someone does it manually. So in the beginning we had a tough time with these skippers to override the computer, and there were delays on their coming in because we could not get enough speed.
>
> Eventually, they say, "No problem. We can override the computer." Two knots is a big help, and we come right in.

On the North Spit, I had a ringside seat at the parade of industrial activity visible from the small county park at the Samoa boat-launching ramp.

There were the sounds, sights, and smells of heavy industry: the clanking of the pulp mills, the belching of steam, and the release of noxious vapors into the air. The fumes were an unhealthy mixture of chlorine gas, sulfur dioxide, formaldehyde, ammonia vapors, and wood dust. But it was the sickly-sweet smell of jobs.

At night the fog swirled around the steel frames of the tall mills, dulled their outlines, and formed small halos around the many pinpricks of lights that festooned their superstructures. It was a tableau of either a futuristic city or the worst of the Industrial Revolution.

The freighter *Forest King,* of Monrovia, made its way slowly up the channel. Fishing boats were outbound. Slowly, very slowly—even tenderly, if that can be imagined—an oceangoing tug pulled a barge with a giant canister secured on top toward a Samoa pulp mill. The tank was loaded with liquid chlorine gas; if ruptured, it would cause more than a few deaths and the evacuation of Eureka. Two additional tugs and a Coast Guard cutter hovered nearby.

The Coast Guard had a number of other duties off this lonely coast, which had seen a fair amount of rum-running during Prohibition. The cutter *Acushnet* was docked nearby. While I was there a military ceremony of leave-taking was conducted for the skipper, who was known as a go-getter. He had been promoted to similar duties in New York harbor. The cutter had recently intercepted a vessel off the coast that carried 1,300 pounds of hashish. Much like a fighter plane, the superstructure of the cutter was adorned with the symbols of victories. There were more than two dozen stylized renderings of a marijuana plant, each representing the seizure of yet another natural resource that had been harvested in vast quantities in this region.

The marijuana-growing boom of the late 1970s and early 1980s also went bust. It became too much of a good thing and had attracted violence, not an unknown commodity in the region. Humboldt County had been touted as the marijuana-growing capital of the nation. Marijuana was ranked as the nation's fourth-largest cash crop behind corn, soybeans, and wheat—these three being supported by government subsidies. In California, the nation's number-one farm state, marijuana was for a time the number-one cash crop, beating out grapes by a few hundred thousand dollars.

The growers armed themselves against "patch pirates" in the Emerald Triangle, as the region of most intense cultivation came to be known. They set booby traps. There were shootings and a few murders. Then federal, state, and local law enforcement agencies joined in what was termed the largest domestic military operation in recent years. They dropped from the air in helicopters, tore out and burned marijuana plants, either destroyed or confiscated private property, and then flew away.

A good thing went bad. The hippies were replaced by commercial dope growers, who hung around bars once frequented by loggers. Ray

Raphael, whose four books, including *Cash Crop,* constitute a contemporary history of the region, wrote:

> From the near extinction of the sea otter in the early 1800s to the depopulation of the salmon today, from peeling the tan oaks to clear-cutting fir and redwood, the depletion of natural resources has been quick and severe since white civilization first arrived in this area. As each new resource was discovered, a small civilization was founded around it; as each resource in turn disappeared, the civilization which depended on it vanished.

In the 1990s the most valuable resource was tourists and the region was reduced to selling mementos of itself: redwood burls, smoked salmon from Alaska, rejuvenated homes turned into bed-and-breakfasts, and gritty logging and farming towns transformed into glitzy replicas of a past that never was. The industry of servicing people who came from elsewhere flourished. And the tour of Humboldt Bay aboard a rebuilt logging vessel never mentioned the Indian massacre. Instead, three little girls walked out onto a dock on Gunther Island and performed a tap dance for the delighted tourists, to the accompaniment of the tune "Lullaby of Broadway."

V

The Great Valley

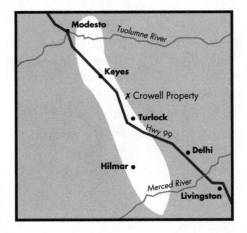

I dropped into the Great Valley by way of Interstate 5, the dividing line between the Land of Fire and the Land of Water. The gently angled descent down the flanks of Mount Shasta, that cyclopean eye that glows from the forehead of California, was made during the winter of an extended drought.

It was an eerie time. There was little snow on the mountain, and the surrounding landscape was dry and crackling when it should have been wet and soggy from winter rains. The giant, serpentine lake behind Shasta Dam, which stored irrigation water for the valley, as do many other reservoirs behind other dams on other rivers, was extremely low. The floating boat ramps extended as far down as they could reach, and grass was growing where there should have been water. A forest fire burned on a ridge to the left, its flickering glow ascending as I drove down the concrete freeway and into the warm valley.

I camped that night amid oak trees in a state recreation area that bordered the Sacramento River. The next morning, before departing, I walked through a grove of apple trees in blossom; green, irrigated fields also spoke to me of the land's great fecundity, not of the decline that I felt was imminent. As it turned out, California would muddle through another year of drought. There had been more serious droughts in the state's distant past; some lasted for a century or two.

As I proceeded south through the valley, there were few visible clues that this water-dependent landscape might be water short or, indeed, might someday wither and dry up, as have all great hydraulic civilizations in the past. The knowledge I had gained while living and working in the valley, however, told me to beware of surface appearances. No matter how many reservoirs, dams, and aqueducts, we lived and prospered in California at the sufferance of the natural elements and the ability of our institutions to maintain huge public works.

I was headed for the center of the valley, the place where I had learned many years ago that California was about land, the water that enriched it, and the people who either benefited or were shunted aside from the riches that accrued from water being applied to the land. The voracious scramble for agricultural wealth had replaced gold mining in the pantheon of wealth and now accounts, in a single year, for more money than all the gold ever mined in this state.

I was searching for a representative segment of the valley—a flat piece of agricultural land—and I was returning to Turlock to find that place. I had served my newspaper apprenticeship in Turlock, then a small farm town in the center of the valley known as the Turkey Capital of the West. After my arrival in the state, I spent six months on a small weekly newspaper in San Carlos, a bedroom community on the Peninsula south of San Francisco. I sold advertising during the day and covered city council and planning commission meetings at night.

It was then time to move on, and I got my first full-time reporting job on the *Turlock Journal,* one of the smallest daily newspapers in the state. It had a circulation of fifty-five hundred. I was the only reporter. I covered police, fire, city government, and anything else that came up. It was also my job to put out the weekly farm page and to take the place of the photographer, the society editor, and the sports editor when they either went on vacation, got drunk, were fired, or quit.

It was at first a traumatic and then, eventually, an intensely rewarding experience for someone who had come to California by way of a youth spent in New Jersey and four college years in Massachusetts. The farm page was the immediate problem, however. I will never forget the farmer who stormed into the newsroom after I had put out a few pitiful efforts and complained about all the freak stories I had run. I thought stories about two-headed calves, walnut trees that grew oranges, and similar curiosities constituted farm news. He laughed derisively. My California education had begun.

It continued in earnest when I made my first visit to a Turlock Irrigation District board meeting. No one covered the irrigation district. We waited for them to tell us the news, but I was curious. Although I had been a political science major in college, I had never heard of this form of government.

On a day when I had nothing else to do after the noon deadline, I drove to the district's office on what was then the outskirts of town. City government was run out of a crumbling downtown building. The irrigation district offices were new and quite substantial. They were built to impress and perhaps to intimidate. The building spoke of longevity, wealth, and power. In time I came to understand that this was the home of a unique and paternalistic institution.

I walked as fearlessly as possible into the boardroom and sat down with notebook in hand. The five directors glanced toward me; perhaps one or two nodded. An old man who sat to their left took one long look and then turned away. They talked about the weather, the crops, and finally the chairman, Abner Crowell, asked my business.

I said I was a reporter from the *Journal*.

Crowell smiled, slightly. There was an extended silence. Papers were shuffled. I began to feel uncomfortable, decided that my time could be better spent elsewhere, and left. But I didn't give up. Whenever I had the time, I returned, and the board slowly got used to my presence. They began to talk about improvement districts, acre-feet of water, and other arcane matters.

I wrote a few harmless and probably inaccurate stories. Crowell took some time to explain a few things to me. If I talked to the old man, Roy Meikle, I would get a terse reply from the chief engineer. Meikle, whom I would come to appreciate as a legend of his time and calling, just barely suffered this fool.

But they needed me, as I also came to understand. They had a new dam to build and needed the approval of the voters in the district to issue bonds to finance the project. So Crowell drove me up to the site for New Don Pedro Dam, whose waters would inundate the old dam of the same name. We got lost the first time. The newspaper publisher, a fellow Rotarian, chided Crowell in a column. We made it the second time. I took pictures and talked with Crowell, who turned out to be amiable and knowledgeable about the mechanics and politics of water.

It was people like Meikle and Crowell who transformed the valley from a marshy, sandy wilderness into the most extensively controlled and richest agricultural landscape in the world. From the moon, only the Great Wall of China and the irrigation canals of the Central Valley were visible. The valley, which resembles a giant cucumber enfolded by coastal and mountain California, was the economic engine that drove the state.

Years later, after I had lived and worked in both rural and urban areas of California, I came across the following passages in two books and thought of my time in the valley. Its importance was underlined by Donald Worster, an environmental historian who specialized in the American West, who wrote in his book *Rivers of Empire*:

> The technological control of water was the basis of a new West. It made possible not only the evolution of a prosperous agriculture but also, to a great extent, the growth of coastal cities like Los Angeles and San Francisco. It eventually made California the leading state in America, and perhaps the single most influential and powerful area in the world for its size.

The prevailing urban view of the valley was one of inconsequence and disdain. John Brinkerhoff Jackson, the specialist in landscapes, described that viewpoint in his 1972 book *American Spaces*. He wrote:

> Disdain for the dull workaday utilitarian countryside (and for the newly rich men who exploited it so effectively) became characteristic of a peculiarly California environmental philosophy: a philosophy which to this day persists in seeing two, and only two, significant aspects of the world: the city and the wilderness.

The valley was a flat, hot landscape to speed across quickly on the way from The City to The Lake, as San Francisco and Lake Tahoe were referred to by the cognoscenti. I always thought that was strange, since such pillars of the San Francisco economic establishment as the Southern Pacific Railroad, the Bank of America, and the Pacific Gas & Electric Company depended on the valley for their inflated existences. In time I came to understand that the Central Valley and Southern California were the most uniquely Californian of all the provinces.

The Great Valley, known also as the Central Valley or the Great Central Valley, consists of the Sacramento Valley in the north and the larger San Joaquin Valley to the south. The dividing line is roughly the Delta, where the Sacramento and San Joaquin Rivers converge in a landscape that was not unlike the bayous and swamps of the South before it was diked and drained, beginning in the 1860s. I have kayaked from Redding to Colusa twice on the Sacramento River and motored through the sloughs of the Delta in a small sailboat. From the water, at least, there is still the feeling of a dense, riverine environment.

The whole valley is one of the great structural depressions in the world. It measures some four hundred fifty miles in length and averages fifty miles in width. The valley was once the floor of an ancient sea. As the Coast Range on the west and the Sierra Nevada in the east rose in recent geologic time, the sea retreated and the valley emerged over the last sixty-five million years. The rivers from the Sierra added their sediment to the vast amounts of primordial ooze that had accumulated under the sea, and eventually the sedimentary sequence reached depths of thirty to forty thousand feet.

Gigantic alluvial fans from the outflow of rivers eventually merged and formed one extensive alluvial plain along the east side of the valley. The land sloped gradually toward the west where water from the individual rivers was collected in the north by the Sacramento River and in the south by the San Joaquin River. The two rivers join in the 750-square-mile Delta and flow to the sea through San Francisco Bay and the Golden Gate.

All the natural ingredients were in place for intensive agriculture. The snow accumulated in the Sierra Nevada; the narrow river canyons that drained the mountains were located in the foothills where dams and reservoirs could be easily constructed; the valley floor was relatively flat and had the proper gradient for canals; and there was plenty of sun. The landscape would be altered—transformed to a greater extent than any similar place on earth—but the elemental factors remained in place and determined what occurred there.

Originally, the valley was a patchwork of aridity and humidity. The Indians rarely traversed the dry plains. They preferred the lush riverbanks and extensive swamps. These marshlands flooded periodically, and tules, other bulrushes, and cattail grew in profusion. The riparian forest was virtually impenetrable: clogged with fallen limbs, berry vines, rose snarls, poison oak, and a general profusion of rank growth. The subtropical prairie was a sea of perennial bunchgrasses, such as purple and nodding needlegrass, interspersed with gracious valley oaks.

Migrating waterfowl blackened the skies. Grizzly bears prowled the forests. Vast herds of deer, pronghorn antelope, and elk, the latter numbering in the thousands, thundered across the plains. The elk were as common as the buffalo farther east, geese stood around as if tame, and "fields of the beautiful blue-flowering lupine" spread all about, remarked John Charles Frémont, the military explorer, as he passed through the valley in the spring of 1845.

The brief spring was, indeed, a glorious time of year. However, at other times of the year and away from the watercourses it was a different valley. William Brewer crossed the valley in the summer of 1864, a drought year. He wrote:

> We came southeast across the plain. The day was hot, as usual, but not so clear. The mountains were invisible through the dusty air; the perfectly level plain stretched away on every side to the horizon, and seemed as boundless and as level as the ocean. It is, in fact, sixty miles wide at this place, and neither tree, nor bush, nor house breaks the monotony. Thus we slowly plod our weary way over it, league after league, day after day. During the entire day we saw beyond us, behind us, sometimes all around us, the deceitful mirage. But it always vanishes as you approach it—heated air, and not cool water, we find in this place.

The large herds of wild cattle and horses who trampled the native grasslands, the exotic plants that slipped into California to replace natives, and the brutal subjugation of Indian tribes were the Spanish and Mexican legacies. Then the Americans displaced the Hispanics and the few remaining Indians in this well-watered valley.

The rapid changeovers of people and cultures and land-use are typical of the California experience. The valley landscape has been superficially altered four times within the last two hundred years: The natural landscape became the livestock landscape around 1800. The livestock landscape became the wheat landscape in 1865. Wheat gave way to irrigated agriculture in 1900. And the urban landscape that crept out from the Bay Area began to replace the rural landscape in 1960. Foreign invaders—Hispanics and the Anglos—were responsible for the first change; all subsequent changes took place within the white-dominated culture.

The Yokut Indians inhabited the San Joaquin Valley when the Hispanics pushed up from Mexico and established a tenuous hold on the coast of Alta California in the second half of the eighteenth century. They were most impressed with the grazing potential of the inland valleys: "The place where we halted was exceedingly beautiful and pleasant, a valley remarkable for its size, adorned with groves of trees, and covered with the finest pasture," commented a member of one of the earliest exploring parties.

The Hispanics let a few of their four-legged beasts loose upon the land in 1769, and they multiplied and multiplied and multiplied. The last official count of mission livestock, made in 1834, totaled two hundred eighty-five thousand cattle, sheep, horses, and mules. But official counts were always low and did not take into account the vast numbers of feral animals. One estimate has placed the number of livestock in Alta California in 1800 at one million head.

In any case, the numbers were too great. Early travelers through the San Joaquin Valley reported that herds of wild horses were swarming over the plains and the numerous cattle were wilder than the deer and elk. Horses were slaughtered at the coastal missions because they were depleting the range needed by the more valuable cattle. The cattle were being slaughtered for their hides. The land was stained red and putrid. An early settler noted, "Thousands of cattle were slain, for their hides only, whilst their carcasses remained to decompose upon the plains."

With the invaders and their goods came the seeds of grasses and weeds from Mexico and Spain. The seeds stuck to the coats of animals and lodged in their intestines, were embedded in packing materials, were found in the ballast of ships, and hid as impurities among more desirous species of exotic plants, such as hay and wheat, that were imported into the country. The noxious weeds were dispersed quickly and widely by the wind and the livestock. Aided by periodic droughts and overgrazing, the native perennials lost their grip upon the landscape and were replaced by the hardier annuals, which were not as desirable for grazing.

No knowledgeable settler would have purposely introduced these species, nor is there any record of such an action. Nevertheless, this incredible transformation of the surface of California did take place within a short period of time. "Few places on earth, if any, have had such a rapid, wholesale replacement of native plants by introduced species," wrote L. T. Burcham in *California Range Land*, a historical survey of the subject made by the California Department of Natural Resources in 1957. In *The Destruction of California*, Raymond F. Dasmann wrote, "The needle grasses and wild rye disappeared and we were left with foxtail and Medusa's head. The June grass, bluegrass, and poppies have gone. We have inherited the poverty grass, tarweed, and thistle." Today 99 percent of the valley landscape has been taken over by these invader species.

The native inhabitants of the Turlock region were also replaced. There were between fifteen and twenty thousand Yokut Indians divided into fifty distinct tribes, each speaking its own dialect, in the San Joaquin Valley at the time the Hispanics appeared on the coast and made a few tentative forays into the interior of California. The Yokuts occupied about 10 percent of the land area of the state, the largest region occupied by a single language group. Their fifty dialects constituted one-third of the languages spoken in the state. They were primarily a river culture, and tribes were spread along the Merced, San Joaquin, and Tuolumne Rivers that encircled Turlock on three sides.

They were also concerned with water. One of their more important ceremonies had to do with rainmaking. Charm stones, later uncovered by farmers leveling their land in preparation for irrigation, were used to lure salmon up the streams.

The Hispanics, accompanied by mission Indians known as neophytes, marched inland in search of converts and slave laborers. The Indians resisted—none more fiercely than Estanislao, whose Christian name, derived from a saint, would one day be bestowed upon Stanislaus County, wherein lies Turlock. Estanislao was an escaped neophyte from Mission San José. He was of mixed blood, tall, slender, and heavily bearded.

Three expeditions were mounted in 1828 and 1829 to capture "dead or alive a certain Estanislao," who had gathered other escaped neophytes around him and was taunting the Mexicans. Among other things, he chided them for being poor shots. It was to be guns and stiff leather jackets against bows and arrows.

The first expedition, led by a sergeant and consisting of fifteen soldiers, engaged in a brief fight with the Indians and then returned to the mission. The second was led by a veteran Indian fighter, who was backed up by forty soldiers, an equal number of Indian auxiliaries, and a cannon. The first shot broke the gun carriage. They approached Estanislao's fortified encampment on the Stanislaus River bottomlands. A Mexican chronicler wrote:

> The village was located in the middle of a willow thicket. These bushes, interlaced one with the other by the great quantity of runners and stems of grapevines, made the area inaccessible even to the rays of the sun, not to speak of affording entry for fighting. When the soldiers approached the war cry resounded through the entire forest, and a few Indians came out to the edge with the sole object of directing at the soldiers such obscene language as they possessed.

The soldiers were enticed into the thicket and bombarded with a barrage of arrows. They were hit in the face, where their leather jackets offered no protection. The heat was enervating. The Mexicans retreated, their casualties being three dead and seven wounded, and returned to Mission San José.

A third expedition, consisting of all the Mexican soldiers in Northern California, was quickly organized. At the age of twenty-one, Ensign Mariano Guadalupe Vallejo, who would rise to be a general and the most influential Hispanic citizen of early California, was put in charge of the force. It consisted of two hundred soldiers and fifty Indian auxiliaries, the latter being ancient enemies of Estanislao's tribe. They too set off for the valley with a cannon.

The "Christian conspirators," as they were referred to by the Mexicans, had erected stockades of oak and cottonwood logs and dug interlocking trenches that were reminiscent of the Modoc Indians' use of their natural landscape. On the first day seven rounds of solid shot and nine of grapeshot were fired from the cannon. Then the weapon was moved to within ten yards of the fortifications, and thirty-one rounds of deadly grapeshot were loosed upon the Indians. The woods were set on fire. The Yokuts were dislodged, and the Indian auxiliaries moved in for the kill. The chronicler, a certain Antonio Maria Osio, wrote in *Historia de California*:

> When a good opportunity presented itself to kill and destroy they burst out like starving hounds and instantly began an atrocious massacre. The woods were burning. They dragged the hidden enemy out of the underground passageways, and finally destroyed everything.

The Mexicans camped that night near the burning forest. A few Indians escaped, including Estanislao, but others were shot down as they attempted to flee the inferno. With no remaining food or ammunition, the expedition returned to Mission San José the next day. Ensign Vallejo reported that three women were killed by bullets, and a short distance away he found seven more corpses. Others were executed, some by hanging. Perhaps a total of thirty Indians were killed. Eleven soldiers and two Indian auxiliaries were wounded. Eighteen stolen horses were recovered. Estanislao sought asylum at the mission and was granted it, over the objections of the general population. He lived out the remainder of his life in peace.

General warfare then broke out in the valley. Sherburne F. Cook of the University of California wrote in *The Conflict Between the California Indians and White Civilization*:

> Almost continuous fighting back and forth resulted, wherein all semblance of real expeditions was lost. At the same time, the expanding economy of the private ranches demanded an increased supply of cheap labor, which was most easily obtained from the adjacent native tribes. Thus punishing stock thieves and capturing farm laborers became almost the same in method.

In this manner a precedent was established. The tremendous productivity of the landscape meant that many people had to work the soil. Farm laborers who would work for low wages were lured from foreign countries, worked under conditions of near slavery, and then disposed of when their presence no longer suited the purposes of those who had imported them. The irony was that the Mexicans, who established this practice in California with the Indians, would themselves become the victims of it.

The Yokuts soon vanished from the scene. Those Indians who lived on the west bank of the San Joaquin River were exterminated by an epidemic in 1833. A traveler reported that "large numbers of their skulls and dead bodies were to be seen under almost every shade tree near water, where uninhabited and deserted villages had been converted into grave yards." To the east, the Indians held up poorly against the encroachment of the Americans during and after the gold rush. An 1881 history of Stanislaus County noted, "Disease has made such inroads upon them that it is but reasonable to expect that the few remaining will soon waste away and perish from the land." The 1910 census, the first to record such matters, counted 533 Yokuts.

With the arrival of the Americans, the emphasis changed from raising cattle for hides to letting the animals loose upon the land to be eventually collected for their quotient of beef. The hides were then wasted. Californians were beef eaters. The California Battalion, a ragtag collection of early settlers and recent emigrants who participated in skirmishes against the Mexicans, marched across the state in 1846; its eating habits were documented by an incredulous observer, Edwin Bryant of Kentucky, who wrote:

> Other provisions being entirely exhausted, beef constitutes the only subsistence for the men, and most of the officers. Under these circumstances, the consumption of beef is astonishing. I do not know that I shall be believed when I state a fact, derived from observation and calculation, that the average consumption per man of fresh beef is at least ten pounds per day. Many of them, I believe, consume much more, and some of them less. Nor does this quantity appear to be injurious to health, or fully satisfy the appetite. I have seen some of the men roast their meat and devour it by the fire from the hour of encamping until late bedtime. They would then sleep until one or two o'clock in the morning, when the cravings of hunger being greater than the desire for repose, the same

occupation would be resumed and continued until the order was given to march. The California beef is generally fat, juicy, and tender, and surpasses in flavor any which I ever tasted elsewhere.

The grazing of livestock and raising of feed for those same four-legged animals—meaning cattle that yielded beef, cows that produced milk and its by-products, and, to a lesser extent, sheep and horses—was by far the most dominant use of land in California and represented the single largest source of gross farm income, not only in the 1840s and 1850s, but also at the present time. In 1990, for instance, when California led the nation for the forty-third consecutive year in cash farm receipts, beef and milk products were the two most valuable commodities, and, when hay and alfalfa were added to the total, accounted for 27 percent of gross farm income. By comparison, grapes were 8 percent and cotton was 6 percent. In addition, approximately 85 percent of the water utilized in California goes to agriculture. Raising feed for livestock is the single largest use of that water.

It could be said without exaggeration that the lowly cow treading back and forth across deserts, mountains, valleys, and coastal hills from Del Norte County in the far northwest to Imperial County in the far southeast—not gold or movies or aerospace or computers—was the single most important ingredient that drove the engine that pushed the California economy for more than two hundred years. Other crops, such as wheat, lettuce, tomatoes, oranges, avocados, almonds, grapes, and eggs—all of which California would produce more of than any other state at one time or another—were, in comparison, minuscule uses of the land and water.

Along with record yields, the agricultural history of California was also replete with accounts of spectacular failures. Extremes of weather, a constant in the state, frequently caused disasters. Huge amounts of rain fell on California during the winter of 1861–1862. The Central Valley again became the inland sea that it once was, and hundreds of thousands of cattle drowned in the rising waters of the Sacramento and San Joaquin Rivers. William Brewer wrote:

Nearly every house and farm over this immense region is gone. There was such a body of water—250 to 300 miles long and 20 to 60 miles wide, the water ice cold and muddy—that the winds made high waves which beat the farm homes in pieces. America has never before seen such desolation by flood as this had been, and seldom has the Old World seen the like.

The rangeland then bloomed in the spring, and what cattle remained feasted on the luxuriant growth of new grasses. Then two years of severe drought followed. The loss from starvation was estimated to lie somewhere between two hundred thousand and one million head of cattle. The land once again stank of rotting flesh, and the few remaining grizzly bears gorged themselves. Brewer rode across the San Joaquin Valley and wrote:

> Where there were green pastures when we camped here two years ago, now all is dry, dusty, bare ground. Three hundred cattle have died by the miserable water hole back of the house, where we get water to drink, and their stench pollutes the air.

Eighty-five years later in his novel *East of Eden,* John Steinbeck described the extremes of weather in the Salinas Valley, another agricultural hothouse just to the west of the San Joaquin Valley:

> I have spoken of the rich years when the rainfall was plentiful. But there were dry years too, and they put a terror on the valley. And it never failed that during the dry years the people forgot about the rich years, and during the wet years they lost all memory of the dry years. It was always that way.

Up until 1867, the region around Turlock was a vast range for cattle. Livestock from ranches located along the Tuolumne and San Joaquin Rivers, where the Indians had also chosen to live, roamed across the swales of the empty prairie. The wild herds trampled the native grasses and anyone who happened to get in their way. A traveler crossing Stanislaus County on foot had a close call in 1863. A herd of longhorns headed toward him on the featureless plain. He said:

> I immediately fell to the ground and crawled on my hands and knees for a long distance until they had lost sight of me. I

afterwards learned that they were infuriated by being caught and branded, and would have killed me had they caught me.

The flood/drought cycle not only brought an end to the livestock boom but also completed the transition from native to exotic grasses, the most widespread of the latter being wheat, which also depended on the natural rainfall.

A vast monoculture of dry-farmed wheat replaced livestock in the Great Valley as the bonanza crop of choice after the disastrous 1860s, and wheat barons replaced cattle kings.

John W. Mitchell, the founder of Turlock, was one of the new land barons. Mitchell, a native of Massachusetts, arrived in California with his brother in 1851 while in his early twenties. He worked as a carpenter in San Francisco until he could acquire the farm implements needed to cut hay and sell it to the teamsters in the Stockton area, who were transporting farm products to the Gold Country. Mitchell and his brother then got in the transportation business themselves.

The young man, against the advice of his friends, began buying and selling land in the San Joaquin Valley. At one time in the late 1850s, he owned half a million acres. Mitchell bought fifty thousand acres one day and sold them the next day for a dollar-an-acre profit. He buried sixty thousand dollars in gold coin near his cabin, and then could not immediately locate it. Mitchell dug frantically, and finally found it.

His modus operandi was described by George H. Tinkham in *History of Stanislaus County*:

> He became the owner of valuable lots along the river, and then he bought lands on the plains; indeed, as fast as he prospered, he bought more lands, thousands of acres, and was extensively engaged in raising sheep, which aided in clearing the lands of brush and weeds, after which he engaged in grain raising on quite as large a scale where once he had had sheep.

Like other speculators, Mitchell used land scrip issued by the federal and state governments to purchase one hundred thousand acres in the Turlock area. The redemption or face value was $1.25, but the market value depended on the going price for the paper notes. The scrip was issued to veterans to promote land settlement or to states to build agricultural colleges. Not all veterans wanted to be frontiersmen and the states were cash hungry, so the scrip was sold for less than its face value. Other more unscrupulous land barons, such as Henry Miller and Charles Lux, whose extensive holdings were located farther south in the valley, also used the notes. Mitchell's holdings were exceeded only by those of Miller and Lux, the state's largest landowners. The two men came to epitomize fraudulent practices of land acquisition.

But Mitchell, according to all surviving accounts, was not of that ilk. He did use the discounted scrip, as was the practice, but he did not abuse the various federal land acquisition laws that Miller and Lux manipulated to their advantage. He did not engage in lawsuits, nor were any brought against him. Mitchell lived quietly and unpretentiously with his wife in the front of his grain warehouse in Turlock. A large man, Mitchell was known to children as "Uncle Johnnie." He was a Mason, he fed hoboes for free, and he donated money to charities. Mitchell left money in his will to four valley towns, including Turlock, to build libraries. "He was looked upon as a very benevolent man in the valley," stated the *San Francisco Chronicle.*

The same year that the Central Pacific announced that it was going to build a railroad through the valley, Mitchell acquired a large block of land north of Turlock. The patent on the first sale of the land that I have chosen to track was issued on May 1, 1868. The going price at that time for public-domain land was seventy-five cents an acre. The price would soon rise, as it always did. (W. W. Robinson wrote fifty years ago in *Land in California* that land prices peaked every twenty years or so, and with each such cycle there was "the same old opportunity to speculate, to make or lose fortunes in the land of California. Buying and selling California, a very old custom, will continue." The cycles now occur at ten-year intervals.)

The land surrounding Turlock was planted with wheat, and in 1872 the *Stanislaus County News* reported that Turlock consisted of a general store, a hotel, a boot and shoe shop, one saloon, and a blacksmith shop. Homes were rising on the stark plain, and work was about to begin on the railroad depot and Mitchell's warehouse. This was at a time when Stanislaus County was heading toward becoming the top wheat-producing county in California and the state was about to lead the nation in wheat production.

California made the jump from the pastoral stage of agriculture to the mechanical stage in a couple of years, and the fruitfulness of the land increased accordingly. The landscape made mechanization possible. It was relatively flat and free of trees and rocks. The soil was light and deep, and giant machines would not bog down in the dry terrain that prevailed for most of the year. Gang plows—six plows, each cutting a furrow one foot wide—were pulled across the vast acreages by teams of eight horses. A harrow and seeder followed.

As the grain turned golden, steam locomotives on wheels were readied to pull threshers. In one day a steam-driven machine threshed 1,618 sacks of grain on a farm outside of Turlock. A greater marvel to the locals were the first combined harvesters that were pulled by two dozen or more horses and moved in huge formations across fields, cutting, threshing, and stacking grain in twenty-four- to thirty-two-foot swaths per machine. In a few years steam-driven combines and tractors appeared.

With the machines came the seasonal farm laborers to do the menial work. At first they were drawn from every available race; then increasing numbers of Chinese were employed as mining and railroad building came to an end. The Chinese lived in their section of Turlock next to the railroad or along the malaria-ridden Merced River.

In 1880, the Chinese represented 85 percent of the agricultural labor force in the counties surrounding Sacramento, and less elsewhere. Chinese laborers were particularly prevalent in hop fields and fruit orchards. A farm publication at the time noted, "It is difficult to see how our annual fruit crop could be harvested and prepared for market without the Chinese." By the time they were driven from agriculture and replaced by the Japanese, the Chinese had participated at the bottom level in every major phase of the California economy.

The end of the Chinese presence in Turlock came in 1886 at the height of the violence that spread throughout the state. Someone set fire to the Chinatown section of town. One of the pioneer Irish settlers, Mrs. William L. Fulkerth, known affectionately as Aunt Abby, recalled:

> Old Purdy, the saloonkeeper, wanted to get rid of Chinatown.
> He pretended to help put out the Chinatown fire but handed
> watercans full of coal oil. Others were setting fire to the Chi-

nese beds. The Chinamen caught on to what was happening and cried, "Godammy, Godammy!"

Most of the Chinese fled to nearby cities, such as Stockton and San Francisco. The Turlock colony declined to one old man. One night his shack also caught fire, and he burned to death. The coroner's jury convened the next day, and after long and thoughtful deliberation, ruled that he had died from burning up. The Chinaman's entire estate consisted of two thirty-pound pigs, so a professional gambler by the name of Cochrane offered to bury him in exchange for the pigs. Eventually, Chinese from elsewhere returned, disinterred the bodies in the Chinese cemetery, and shipped them home to China. That ended the Chinese presence in Turlock.

The Turlock of the 1880s and early 1890s reflected the prosperity of the wheat era. There were now three hotels, three general stores, one drugstore, one tin shop, one boot and shoe shop, sixteen saloons, and five grain warehouses, including Mitchell's on Olive Street. Teamsters would start for town from outlying farms at three A.M. with huge wagons loaded with six to nine tons of grain that were pulled by teams of six, eight, or ten horses. They lined up to unload, and sometimes the line stretched far out of town. One day six hundred horses and mules from such teams were counted in town.

What the town possessed in terms of prosperity, it lacked in aesthetics. A traveler passed through Turlock around 1880 and noted in a book, under the subhead of "Horrible Turlock":

> In all California, I saw no other town or village like Turlock. Located on a branch of the Southern Pacific Railroad, it was the market town for a large, and just then, prosperous farming district, and was in the height of its initial prosperity. In its buildings, it presented that strange contrast of huts and large store-houses, often seen in the new West. But oh! the dirt! the dust! the tin cans! the frying, sizzling heat! the sweat and dust! the dust and sweat!

On October 4, 1882, the piece of land north of town changed hands. Roger M. Williams purchased 325.39 acres—slightly more than a half section—from Mitchell for two dollars an acre, or a total of $650. Ten years earlier Mitchell had mortgaged the property for $125, and one year after the sale Williams mortgaged the land for $3,650, or $11.22 an acre. For the first time the property was given a legal description. It was known as the "West ½ of Section 3 in Township 5 South Range 10 East."

The land needed to be identified with exactness because it had become a commodity. So the valley was divided into small grids based on the points of the compass that were derived from the known location of the Mount Diablo Base Line and Meridian. The sense of a natural landscape was lost in the process. The artificial divisions fostered the illusion of uniformity as land became property. The farmer or land speculator tended to focus only on what was contained within property lines and not on the natural processes that bound them together.

Williams, who would play an important role in the next phase of California agriculture, immediately began using his two sections as leverage for loans to purchase more land. He paid off the loans within one year, the interest varying from nothing to 10 percent. Late in the decade the county paid Williams $36.30 for the right-of-way for Monte Vista Road, indicating that settlement was taking place north of Turlock.

Turlock's first flush of prosperity was short-lived. On the night of October 3, 1893, fire—that scourge of so many early California settlements—roared through the downtown business section and left gaping black holes in its wake that would not be filled for another ten years or so. A saloonkeeper who lost his bar killed himself by drinking a jug of coal tar. Another coroner's jury was convened and did not render a verdict until it had consumed his inventory of liquor.

There was another fire in 1894, the year wheat prices hit rock bottom. The land had been exhausted by too much of a good thing. Other states and foreign countries had usurped California's predominance in grain production. Valley farmers began to switch to the greater certainty of irrigated crops.

The search for more water took a variety of forms. Gunpowder was exploded and giant bonfires were lit by rainmakers. Rain supposedly followed the plow, but when land was plowed it didn't necessarily rain. Trees attracted rain, it was thought. Trees were then planted throughout the valley, one man in Modesto becoming known as the "Eucalyptus

King." Flood an area, and it was thought that it would rain. That failed too. Irrigation was another possibility, but there were opponents to irrigation. Standing water caused the "general prevalence of chills and fever and other bilious diseases," said one doctor. When the temperature rose higher than 60 degrees, a not uncommon occurrence, "poisonous fermentation" of the water supposedly resulted.

There were also proponents of irrigation, the main argument in favor being that both the farmer and the land speculator would benefit quickly from the promise or actual application of a steady supply of water upon land. That recurring evil—drought—would be banished forever. Irrigation became a movement, then swelled into one of those panaceas that periodically swept California: Irrigation was blessed by God, strengthened the family, evened out the disparity between poverty and wealth, and would put food on the table for the nation.

The novel idea of an irrigation district, a special governmental entity to develop and manage the distribution of water, took hold first in Stanislaus County. The Wright Act, named after the Stanislaus County legislator who authored it, unanimously passed the legislature in 1887 at the height of the irrigation crusade. Carefully drafted, conservative in tone, thorough, and a paean to local control, the act found its embodiment in the Turlock Irrigation District, the first such entity to be formed under its provisions.

Such districts taxed everyone within their boundaries, regardless of their need for water. It was, in the words of historian Donald Worster, "a means of getting more income out of a river without surrendering to urban capitalists" or, as would be the case in fifteen years when the Reclamation Act was passed, surrendering control to the federal government.

As with other crusades, California exported the idea, and it caught on throughout the West with mixed results. Donald J. Pisani wrote in his history of the irrigation movement: "Only a few well-managed ventures, such as the Modesto and Turlock districts, weathered the financial storms of the 1890s; the remainder died at birth or within a few years. And as the speculative balloons burst, even the soundest districts suffered from the aftershocks." Pisani added in a footnote, "Even in the darkest days of the 1890s, the Turlock District remained in relatively good health."

Irrigation and the introduction of the refrigerated railroad car provided the means for the next bonanza. Worster wrote, "Artificial water, mixed with human enterprise and labor, would eventually make California the premier agricultural center of the entire earth, a modern kingdom by the sea." It was the dependency of the farmer upon the local irrigation district, or in the case of reclamation projects, the federal government—not the lonely Jeffersonian yeoman plodding along behind his single-bladed plow—that created this kingdom.

On a more immediate level, water rights to the Tuolumne River had to be obtained and the water had to be moved to Turlock.

When Assemblyman C. C. Wright returned to Stanislaus County after the governor signed his bill, it was the Fourth of July in March. A large, cheering crowd and the Turlock band greeted him at the train station. An old signal cannon boomed out. There were huzzahs, hip hip hoorays, and speeches. Two months later the vote to form the Turlock district passed by a four-to-one margin, and one of the five farmers elected to the board of directors was Roger Williams, the owner of the flat piece of land north of Turlock.

Williams was one of the first movers and shakers of the Stanislaus County water establishment, and for a brief moment achieved statewide prominence. A resident of Ceres on the south bank of the Tuolumne River, just south of Modesto, he was one of the community leaders who financed construction of Grange Hall, the town's first community center, polling place, and high school. In addition to being a director, he was the district's secretary. As such he was the first and for a few years the only employee. The Turlock Irrigation District office was located for a time in Ceres, nine miles north of Turlock. Whether Williams farmed his Turlock property or, as is more likely considering his many other duties, leased it to a tenant is not known. He seems to have been a combination farmer, land speculator, and water district functionary.

As the latter, it was Williams who found office space, purchased supplies, took notes and wrote up the board's minutes, sought advice from Fresno farmers on how to construct canals, hired an engineer, kept the books, and made arrangements for a land survey and maps. Most important, it was Williams, functioning as a committee of one, who obtained the district's water rights to the Tuolumne River. Turlock secured the first rights. Under the doctrine of "first in time first in right," the small irrigation district had a great deal of leverage over those who would eventually want to share that water.

It was also Williams who made the preparations for a bond election and carried out the successful sale of $50,000 in bonds in November 1887 and another $550,000 worth of bonds the following year. When it came time to construct La Grange Dam, the first irrigation dam on the Tuolumne River, it was Williams who was Turlock's representative on the joint Turlock-Modesto committee that oversaw construction. A history of the county noted that Williams, along with his counterpart from the Modesto Irrigation District, "performed Herculean work." The account added, "The stability and successful workmanship of the dam are due largely to their labor of constant superintendence."

It was also Williams who prepared the tribute for his fellow director and the man from whom he had purchased the property north of Turlock when Mitchell died in 1893. The board's resolution that was "spread upon the minutes" said that Mitchell's death was the most severe blow yet dealt the district and cited his "untiring energy, constant devotion to the interests of the District, even temper, and courteous demeanor toward his fellow directors and all with whom he came in contact." As the years passed and other employees were hired, Williams gave up his paid position and became president of the board. He outlasted all the other original directors and served until 1905. He died in Ceres in 1919.

La Grange Dam

Williams's brief moment of statewide prominence came when, acting in the interests of the district, he filed a friendly lawsuit challenging the constitutionality of the Wright Act. The State Supreme Court ruled the act constitutional in *Turlock Irrigation District* v. *Williams,* a court case mentioned in all thorough discussions of California water. What it meant was that the district had effectively diffused all potential unfriendly suits, and, along with other districts in the state, it could now go about its business. When the court's decision was announced in May 1888, guns were fired, bonfires were lit, and a county newspaper declared, "It was a memorable event for Stanislaus County."

Abner Crowell, Sr., the next owner of that plot of land, came around the Horn in a sailing ship in the 1850s. Instead of making for the gold fields, the young New Englander settled in Hanford in the San Joaquin Valley. Crowell taught school for a time in the small farm community in Fresno County. He met Mary Konoyer, who had come to California in a wagon train, and they married. From school teaching, Crowell moved on to horse trading. Later he bought cows and built a cheese plant. Crowell was an aggressive businessman, and he prospered. The couple had six sons and three daughters, which was a help. By the age of ten, the boys were milking a dozen cows apiece. When in their teens, they were sent out in pairs in horse-drawn wagons and peddled cheese up and down the valley. Arthur and Charlie made up one such team.

Crowell was an austere, frugal man. He did not favor the use of tobacco. He told his children that if they did not smoke by the age of eighteen, he would give them the pick of his horses. Arthur made his pick. Later in life he took up smoking; and when his father visited, the middle-aged man hid his cigarettes. Crowell was that kind of father.

The elder Crowell worked in the same suit and hat that he wore to visit his banker in San Francisco. One time the doorman, not recognizing Crowell, would not let him into the bank until the manager was called and identified his valued customer. His grandchildren recalled being forced to pick blackberries in their grandfather's yard. The old man then took the children out in his touring car to sell the fruit.

Irrigation would be the salvation of the West, Crowell believed, and he bought bonds sold by irrigation districts throughout the western states. Among his investments were Turlock Irrigation District bonds, which he purchased from Roger Williams. He also bought Williams's half-section north of Turlock at a time when land prices were depressed in the 1890s. One quarter-section was put in his name and another in his wife's name. Charlie acquired eighty acres of his father's land in 1898. Arthur was slated to operate the cheese plant, but when that burned to the ground, he bought eighty acres from his mother at fifty dollars an acre. The remaining one hundred sixty acres were sold to persons outside the family.

Arthur's one-eighth section was just to the west of his brother's property. Both fronted on Monte Vista Road, then a sandy wagon track. Neither piece of property had ever been irrigated. It was worn-out grazing and wheat land waiting to be rejuvenated. The brothers were minimally equipped for the job. Each had five horses, one more than was needed to pull a four-horse Fresno scraper, and they shared a milk cow. Charlie had built a small house, barn, and corral by the time Arthur arrived in 1900. Arthur quickly put up a small bunkhouse on the southeast corner of his property. He ate his meals with Charlie, who was married. Then Arthur set about leveling his land in preparation for irrigation.

Land in its natural state in the San Joaquin Valley was only relatively flat. It was not flat enough, or flat enough in the right way, to run water across it without ponding or missing areas entirely. For instance, the swales on Arthur Crowell's property amounted to four-foot rises and drops. The art of leveling land for irrigation was exactly that, an art that developed over the years into a science. Gradually, the valley, following the natural slope, was terraced in a series of fields that fell incrementally toward the San Joaquin River.

The Fresno scraper was the mechanical device that won the agricultural West and remade the valley floor. Any native vegetation missed by livestock or the wheat farmers was carried away by the scrapers. The Fresno scraper was a long, scoop-shaped device pulled by one, two, or four horses. With the reins in one hand and the other hand on a lever, the farmer could guide the team while at the same time scoop, carry, or dump the dirt, depending on which way he flipped the lever. Most farmers leveled the land by trial and error, using the eyeball method. Only a few could afford a surveyor. The land was irrigated, mistakes were noted, and then the farmer attempted to rectify them.

Arthur Crowell worked through the blistering heat constructing check levees and lateral ditches with a scraper. It was hard, grueling work, and the brothers helped each other out. The water arrived in the district between 1900 and 1903, depending on the location of the farm. The problem when the first water arrived was that few farmers knew how much to use. So, to be safe, they poured copious amounts of water on their crops. Frank Adams, an outside irrigation expert, commented, "There is no doubt that too much water is used by a majority of the irrigators in both the Modesto and Turlock districts." The level of the ground-water rose, ponded, and stifled crops until farmers learned better irrigation practices and pumps were installed to get rid of the excess water.

Alfalfa, a highly water-dependent crop, and dairy cows proliferated from the start. They had a symbiotic relationship: Alfalfa yielded hay that was fed to cows, and both produced immediate income. In 1904, when the first crop census was taken in the Turlock district, 98 percent of the irrigated acreage was planted in alfalfa. Alfalfa replaced wheat as the crop of choice, and the raising of different feeds for cows would remain dominant in the Turlock district for many years to come. California common alfalfa, or Chilean alfalfa, as it was also known, was the most popular variety—a blue-flowered herb whose place of origin was southwestern Asia.

Unirrigated and irrigated land

A land boom, not unlike what took place elsewhere in California when the promise of water loomed on the horizon or was actually imminent, swept Stanislaus County. Huge tracts of land were advertised in glowing terms: Prices will increase 10 percent a month; orchards, vineyards, orange groves, alfalfa fields, and "pretty, well-built homes" will sprout from the floor of the desert. The advertisement for the Hickman Ranch, located in one of the more marginal growing areas, continued, "Make your selection now, only a small deposit is required."

Cut-rate excursion fares to inspect farm properties were offered to people of proven means by the Southern Pacific Railroad, which paid for the printing of three thousand postcards extolling the beauty and fecundity of Stanislaus County. An immense model of the Turlock and Modesto irrigation systems was displayed at the annual conclave of the Knights of Pythias in San Francisco. A lecturer was hired, and he traveled to Los Angeles. The speaker gave twenty-four lectures a month, addressed twenty-one thousand persons in one year, and was credited with fifty land sales.

The most successful promotion took place on sandy land just to the south of Turlock. It seemed as though the Swedes arrived overnight. Lured to Turlock by the deceptive advertising of a fellow Swede in a church publication, they quickly displaced those whites of English, Irish, and German descent who had come before them. Within a few years the Swedes dominated the town. Their language was spoken in stores, in the schools, and in most of the churches; and they became the social, economic, and political elite of the town. Others grumbled at the takeover by the newcomers.

The influx began when the promoter, Nels O. Hultberg, who could not speak Swedish, put an ad in a Bay Area newspaper for a Swedish-speaking secretary. Esther Hall, a young woman of Swedish descent who had just graduated from secretarial school, was hired; she moved to Turlock in December 1902, just as Hultberg's land-promotion scheme was getting under way.

Hultberg had been a missionary for the Swedish Mission Church and a gold miner in Alaska. With the little money he had made from mining and the goodwill he had accumulated from missionary work, Hultberg traveled to California. He became a real estate salesman for the Fin de

Siècle Investment Company, which was composed of Mitchell's heirs. They were having a hard time liquidating the estate. Hultberg set about selling barren land, using pictures of thriving orchards in San Jose and elsewhere as bait for his fellow Swedes.

The promoter maintained that his goal was to establish the largest Swedish colony in California, if not the country. He targeted fellow members of the Swedish Mission Church and stated in an advertisement in the *Mission Friend*, the church publication:

> We offer the cheapest and best land, with the most plentiful and cheapest water to be found anywhere in California. Here we have good land suitable for the cultivation of all sorts of fruit, an ample supply of water, the most delightful climate, a market near at hand, excellent communications.

Of these advertisements, Miss Hall later said:

> You know the Swedes believe everything they read, especially in their Church papers, so when they read about this wonderful place in California, many of them decided they wanted to get away from where they were and come to this paradise where every man could sit under his own fig tree and eat his own grapes.

Hultberg, like so many other land hustlers in California, was selling the promise of a Mediterranean utopia to people from a cold climate. No such utopia existed at the time, nor was there any guarantee it would materialize in the future. But people bought, and they suffered or prospered accordingly. Swedes from the eastern and midwestern states put down their money and boarded trains. The men rode in the boxcars with the farm implements, livestock, and family belongings; the women and children rode in the passenger cars. Over a period of fifteen months, nearly one thousand Swedish settlers arrived in Turlock.

When they unloaded their belongings at the railroad station and looked around, their hearts broke. Many cried. The rude hamlet, still showing the scars from the fires of the previous decade, was in an advanced stage of decrepitude. There was barely a tree growing. What houses and business establishments there were stood as stark silhouettes against the barren landscape. Front Street was not paved with gold but with straw, so that the large grain wagons would not bog down in the deep sand.

Hultberg hid when the emigrants arrived. It was Miss Hall's job to meet the newcomers at the station. She recalled:

> Oh, those were awful hard times. Those that had money got out but most of them had sunk all they had and they were stuck. I worked at the land office at this time and was sending out literature telling people how grand it was, and many times I asked God to forgive me for helping to fool those poor people.

Hilmar, the Swedish colony a few miles to the south of Turlock that was named by Hultberg for his son, had sandier soil and less to recommend it than even Turlock. There were blinding, suffocating sandstorms; coyotes that howled in the night and preyed upon livestock; grasshoppers and jackrabbits that stripped newly planted crops to their bare stalks; gophers who started holes in sandy irrigation ditches that widened unless a wife or a child was inserted as a human plug until the farmer could repair the levee; and always—from mid-spring through mid-fall—the heat, oftentimes over 100 degrees.

As houses were erected, they provided the only shade. In front of each house and on either side of the drive the two obligatory palm trees—the California symbol of the good life—were planted and carefully tended by the Swedes, many of whom had come from the eastern coal mines and were bothered until their deaths by fits of coughing brought on by the dust.

Single Swedish women were much in demand. Miss Hall's long, dark hair was piled stylishly above her high forehead. She had a strong face with direct eyes. Around her neck was a high collar that buttoned in front. It was topped with a lace fringe. Women wore long skirts with long-sleeved shirtwaists of a different color, and men donned dark suits for church, the social event of the long week. Miss Hall described the process of building a community in the following manner:

> Many of the young men thought the Swedish girls looked pretty good and they used to come to the Church services and sit through long sermons, and those old Swedish preachers were pretty long winded. They didn't understand a word but

they didn't fall asleep. Since then I have seen those same fellows fall asleep in a Church where they understood every word. Well, after Church Sunday evenings the young men would line up on either side of the path leading from the Church to the gate, then when the young lady he liked came along he would ask to see her home. Let me tell you many of the Turlock families started right at that church.

One of the young men who sat uncomprehendingly through the sermons in Swedish was Arthur Crowell. He asked to see Esther Hall home, and soon they were married. Esther Crowell moved out to the Monte Vista Road property. Arthur put the bunkhouse on skids and pulled it over to the southwestern edge of the farm, where it became housing for the hired hands. Then he set about building a small house nearby for $430. Rooms were added as the family grew.

It was a hard life for the wife of a farmer, and Esther was sickly. She may have had tuberculosis, valley fever, or malaria. She cooked, did chores, drove the teams of horses, tended the vegetable garden, grew roses around the house, and bore and raised three children. Her son, Verne, recalled:

> One time she took me out and we walked down Monte Vista where there was a row of eucalyptus trees and we sat in the shade and she said, "Verne, I don't think I'll be with you very long. You've got to learn to get along without me." But she died at the age of ninety-six as healthy as a horse up to the last few years of her life. She was just overworked.

Looking back some sixty years from the vantage point of the age of television and astronauts in 1962, Mrs. Crowell reminisced about a simpler, more rudimentary time: There was no running water in the house. An outhouse was the toilet. Water for cooking and bathing was brought in from the well and heated on the woodstove, even in summer. Baths were taken in a large pan. Refrigeration was a cellar or a breeze on wet burlap that covered a box. Air-conditioning was damp sheets.

People walked everywhere on paths beaten into the weeds. They walked ankle-deep in the hot sand, and the women hiked up their skirts. For longer journeys, there was the wagon. They drove to the Merced

River to swim and fish. Salmon were caught in nets and canned because of the scarcity of fresh meat. The battery-operated telephone came before electricity, and the first automobiles arrived in 1906. "They were a nuisance because they scared the horses and caused a lot of accidents," Mrs. Crowell recalled.

Arthur Crowell planted alfalfa, pastured other farmers' cows, and operated a small dairy. The cows—named Bessie, Whitey, Peggie, Spottie, Fessie, Alice, and Maud—and horses were given the feed that was raised on the farm, and the pigs were fed the skimmed milk the cows produced. For a few years Crowell raised turkeys, selling $351.60 worth of birds in 1909. By 1916 he was experimenting with crops: There was some spoilage among the blackeye beans, and cranberries were a total failure.

Melons were the big cash crop for a few years. One year Crowell netted $4,430 on cantaloupes. He also raised casabas and watermelons. The trick was to take advantage of the rising water table and plant the melons in a field where they did not need to be irrigated. The melons were packed in a small shed. The actual packing had to be done by an expert, but the Crowells' two daughters and one son nailed the crates together and were paid a few pennies per crate. As they got older the children sorted the melons, mixed the paste, and affixed such brand labels as O. G. Olsen and Turlock Fruit. There was one hired hand to help. Seasonal laborers were mostly Greek and a few Filipinos.

They began to prosper in a small way. Relatives and friends visited. There were always a dozen or so people sitting down for a meal at the long table on the back porch. The first meal was at seven A.M. Winter and summer there was mush (oatmeal), fried eggs, cornbread, ham, and bacon. There was a big lunch at ten A.M. featuring meat, mashed potatoes, gravy, fresh vegetables in season, pies, cakes, iced tea, and coffee. The four P.M. meal was lighter. On hot days between meals and chores, the kids went swimming across the street in the irrigation canal, where a drop formed a deep pool and a plank served as a diving board. There was a small elementary school to the west on Monte Vista Road, right where the freeway cloverleaf is now located. Three rooms housed eight grades and between fifty and sixty pupils.

The town also grew as irrigated agriculture took hold and the money started rolling in. Once again there was a boomtown atmosphere. Two-story brick buildings were built on Main Street, sidewalks were constructed, and in 1909 the first Turlock auto show featured motorists from the San Francisco Chamber of Commerce, who drove out to demonstrate their automobiles. Self-important men posed for the photographer in their machines parked on the sandy main street, surrounded by small boys who imitated their poses. The next year the Turlock Irrigation District moved its office into the small First National Bank building.

By 1915 the population had jumped to over three thousand, and Turlock was known by the first of a series of promotional slogans that sought to give the town its place in the California sun. The town was the self-proclaimed "Melon Capital of the World," complete with an annual melon festival, queen, and song—"Way Down in Turlock Where the Watermelons Grow." One verse went:

> Come on down to Turlock,
> We'll meet you at the train.
> Look our district over,
> You'll have everything to gain.
> Here's where the crops are bumper,
> You'll have to be a humper,
> To keep in line with Turlock,
> Where the watermelons grow.

Turlock gained another distinction during this period. In a valley that took its religion quite seriously, Turlock stood out. Supposedly it had more churches per capita than any other town in the United States and was listed for holding such a record in *Ripley's Believe It or Not* during the 1930s. The Swedes brought their versions of religion to Turlock. Later immigrant groups did the same, as did emigrants from within the country. Religion flourished in the fertile spiritual soil of Turlock, much like crops did when irrigated. One book on irrigation and religion said of Turlock that churches "multiply like Amoeba." Factions begat factions. Saloons were driven out of town for a while, and they never regained their earlier prominence.

The town was dominated by the conservative, evangelical nature of the dominant religions. The 1958 Turlock Golden Jubilee pamphlet featured a photograph of Billy Graham and his message: "It is indeed a privilege to be included on a program which recognizes Christianity and the Church as having had the greatest part in our heritage." Graham addressed a crowd of fifteen thousand—a number that was twice the city's population—on the high school football field. Nearly thirteen hundred sang in the massed choir. Ten years later the population was thirteen thousand, and there were thirty-seven churches within the city limits. However, eighty churches were located within a radius of ten miles of city hall, an area that included both the city and the surrounding unincorporated areas.

Actually the Turlock Irrigation District and one man in particular, Roy Meikle, were more responsible for the fact that Turlock had a heritage than anything else. Alan M. Paterson stated in his history of the district, *Land, Water, and Power*: "The impact of the irrigation revolution on every level—economic, social, environmental—was so profound that irrigation easily became the most important single fact in the history of the Turlock region." And Meikle was the embodiment of irrigation for Turlock.

The quiet, self-effacing man who would serve his district so well was born in Ohio and attended Stanford University. He quit after completing the engineering curriculum but before he got his degree. Meikle worked for three water agencies in the northwest before joining the Turlock district in 1912. He became chief engineer in 1914, a position he would hold until he retired at the age of eighty-seven in 1971. During his nearly sixty years with the district, Meikle oversaw the construction of two major dams and countless lesser projects with frugality and great tenacity. What he accomplished was the supplying of cheap water for irrigation and the generation of electricity that would pay for the dams and canals and drive the hundreds of pumps needed to keep the district from drowning in its own water.

Meikle's career spanned the era of great public works in California. Near the end of his time, Meikle began to encounter opposition to the district's goals by agencies and people who had opposing objectives, such as saving the salmon run on the Tuolumne River. No longer could an entity like the Turlock Irrigation District exist in isolation, no matter how much it might want to. As California's great natural resources—such as its timber, oil, and water—were sought more avidly and became scarcer in the process, other claims for their use—or the novel argument that they be left untouched—infringed upon the single-minded purposes of special districts. Compromises needed to be struck.

It was a process that Meikle didn't quite understand as he aged. At the height of his confrontation with the federal and state governments over construction of New Don Pedro Dam, Meikle asked in a memo to himself in 1963: "Why should Turlock build a fish industry and recreation facilities for the Government?" The district's duty was to supply irrigation water to its customers, and that was that. Meikle retired before he had to deal fully with the painful reality that the political power that governed the disposition of water had shifted in favor of urban uses over rural agriculture.

Although his job title never was any grander than "chief engineer," Meikle was totally in control over the years. He was careful to seem to defer to the five farmers who were the directors. In an early photograph the short man with the level gaze and thin lips stood behind the blustery-looking directors, who had to lean forward to catch his softly spoken words. The district employees both venerated and feared "Mr. Meikle," as they called him. Years after his death in 1975, an older employee said, "His word was as good as God's word. Now you need everything in writing to cover your ass."

Meikle was old-fashioned. He believed in serving the public. In his 1935 "Suggestions for Employees Contacting the Public," the chief engineer said, "Remember above all else, that we are here to help the public; *not* tell it what *we* want it to do. Put that thought into all your transactions." He suggested that those employees who had contact with the public, such as cashiers, say "Good morning" or "Good afternoon," rather than the more abrupt "Who's next?" In 1952, using the argot of the region, Meikle issued "Four Commandments" for obtaining quick answers to public complaints.

The chief engineer did not lash out publicly at critics, but he excoriated them in private notes. Of the editor of the *Modesto Bee* who criticized the district for the cost of promotional literature, Meikle noted, "This man is either ignorant or a liar or both." Of a 1953 story in the *Los Angeles Times* on the district, Meikle noted, "The statement that 'water is so abundant that wells must pump it back from the ground into the canals' is a new way to describe the drainage problem."

Being as politic as he could, since he was dealing with customers, Meikle was constantly critical of the farmers' overuse of water. He could not tolerate waste in any form. In one private note after another to himself, such as this 1962 memo when Meikle was in the midst of negotiations for New Don Pedro, he outlined the issue: "In other words, the use of too much water by the irrigators encouraged by its low cost is probably the District's No. 1 problem." During his time the groundwater level had risen from sixty to four feet beneath the surface, and in many places ponds and lakes had reappeared.

His private notes and the histories of the district that he wrote showed again and again that alfalfa, feed grains, clover, and irrigated pasturelands constituted by far the greatest acreage in the district and took the most water. In a look backward at the history of the district in 1950, Meikle wrote:

> The irrigation of permanent pasture on sandy soil requires frequent irrigation and a large amount of the water applied passes through the soil into the ground water. In some areas this causes a serious drainage problem requiring more drainage pumping. Some lands are not leveled, checked and irrigated with any idea of conserving water.

Meikle was not universally loved. In 1945 an old man who disliked his policies broke into a board meeting, took careful aim with a .32-caliber revolver, and fired one shot. Meikle put his hand up to stop the bullet. His hand deflected the bullet, which lodged in his stomach. When his son, Jim, came home from World War II, Meikle asked to be instructed in the use of his army-issue .45-caliber pistol. He kept the pistol in a desk drawer in his office.

Meikle's first priority was his work. He never took a vacation. He had no hobbies. He joined friends at a Saturday luncheon gathering called the Knife and Fork Club. His wife was devoted; she played a lot of bridge. His friends took Jim hunting and fishing. There were two other children, a daughter and another son who was institutionalized at an early age because of an accident at birth. Meikle never raised his voice at home or uttered a profane word. In fact, he spoke very little. When the family drove to San Francisco and back, there was no conversation, nor was there any talk at the dinner table. In a church town, Meikle rarely went to church. Meikle worked, and it seemed like Turlock prospered.

But not all of Turlock prospered. For some a place in the California sun with a piece of land and an adequate water supply was a more difficult, if not impossible, goal to attain. The unofficial policy was to keep the town core and the region around it out to an undefined distance predominately white. By subtle pressures, people of other colors were pushed outside. There were some exceptions, such as the Mexican bars along South First Street and the Filipino and Mexican housing areas on the southwest side of town.

There literally was a right and a wrong side of the railroad tracks in Turlock and other valley towns: right was to the east and wrong was to the west. Gerald Haslam, a native of the valley and a California state university professor, noted, "There has tended to be a direct link between centralized irrigation systems and centralized political and economic power; and that in turn has created a paternalistic, class-ridden society with nonwhites on the bottom."

In their time Chinese, Japanese, East Indian, Filipino, Mexican, Okie, and more Mexican farm laborers surged across the valley, one group being replaced by the next when they sought to settle in threatening numbers. Carey McWilliams wrote in his groundbreaking book *Factories in the Field*, "Here the practice has been to use a race for a purpose and then to kick it out, in preference for some weaker racial unit. In each instance the shift in racial units has been accomplished by a determined effort to drive the offending race from the scene."

Only the Swedes, who came to dominate the Turlock region, and the Portuguese and Assyrians, who had to battle for their respective niches, made a dent in the social and economic hierarchy of the community. In a study of the Assyrian community and class and ethnic differences in Turlock, Arian Beit Ishaya wrote:

> In the history of the Turlock region the Anglo-American and the Nordic people have been considered superior to other ethnics of Portuguese, Assyrian, Chinese, Japanese, and Mexican-American backgrounds. In the 1920s and 1930s the

propaganda against the "Yellow Race" in Turlock reached hysterical proportions.

One way to ensure the continued whiteness of Turlock was to create a buffer between it and the expanding Japanese colony just to the south of the Merced River in the Livingston area. That buffer was the unsuccessful Delhi Colony.

Agricultural colonies that operated as cooperatives were the idea of Elwood Mead, a professor of rural institutions at the University of California School of Agriculture who later headed the federal Bureau of Reclamation. Mead was an expert in irrigation and a promoter of the Jeffersonian ideal of small farmers owning small plots of land. Mead also was an outspoken advocate of white dominance, which reached a peak during the mass anti-Japanese hysteria of 1920 and 1921. The Progressives, under Governor Hiram Johnson, incorporated Mead's ideas into their rural program, which was twofold: break up land monopolies and restore the countryside to the whites, who would occupy the colonies.

Mead and other white Californians feared southern European and Japanese domination of rural areas. The State Land Settlement Board, headed by Mead and headquartered in Agriculture Hall at the University of California, published a pamphlet in 1920 containing laudatory articles about the colonies from *Sunset* and *Country Gentleman* magazines. The *Sunset* article pointed out that Athens and Rome fell after "imported slaves" took over the work. It went on to quote Mead as stating:

> There are sections [of rural California] now which include people from every country in southern Europe and nearly all of Asia. If they intermarry what will the mongrel descendants be like? They have been well described as "unhappy beings every cell of whose bodies is a battle ground of joining heredities who express their souls in acts of hectic violence and aimless instability." If they do not intermarry, then each of our great valleys will be the home of racial friction which will make the Balkins seem like a prayer meeting.

These aliens, wrote Mead, "were in sorry contrast to the State's first settlers, who were the finest type of American citizen this nation had produced." Mead continued in his 1920 book *Helping Men Own Farms*:

The California pioneer had been a citizen first, a money maker second. He was generous and public spirited to a fault. In contrast, the alien renter had no interest in rural welfare. He had a racial aloofness and he farmed the land to get all he could out of it in the period of his lease. Wherever he displaced the American, he put rural life on the down grade.

World War I veterans were recruited for the two colonies, one at Durham near Chico and the other in Delhi, just south of Turlock. The Delhi Colony attracted one hundred thirty white families who were given low-interest loans from the state to build farmhouses and make other improvements to their small parcels. A 1923 state legislative report on the two colonies cited the danger of Japanese dominance in certain crops and added, "The more Durhams and Delhis there are, the more certain it is that rural California will in the next half century remain the frontier of the white man's world."

Despite being supplied with irrigation water by the Turlock Irrigation District, the cooperative venture failed. Irrigation district historian Paterson wrote that Delhi "became the irrigation ideal gone haywire." The colonists faced the same hardships other settlers encountered, except they had the financial and technical backing of the state.

There were problems. Some of the veterans were disabled, and most had no farming experience. The soil was exceptionally poor, and the university professors did an inadequate job in laying out parcels and the irrigation system. The bureaucracy that ran the project was heavily staffed. Mead was hung in effigy by the Ku Klux Klan, and a defective bomb was planted with the intent of killing the project superintendent. The state eventually lost $2 million in defaulted colony loans.

On the other side of the Delhi Colony from Turlock, the Yamato Colony, a Japanese cooperative venture, encountered the same natural conditions plus racial hostility in Merced County; but it survived, eventually flourished, and upgraded rural life in the process.

The Japanese were brought to California to replace the Chinese after the Chinese Exclusion Act of 1882 was passed. They experienced the gamut of racial hatred. The Japanese were snubbed, called names, spat upon, shot at, beaten up, kidnapped, sent to segregated schools, denied citizenship, and denied the right to own land. Further numbers of

Japanese were denied entry into the country by exclusionist legislation, and many who had entered were incarcerated during World War II.

Utilizing ethnic solidarity in their cooperative economic ventures, the Japanese moved quickly from being farm laborers to farm renters and owners. This caused problems, as Ronald Takaki explained in *Strangers from a Different Shore*: "But the very success of the Japanese in enterprise further aroused waves of exclusionist agitation, and their very withdrawal into their self-contained ethnic communities for survival and protection reinforced hostile claims of their unassimilability and their condition as 'strangers.'"

Although ostensibly barred from owning farmland by the 1913 California Alien Land Act, the Japanese either rented or found ways around the restrictions. They were knowledgeable and hardworking farmers. By 1920, when anti-Japanese feelings were at fever pitch, tenants and owners of lands farmed by persons of Japanese descent outnumbered the combined statewide totals of those of Italian, Portuguese, and German descent.

A certain ugliness reared its head in Turlock and its environs. Elbert Adams, owner and editor of the *Livingston Chronicle* to the south, wrote, "We could not blind our eyes or deaden our senses to the fact that more Japanese were coming in here." Adams was elected president of the Merced County Anti-Japanese Association. First the Turlock Board of Trade and then its Livingston counterpart passed resolutions stating the Japanese were "unassimilated and unassimilable." An unsuccessful attempt was made to halt further sales of land to those of Japanese descent.

A mass meeting was held, and a unanimous vote authorized white business and farm organizations to erect signs on the highway to the north and south of Livingston stating NO MORE JAPANESE WANTED HERE. The plain, block-letter signs were erected, but it wasn't long before the wording was changed to LIVINGSTON, THE COMMUNITY WITH A DESTINY. Business had suffered.

There were public and private threats against the Japanese. Referring to a Japanese-financed campaign against passage of a 1920 initiative measure designed to plug loopholes in the 1913 Alien Land Act, Adams warned:

> If any of them ["our" Livingston Japanese] are contributing to the "cause" financially, we advise that they stop it forthwith.

Your homes here, neighbors, are secured and safe. Let well enough alone. Perpetuate the good feelings that exist here if you can. Please don't create a necessity for unpleasant words or actions.

Words escalated into action in Turlock in July 1921. Since the beginning of the melon-picking season, the streets of the small town had been seething. Japanese labor contractors, barred by whites from formal negotiations between pickers and growers, had countered by providing cheaper labor. White fruit pickers were wandering the streets while the local restaurants "were filled with Japanese eating expensive breakfasts," according to the *Modesto Evening News.*

The Turlock establishment girded itself. A petition with the names of eighty merchants and businessmen was presented to the directors of the chamber of commerce, who quickly went on record as being "strictly against the Japanese." They added in a formal resolution that "This district should be a district for white people and for Americans." Preference should be given to white over Oriental laborers. The local American Legion post passed a similar resolution, adding that it was against the "Japanese invasion" of the greater Turlock area. The local newspapers referred to the Japanese as "Asiatics," "little brown men," "Nipponese," and, most frequently, as "Japs." They were filled with news about the Jap menace.

Approval for action seemed to have been given by the establishment. Shortly after midnight on July 20, a small fleet of trucks pulled up before the Tokio Hotel on Front Street. A mob armed with clubs, variously estimated at between sixty and one hundred fifty men, yelled obscenities and threatened to burn down the hotel. The doors opened, and a delegation went inside and gave the Japanese laborers ten minutes to pack their belongings and get in the trucks. The vehicles proceeded to two nearby bunkhouses on Center Street and several ranches south of town.

Between fifty-eight and one hundred twenty Japanese, the numbers again varying with the account, were driven to nearby Keyes, where the northbound freight was flagged at 2:55 A.M. They were put on board and warned never to return. There was a general exodus of the remaining Japanese laborers from town the next day. One of the anonymous kidnappers told a reporter, "This action is intended to apply to future work here as well as the present, and it is also to let the melon people know that white labor is to be respected in this district." The two Turlock police officers who were on duty that night knew in advance about the raids, yet did nothing about them.

There was one day of self-congratulation, and then the reaction set in. The Harding administration was in the midst of delicate negotiations with the Japanese government over the terms for the latter's participation in an international disarmament conference. Suddenly Turlock burst upon the world via the front pages. A Japanese newspaper commented, at a time of great revolutionary unrest in Russia, "If the American government cannot be held responsible for this outrage, it is no better than the Russian government." Other newspapers in Japan reacted angrily to the Turlock incident and demanded an indemnity.

There was a lot of official backpedaling. Everyone from the directors of the Turlock Chamber of Commerce, to the publisher of the *Farmers' Daily Journal* of Turlock, to the Stanislaus County district attorney, to the California Exclusionist League, to the California governor, and to California's two senators in Washington, D.C., said that, although they had no liking of the Japanese, they should be dealt with in a legal manner. This was an economic issue, not a racial incident, they maintained.

The first person arrested was, predictably, the head of the fruit pickers' union. James L. Shea, a former semipro baseball player, was reported to have said, "Pinched, by God. So I'm the goat!" The sheriff also tried to pin it on the radical Wobblies, but no members of the militant union could be found. (The Republican *San Francisco Chronicle* professed greater love for the Japanese than for the Wobblies.) Eventually, nearly ten men were arrested on kidnapping charges. They were truck drivers, farmers, a store clerk, fruit pickers, and a policeman.

All the out-of-town newspaper correspondents had departed from Turlock by the time masked, white bandits began terrorizing Japanese farmers living in outlying areas over the next few months. No traces or clues to their identities could be found. There were similar incidents elsewhere in the valley. According to the Japanese consul general of San Francisco, who investigated one incident, "low-class immigrants from southern Europe" who felt their jobs were endangered forced Japanese workers to board a train at the Livingston station. Such incidents, combined with restrictive land and immigration legislation passed at the state and federal levels, stuck in the craw of the Japanese government.

World War II fed the latent racial hysteria. The Japanese were taken to an assembly center hastily constructed at the Merced County Fairgrounds; from there they were sent to such relocation camps as Amache in Colorado and Tule Lake in northeastern California. They put their lands in trusteeships run, for the most part honestly, by white managers. Some of their homes were burglarized during their absence.

Unlike other Japanese, they had homes and farmlands to return to after World War II, although their reappearance on the scene was not welcome. The *Livingston Chronicle* opposed their return. Shots were fired into their churches and homes, and mattresses and sheet metal were piled against the walls to stop the bullets. The Japanese appealed for protection. None was forthcoming. A county supervisor, explaining why he was against such aid, asked, "There hasn't been any murder yet, has there?"

While the Japanese were viewed as different, clannish, and an economic threat, the Filipinos who followed them were depicted mainly as a sexual threat. V. S. McClatchy of the Central Valley newspaper family testified before a congressional committee: "You can realize, with the declared preference of the Filipino for white women and the willingness on the part of some white females to yield to that preference, the situation which arises." The Filipinos were confined, for the most part, to labor camps south of Turlock. Few owned or rented land, and for that reason they were less of a target.

After selling off twenty acres in 1910 and renting out ten of their acres, the Crowells farmed the remaining fifty acres and raised three children during the Depression. Of the Turlock area during the 1930s, *The WPA Guide to California* noted:

> U.S. 99 crosses and recrosses a network of canals on its way to TURLOCK (101 alt., 4,276 pop.), in the midst of hundreds of small farms—all depending on the vast Turlock Irrigation District launched in 1887. The farms produce dairy and poultry products and a variety of crops: alfalfa, sweet potatoes, watermelons, peaches, apricots, grapes, and grain. The town celebrates its watermelon crop at an August melon carnival.

The Depression was not an easy time for most of the people in Stanislaus County. A Farm Housing Survey was taken in 1934; and the enumerators noted the poor condition of farmhouses. Most were drafty, needed new roofs, and had no indoor plumbing. Some had been abandoned only to be reoccupied by families fleeing worse conditions in the cities. One enumerator gave the following breakdown for her district:

> I would say there are at least three-fourths of the farmers who need help. One-fourth, perhaps, of all the farm owners are making a living, sending children to college, and are not in debt. Have lovely homes, can pay their taxes and keep their places in repair. Of the other three-fourths, some of them own their homes, do some farming, have cows, fruit, but the market value for their produce is so low they cannot keep even. Taxes and repairs and the things they must buy are still high. They say, "Give us a fair price for our produce, we can get along."

The enumerators, with clipboards in hand, approached the homes and knocked on the doors. They reported on what they saw inside in terms of racial categories. Homes inhabited by Portuguese were either in poor condition or "nicely improved," Italians were "comfortably fixed," Japanese were poor but "very clean," and Swedish homes were in "fine condition" and were "clean and comfortable." Racial characteristics, as described by one enumerator with a Scottish surname, were as follows:

> *The Swedes* are like the Scotch—doubting Thomases. But once convinced they capitulate handsomely. Out comes the coffee and cake, no matter what the hour and when the schedule is finished friend invites friend to visit "and we'll finish our chat."
>
> *The Japanese* meet you with an ingratiating smile and "No Splik Inglais." That's their story and they stick to it despite all comers.
>
> *The Filipino* smiles and while you look at a tumble-down shack he assures you that everything is in perfect condition. He knows you know he is lying, and he smiles and keeps right on—smooth voiced—maddening!

The Portuguese is either slow to answer or knows each one as fast as you can read and check. He doesn't doubt you and inclines to a friendly camaraderie.

The Armenian is subservient, nevertheless he says "yes" to everything lest he should miss something. He wants everything he can get and is taking no chances.

The American meets you as an equal. He's just like us and needs no description.

The Crowells were relatively well off, in terms of Depression-era standards. It was a time of transition for the family. Arthur, whose interests were veering in the direction of golf and banking, was in the process of turning the Monte Vista Road farm over to his son, Verne. The father was a charter member of the Turlock Country Club, where he made friends with the owner of the local Ford agency and the owner of a Hilmar warehouse. The three men took over the Stanislaus-Merced Savings and Loan Association, and Crowell became vice president and then president. He helped his son milk in the early mornings, and then drove into town to conduct business and play golf.

Verne Crowell, a cousin to Abner, who I was acquainted with, was born and bred to farming, and it stuck to him like fresh cow manure to new boots. He was born in the front bedroom of the farmhouse that faced onto the sandy road on a hot summer night in 1916. Verne's greatest childhood passion was toy farm animals. When he had a nickel he would spend it at Woolworth's on celluloid horses, cows, pigs, and poultry. Verne constructed a whole farm in the backyard, complete with miniature irrigation ditches into which he turned water. He watched his father irrigate and learned from him. He also learned from the ditch tender, who was the embodiment of the irrigation district to the farmer.

There was one ditch tender who took care of the whole of Lateral 3, using a horse and buggy at first to get around. Crowell recalled:

> In those days we left a note up on Lateral 3, and he'd see it when he drove by in his little buckboard. We didn't have telephones yet. So he would duck on down here and tell my dad when he could have the water. As my dad wanted him on his side, he would take him outside, because my mother was strictly against liquor, to the tank house; and they would gen-

erally have a jug of wine or a drink or two out there. Sometimes this ditch tender would get in with the Portuguese and Italians and they would let him drink all he wanted. He'd get so drunk he couldn't even walk. But he always remembered what the deal was on water. He was fabulous.

The ditch tender graduated from a horse and buggy to a Model T Ford:

He still had the same problem. He'd get drunk and run right into the canal, you know; and we'd take a team of horses and pull his Ford out. But you never criticized a ditch tender because if they didn't like you they would just murder you. You'd do all your irrigating at night.

Crowell went to the school on Monte Vista Road and then to Turlock High School in town. After graduating in 1934, he went to work full-time for his father. The young farmer saved his money and bought a used 1933 Plymouth. Crowell got a friend to buy two gallons of gas, and they went out to visit the Whitaker sisters. "George got in the rumble seat with Barbara, and I got in the front seat with her sister, Margaret. We were gone

Lateral 3, with the Crowell property in the distance

for three or four hours chasing around the country, and somehow when we came home Barbara was in the front seat with me."

After one year at the University of California at Davis, then primarily an agricultural school, Verne returned home and he and Barbara were married. Young Crowell worked for his father as the hired hand at the standard wage of thirty dollars a month and a room for himself and his bride. When his parents moved to town, Verne took over managing the farm. He could not afford a hired hand. Verne and his father agreed on some matters and clashed on others:

> We had six horses and were milking by hand. The first thing I did was I said I wanted to buy a milking machine. He didn't kick too much at that because he wanted to get away from milking those cows, too. So I put in an automatic milking machine. A year and one-half later I was working myself to death. I was as skinny as a rail. I talked to the tractor dealer. Dad really got upset. He said, "I don't ever want to see a tractor on this place. If you are driving a tractor out in the field and it leaks oil, there will be no crops. If a horse stops and takes a crap, you get a good crop there." He just disliked change.
>
> My mother called me that night and said, "Verne, your dad is really upset and he says if you buy a tractor he is not ever going to come back out on the ranch again." I said the hell with it. Six months later Dad came out. He had been traveling around the country for the bank. He said, "It seems like the fellows who are doing best are the ones with tractors. So if you want to buy a tractor, I'll sign your note." I said fine.

The elder Crowell had rented out the north ten acres some years previously, but Verne farmed more aggressively; so he took those ten acres back, bringing the total acreage of the home ranch to sixty. He rented additional acreage in amounts that varied year to year from sixty to two hundred acres. The total number of acres that Crowell farmed put him close to the average size of a California farm. As to crops, Crowell stayed away from melons, whose prices were too volatile. He preferred beans, which were steadier, the customary alfalfa for feed, and dairy cows. A hired man milked the cows and spent a half day in the fields.

During World War II, Crowell milked his own cows. He bought twenty acres on the south side of Monte Vista Road for $325 an acre. The price of beans rose from two cents a pound to fifteen cents. One crop of blackeye beans paid for the land. "I was in hog heaven," said Crowell. "I like black-eyes. I got some tremendous crops of beans." After World War II, Crowell kept adding to his livestock portfolio, bought more farm machinery, and rented additional acreage. He planted one crop of casaba melons but made no money on his only venture into Turlock's claim to fame.

Barbara and Verne raised two sons and a daughter in the ranch house where Verne was born. Except for the payments on the tractor, they owed no money. He explained:

> We were from the old school. We didn't want to owe anybody any money. We got one milk check a month and that was practically all our income, except for when we sold our beans. The way we did it then was that we did not pay our bills by mail. My wife and I went around to each place where we had charged up something and paid them off from the money from the milk check. Whatever we had left after that was what we had. If we had $50 maybe we would buy some new clothes for Barbara and the kids. If it was $10 or $12, we would just buy groceries. Later we found out that we had absolutely no credit rating.

Turlock went searching for a new identity in the postwar years. Melon production had peaked and then dropped in the 1930s, so the town no longer qualified as the Melon Capital of the World. The Turlock Chamber of Commerce, which had replaced the Turlock Board of Trade, headed the search. The chamber was the single most important organization in town, outside of the irrigation district. Local historian and college administrator John E. Caswell, when asked who ran the town, replied, "It is important to have a Scandinavian name. It is important to be a churchman, particularly if you are affiliated with one or more of the larger and more conservative churches. It is important to have held office in the Chamber of Commerce."

The chamber looked around for one of those dearly loved, innocuous slogans that would promote the commercial vitality of the community.

For a time after the war, Turlock was known as "The Thrifty City." With the Far West Turkey Show being held annually in Turlock from 1946 to 1965, Turlock took on the moniker "Turkey Capital of the West." At the time of the golden jubilee in 1958, the slogan was "Dust to Destiny." When Turlock was desperately seeking to become something more than a farming town and was selected as the site for a state college in 1959, it became "Home of Stanislaus State College." The college was known colloquially as "Turkey Tech."

A college wasn't enough, so in 1964 it became "Growing with Agriculture, Industry, Education." As Turlock became a bedroom community for commuters escaping more densely packed urban areas in the valley and the Bay Area, the emphasis was placed on its bucolic image. Signs on the freeway that now bypasses the small city (101 altitude, 42,200 population) declared TURLOCK. KEY TO THE VALLEY. A NATURAL PLACE TO GROW.

I arrived in Turlock during the transition from turkeys to college students. During my two years in Turlock, there were two major issues that dominated my reportorial time. One was the effort to establish the college on its permanent campus at Monte Vista and Geer roads, and the other was the negotiations leading to construction of New Don Pedro Dam.

As is usual with newspapers (at least I have found it so in California, whether it be in Turlock or Los Angeles), the publisher was actively involved in both issues because they would have a great impact on the growth of the community and, thus, on the growth of his property. The *Journal's* publisher, Stan Wilson, headed the effort to obtain the college for Turlock. He and his friends in the chamber of commerce and the Rotary Club wanted that dam built, as did an "overwhelming" (the adjective I used in the story) number of voters in the irrigation district, who approved the bond issue for its construction by a vote of 5,745 to 126.

To approve such a bond issue by majorities of 98 percent in the Turlock, 97 percent in the Modesto, and 92 percent in the San Francisco water districts would be impossible today. But in the early 1960s, the concept of growth was sweeping California. The Turlock district needed the new dam for holdover storage during the dry years that were to come and for additional electrical generation. Turlock's partners in the dam, the Modesto Irrigation District and the City of San Francisco, needed it to accommodate and promote growth. Crowell and Meikle, among others, were quite pleased with the vote, although they foresaw that it would be some time before actual construction could start.

The sticking point was the amount of water that would be released for the downstream salmon. The state Department of Fish and Game and the federal Department of the Interior under Secretary Stewart Udall pushed for more water than Meikle wanted to surrender for that purpose, which was none. An attorney retained by the district opined that "the Districts can not legally dispose of any irrigation water unless it is surplus." Fish simply were not within the purview of an irrigation district's operation. "To consider any compromise at this time which could reduce the District's water supply in a dry year is unthinkable," wrote Meikle to himself.

No such compromise was considered. District historian Paterson wrote of the final 1966 agreement, "By threatening to scuttle the project if their terms were not met, Meikle and Crowell had won important concessions insuring that the project would be profitable for the district."

When the new reservoir behind the stark, earth-filled dam began filling in 1971, Meikle, who had an abiding love for the old concrete-arch dam that would be inundated by the rising waters, refused to watch. He said that he did not like to go to funerals. The chief engineer was not an emotional person, but his voice broke when talking about the loss of the old dam, whose construction he had overseen fifty years earlier.

Even prior to Meikle's death in 1975, the rules of the game had changed. Such dams on the Tuolumne River could not be built today because of environmental, seismic, and economic constraints. A new dam on the neighboring Stanislaus River just squeaked by in the mid-1970s. The Turlock district, in conjunction with Modesto, was unsuccessful in its effort to build a nuclear power plant; the same fate is probably in store for a proposed hydroelectric facility on a tributary of the Tuolumne.

With increased urbanization, the supplying of electrical power became more important than irrigation water to the district. After peaking in 1948, irrigated acreage within the district's boundaries dropped in the latter half of the twentieth century. But farmers still controlled the district, despite their lessening numbers and importance. Raising feed for livestock still accounted for the most acreage and by far the greatest use of water. Ninety-four percent of the district's water went to agricultural use.

The severe droughts of the two years in the mid-1970s and the stretch of years beginning in the late 1980s and extending into the early 1990s caused losses in revenue from electrical generation. The district began to lose its luster, despite construction of a magnificent new headquarters building. There were poor management decisions, contention within the district, fear of economic ruin, and the possible loss of water to those who could afford to pay more for it, meaning the cities.

To survive, the district and the farmers began to suck the valley dry. They pumped more and more water from the savings bank of the underground water table, which sank farther and farther, thus necessitating the construction of deeper and deeper wells and the use of additional electricity to pump the water to the surface.

Like all great irrigation societies that had preceded it, the Central Valley would eventually choke on the wastes it produced. When the infrastructure became bankrupt and when all affordable engineering solutions had been exhausted, the valley would revert to the sandy wastes that had once been crossed by Indians, soldiers, and settlers. The only question was when. In his history of California agriculture *Harvest Empire*, Lawrence Jelinek wrote:

> Concern has also increased over the long-term impact of intensive cultivation practices on the soils of the San Joaquin and Imperial Valleys. According to a federal-state report issued in January, 1979, the use of irrigation, fertilizers, and pesticides is threatening to turn substantial tracts of the San Joaquin Valley into desert while materially reducing the fertility of still larger tracts.

Below the last dam, the Tuolumne River was a turgid stream that carried selenium, manganese, and boron from the surrounding fields into the cesspool that the San Joaquin River had become. The native grasses and oak trees were all but gone. The wildlife was being poisoned by the toxic chemicals used to grow feed and food.

Humans were next. All the drinking water came from wells. One-third of the wells supplying water for domestic consumption within the Turlock Irrigation District were contaminated. The area had the second largest cluster of wells in which pesticides had been detected. (The largest cluster was in the Fresno area.) The two agricultural chemicals that caused the most problems were nitrates contained in fertilizers, which sickened infants, and DBCP (dibromochloropropane), a fumigant used for nematode control until it was found to cause sterility in men. Contaminated wells were abandoned in the district, and the irrigation district board was thinking about furnishing drinking water from the Tuolumne River. Stockton, Modesto, and Fresno were confronted with the same problem.

The soil on Verne Crowell's land was sandy loam. It was easy to work but needed fertilizer to make it productive. Manure was the first enrichener used on it, and then fertilizing became more scientific. A soil analysis was made and a commercial product that would correct the deficiencies was recommended. Fertilizer made all the difference. Crowell said:

> When I first started farming no one would rent a sandy piece of ground because it was so unproductive. Now sandy ground rents as good as anywhere else because you can put fertilizer on it and get a good crop. It does take more water, but if you got enough water and enough fertilizer you can grow anything on sandy ground.

Crowell kept expanding his operation. He had his daughter and two sons to help him. He treated the boys as his father had treated him. They worked hard and were paid a wage. "Shoot, a little kid twelve years old on a tractor could take the place of a man who was costing you three dollars or four dollars a day," said Crowell. His daughter helped her mother and ran the hay rake. One night she put water instead of gas in the tank of the hay baler. "I was so tired and had so much work to do I think I cried," said Crowell. He didn't trust himself to scold her, so he asked Barbara to do it. "Her mind was just elsewhere," said Crowell, who lost two days of hay baling.

They built a modern, one-story ranch house on the southeast corner of the property and moved out of the old farmhouse. Crowell didn't like artificial air, but as a concession to his wife, the new house was built with the ducts but without an air-conditioning unit. "We lived that way for probably eight or ten years and she was complaining all the time and I liked it that way. Any friends we had out thought it was hotter than hell." Crowell gave in, and his friends donned their heaviest coats to help the couple celebrate the installation of a cooling unit.

In the 1950s the price of land began to rise precipitously. Dairy farmers from Los Angeles County who had been forced out by encroaching suburbs took the cash from the sale of their farms and headed north to buy cheaper farmland in the valley. "And, of course, it caused the price of land to skyrocket," said Crowell. The location of the college north of Turlock added to the boom. Crowell sensed change:

> Land had been something to farm to me. It was not something to turn over and make money on. In fact, the first piece of land I traded off, it was just like to kill me because I couldn't bear the thought of parting with a piece of land.

But part he did. When Crowell's father died, the son was made a director of the bank; and he began to learn the intricacies of buying and selling land. When the college was sited next door, he became knowledgeable about real estate out of immediate necessity. His land was rapidly increasing in value, and the taxes were killing him. "All of a sudden I was paying $100 an acre in taxes on land that I couldn't rent for $45. That was why I started in on the apartments. I had to do something." Ten acres opposite the college were sold off in two-and-a-half-acre chunks for student apartments. Crowell kept a half-interest in the development and eventually realized more than $1.5 million from it.

> I started trading, and that was the first real money I ever had, too. When you are farming and expanding, you never get any money. There is always some new machinery to pay for or new barn to build or cattle or some damn thing and you never seem to have any money and when we built those apartments, then we started getting some income from them. And when we finally sold them, that was the first real money I ever had.

The price of land spurted upward in the late 1980s. Crowell bought back the north twenty acres that his father had sold seventy years earlier for $12,000 an acre in 1982 and sold it six years later for $45,000 an acre. The developer, Bob Florsheim of Stockton, devised a subdivision plan that called for five single-family homes per acre. Ninety-four such homes would produce an estimated population of three hundred in what was now called Rolling Hills Estates, although there were no hills in sight and five homes per acre hardly qualified as an estate.

The loss of farmland was not a factor. The removal of twenty acres from crops would not amount to a "significant reduction" in Stanislaus County agriculture, it was determined by planners. There was no consideration of the cumulative effect of a number of such twenty-acre withdrawals. The Turlock Irrigation District attached a minor caveat to the development: "Since the land will no longer irrigate, an agreement to Abandon Use of Ditch must be requested by the owner before the final map is approved."

The Turlock City Council approved the plans in August 1989, but Florsheim never developed the property. He sold it, complete with approved plan, to Brooks & Sons of Turlock, a father-and-son real estate development and construction firm. The price was $2.3 million for what had cost $900,000 a year earlier. That worked out to $115,000 an acre—an increase of 153,333 percent over what Mitchell had paid for an acre 121 years ago.

George Kapor of Hilmar tore out the irrigation system and replaced it with a storm-drain system for Rolling Hills Estates. Kapor Brothers had been in the land-leveling business for fifty years. George Kapor, the

Farms and subdivisions near the Crowell property

father, started the business with a single dozer and a drag scraper. He had leveled Verne Crowell's land for irrigation purposes; and now his son, also named George, constructed storm drains for subdivisions on that same land.

The recession of the early 1990s dealt a temporary setback to the California real estate market. David Brooks, the son, explained what that meant: "We bought ninety-four paper lots. They made the money. If things hadn't slowed down we would have made the money." The marketing idea behind Rolling Hills was to get people to upgrade. "Move *up* to Rolling Hills Estates" was the sales pitch. Brooks ticked off the amenities: garage-door opener, trash compactor, fully landscaped yard. The homes sold for an average of $140,000 in 1992, bringing the value of a totally developed acre to $700,000 (a 933,333 percent increase).

A few months later Crowell's next sale consisted of twenty-seven acres in the center of the old farm for $1,620,000, which worked out to $60,000 per acre for raw land without an approved subdivision plan, an increase of $15,000 over the previous sale. Ray Franco of Heartland Property, a Turlock developer, submitted a plan for a "conventional subdivision" of 129 homes, again averaging five per acre, that would serve a population of nearly four hundred people and represent an estimated investment of $5 million. The subdivision was given the enticing name of North View Meadows II, although the view to the north was blocked by the other subdivision and no longer was any meadow in sight.

For Crowell, the sale represented the chance to increase his farm holdings. He said:

> When it really became apparent that the land was worth more to somebody else than me, was when this land got up to $60,000 an acre. That was when I sold the land they are putting these houses on. These builders were making buckets of money on developments and were running out of good areas to develop.
>
> At about the same time this other ranch, this 300 acres to the west of Turlock, came up for sale for $1.3 million, so I decided it was a good time to move. I traded twenty-seven acres for three hundred. That's what I did. I got out of taxes, sold at the peak of the market, made a few hundred thousand dollars, and got much more land.

Crowell had one foot in the future of California farming and one in the past. The outfall from the Turlock sewer plant ran through his new property, and he had the riparian rights to the treated water. The recycled water was rich in nitrogen, so he used less fertilizer. The drought did not affect the operation on this dairy ranch, which was run by his son Mike, who was on the board of the Turlock Irrigation District. Of his new purchase, Crowell said, "Fussing with cattle is my joy in life."

Crowell also leased four hundred acres of grazing land to the east of Turlock at the edge of the Sierra foothills. To flood-irrigate that pastureland took massive amounts of water. "It's a monster as far as using water goes," said Crowell, a vigorous three-quarters of a century old. He added, "The way people are coming into California, I don't know how you are going to farm in thirty years. You are not going to be growing much cow feed anymore. Cow feed takes lots of water."

Of the original purchase of eighty acres made in the name of his grandmother nearly one hundred years ago, Crowell's holdings in the early 1990s were a little over twenty-two acres. He retained ownership of the 1.43 acres on which his air-conditioned home stood; of the 11.95 adjacent acres where there was a corral for Barbara's horses, the original farmhouse (now occupied by Crowell's herdsman), the old dairy buildings, and scattered farm

The old farmhouse

equipment; and 7.95 acres facing the college, now called a state university. The latter two undeveloped parcels were designated as storm drains by the city, meaning the storm runoff from the adjacent, asphalt-covered subdivisions would empty into the sandy loam on which crops were once raised. Son Gary's house was on 0.93 of an acre at the northwest corner.

The old farmhouse, tree-shaded and shuttered on the street side, was the past to Crowell and an eyesore to his neighbors. These neighbors tended to be new to the area and overwhelmingly white. Most of these residents, who lived behind the walls of their respective subdivisions, were married, in their late twenties, and parents of at least one child. They had inherited a very small piece of flat land with a long history; but few, if any, had knowledge of that legacy.

In a place where recorded history is so short and the people so transient, the few remaining valley oak trees stand out with their deep-rooted pasts. Some are six hundred years old. Their trunks, which reach diameters of nine feet, are strong enough, considering the sail effect of their large canopies, to withstand the spring winds that sweep across the valley. The trees are huge and many-limbed in a graceful, drooping sort of way.

In their time, they had supplied the Indians with their staple food, the acorn. Early travelers from the East, who were acquainted with different species of oaks, were taken aback by the variety, size, and widespread presence of the California oaks. William Brewer wrote wistfully of the valley oaks that he encountered:

> First I passed through a wild canyon, then over hills covered
> with oats, with here and there trees—oaks and pines. Some of
> these oaks were noble ones indeed. How I wish one stood in
> our yard at home.

Brewer measured the trees and was impressed enough to write the measurements in italics: one had a *diameter* of over six feet and the branches spread *over seventy-five feet each way*. "I lay in its shade a little while before going on," said Brewer.

To early valley settlers, who had few boulders to clear from their fields, oaks were the main obstacles. Oak trees made good firewood. A rancher near present-day Caswell State Park on the Stanislaus River cut

Valley oaks in a state park

three hundred cords of wood one year. Much of that wood was sold. Some was kept and burned for heat and for cooking. Down came the oak trees to fuel the steam engines that drove the huge wheat combines, sawmills, trains, and steamboats. Oak was also used in construction, whether it be parquet floors or railroad trestles.

A history of the Chinese in San Joaquin County stated:

> In the 1880 census alone eighty-five Chinese were employed as woodcutters. Almost singlehanded in this endeavor, they changed the face of the county from the heavily-timbered region of early days to one more closely resembling a savannah. Chinese woodcutters cleared the riverbottoms and tenaciously grubbed stumps from the flat farmlands.

In this manner the valley was shaved clean, except for a narrow fringe along the rivers. Where once there had been riparian woodlands ten miles wide, now they do not exceed a few hundred feet and are confined to preserves. Oaks once covered a million acres, but only 1 percent of that acreage is now considered pristine oak woodlands. The king of trees has been reduced to a few solitary mendicants.

VI

The Fractured Province

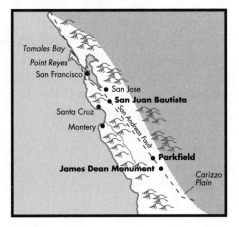

I drove west from the valley and crossed the Diablo Range on a bucolic spring day. Pastoral California was in all its benign glory. Robert Louis Stevenson had written of just such a time in *The Silverado Squatters*: "It was then that favoured moment in the Californian year, when the rains are over and the dusty summer has not yet set in; often visited by fresh airs, now from the mountain, now across Sonoma from the sea; very quiet, very idle, very silent but for the breezes and the cattle-bells afield." Except for the "eternal poison oak," he added, all that grew was "beautiful, healthy, scented, and rich in bloom."

The soft, green hills were splashed with swatches of yellow, orange, and violet wildflowers. The roadside boulders were etched with marks proclaiming Bunny, Shadow, Randy, J.J., and Connie. Light and shadow alternated on that partly cloudy day. The higher the road ascended, however, the more barren and scratchy the land became. The desolation peaked at the crest, where the blackened remains of fire-scarred trees and chaparral stood out like so many contorted skeletons.

On impulse, I took a right turn at a fork in the road and quite unexpectedly encountered Lee's Field, where paint-ball war games were being conducted on this Sunday with all the enthusiasm of men and boys who had never experienced actual combat. The next unit to take the field was readying for the attack under the metal roof of an open shed. A small boy was given a loaded pistol by his father, who warned him not to pull the trigger in the shed. I flinched every time he waved it around, although I knew from a conversation I had overheard that the worst I could expect from a paint-ball wound was a hickey.

The shed contained the barely restrained excitement of two football teams about to take the field. Scattered about on long tables were all the paraphernalia and weapons—camouflage clothing and flak jackets and Uzi submachine guns and 9mm pistol look-alikes—that could be seen on the streets of any self-respecting California city.

I drove up the road and parked where I could overlook the pine-and-rock-dotted ravine where a major battle was being waged. I made my lunch in the van, took out a chair and binoculars, and watched the warfare unfold below me while I ate. I felt like a god. Below me the struggle of mortals proceeded in the following manner:

> *(Ping, ping, and then a whole string of pings.)*
> Where's the guy on your left? Is he still up there?
> Mow 'em down.
> *(A whole string of pings.)*
> He's over there to the left, to the left.
> *(Five or six scattered pings.)*
> Hit the ground, hit the . . .
> *(The words were obliterated by the noise of intense firing.)*
> Hit.
> *(A dead soldier declared himself dead.)*
> I got one.
> Good shot.
> Move up high. Move up.
> Okay, there's one more to your left. Behind the rock, the rock. Move around to the right.
> *(I could see the hunted man crouch lower behind the rock. His attacker moved forward. Could I save him by yelling? A string of pings. Too late.)*

He's hit. Dead man!
Okay, that's it. Game. Change sides.

I drove back to the road that led westward to Mount Hoffman. The domed structures of Lick Observatory rose above the pines on a ridge reached by a narrow mountain road. The aging facility had seen better days: It had been patched after numerous earthquakes. Rust and chipped paint indicated that recent cuts in the University of California's budget had taken a toll on maintenance. Light and smog pollution from nearby metropolitan San Jose lessened the effectiveness of the telescopes. Yet, to go in one hour from mock weapons pointed at humans to telescopes pointed at the universe seemed to me very California.

The clouds blew past on either side and below the observatory, whose late-nineteenth-century architecture suggested a time warp in the sky. When the clouds parted, I could place myself within a context. There were distant views of the Sierra Nevada, the Central Valley, the Coast Ranges, and the Pacific Ocean—a cross-section and sampling at mid-state of the tremendous variety that constituted the landscapes of California. I descended to follow the parallel ranks of mountains and valleys and coastline that constitute the Fractured Province.

The San Andreas Fault slices ashore just north of Point Arena and, in a fairly straight line, cleaves the low mountain peaks and ridges known collectively as the Coast Range, the central coast, and interior valleys for three hundred fifty miles in a southeasterly direction to Big Bend; there it veers more easterly in keeping with the direction of the Transverse Ranges, whose ramparts seal off Southern California from the rest of the state.

This landscape can be the most beguiling and lyrically beautiful part of California. It is where vibrant-colored wildflowers and green velvet hillsides dotted with lollipop oaks mock the concept of winter. The verdancy of early spring, however, is fleeting. During the longer summer and fall months, the dry, golden grasses—the true gold of California—predominate; and these interior valleys become crackling hot.

It is here that San Francisco, everyone's favorite city (actually more eastern than indigenous, and for that reason unreal and favored by the many transients), is located. The *San Francisco Chronicle,* the one newspaper that circulates throughout the region, publishes a "Geo Watch" feature on the weather page each Friday. Here is the report for the past week:

> Seismic activity increased to a near-normal level with 27 earth movements measuring at least magnitude 1.0 in the greater Bay Area. The week's most powerful quake registered 3.8 on the Richter scale from an epicenter near The Geysers geothermal field. It awakened many people in parts of Lake, Sonoma, and Napa counties during the early morning hours on Monday, and shook some items off shelves. A slightly weaker and less troublesome tremor occurred near San Juan Bautista at a more respectable hour on the previous Thursday. There was the usual scattering of small slips along the southern stretch of the San Andreas, and weak quakes near Pacifica and San Leandro.

This seismic activity did not take place along a single line demarcating the Pacific and North American Plates. The plates grind against each other in a much more complex manner. The widest expression of such movement is the San Andreas Fault System, a network of parallel faults some fifty miles wide at the latitude of San Francisco. Here the system includes, from west to east, the San Gregorio–Hosgri, San Andreas, Hayward–Rogers Creek, Concord–Green Valley, Calaveras, and Greenville fault zones. The sheared rock in the fault zone is anywhere from one-third of a mile to one mile wide. Finally, there is the narrow fault line itself—the most recent fracture within the fault zone.

I have lived about as close as one can to the San Andreas Fault in the area where the maximum displacement of twenty feet was experienced during the 1906 San Francisco earthquake. For the first eleven years I lived just to the west of the fault on the Pacific Plate, which gradually is making its way toward Alaska. Since then I have lived just to the east of the fault on the more stable North American Plate. The Pacific Plate, on which most residents of the state live, is suspect terrain; it has come from elsewhere and is of uncertain origin. The North American Plate is mainland U.S.A.

My house sits on the edge of a one-hundred-foot-high bluff overlooking a flat pasture. On the other side of the pasture—three-quarters of a mile wide at this point—is the steep Inverness Ridge. The flat land at the head of Tomales Bay was formed by sediment transported downstream from the local extension of the Coast Range, which is within the San Andreas Fault System. The pasture, a wetland before a local rancher constructed dikes, is the San Andreas Fault Zone. The last mile of Lagunitas Creek follows the rupture of the 1906 earthquake and marks the actual San Andreas Fault at this point. With dairy cows grazing below in the summer months and migrating waterfowl splashing about in shallow ponds formed by the winter rains, the scenery is quite idyllic. But that is deceptive.

The landscape is unstable and chaotic. To the west of the fault, the land is slowly and inexorably moving at an average of two inches a year toward the northwest; in thirty million years it will be 350 miles closer to Alaska, and thirty million years ago it was an equivalent distance to the south. From 1769 to the beginning of the present decade, a period of years representing the bulk of California's recorded history, there were 117 major earthquakes within the San Andreas Fault System. "Major" meant that the quakes were felt and recorded in written documents or, after the invention of the seismograph, measured as moderate or higher on the Richter scale.

The landscape lay still, crept, or jerked during repetitive cycles of strain, accumulation, and release over millions of years. The stages of the cycle, as described in a U.S. Geological Survey professional paper, include "a low level of seismicity in the first part of the cycle once aftershocks of the latest event subside, a rise in regional activity as strain accumulates, and ultimately the occurrence of another earthquake with its attendant foreshocks and aftershocks, which initiates the next cycle." A tremendous amount of energy builds up when the land strains against itself. It is released when there is a rupture.

California is being pushed and shoved about. Two long-term movements of the earth are taking place. To the north of Cape Mendocino, the subduction that has given rise to the volcanoes and lava flows of the Cascade Range is from west to east. To the south, where the San Andreas Fault is the dominant force, the movement is toward the northwest. California's complex history was shaped and given its episodic rhythm by these cataclysmic cycles and contradictory movements.

In the nineteenth century there were various theories to explain the cause of earthquakes, among them the electrical theory, the planetary theory, the underground explosion or wind theory, the volcanic theory, and the general-wickedness-of-humans theory. The scientific concept of there being a long, continuous fault along which earthquakes occurred was slow to emerge, not coming to the forefront until after the 1906 quake.

William Brewer, a close observer of natural phenomena, and his boss, Josiah Dwight Whitney, the state geologist, experienced a number of earthquakes and crossed and recrossed what would come to be recognized and called the San Andreas Fault. Yet they never noticed a pattern in the occurrences of earthquakes or the extended dent in the earth that a fault left upon the land. While camped along the fault at San Juan Bautista, east of Monterey, Brewer received a letter from Whitney, who was in San Francisco in early July 1861. Brewer wrote in his journal:

> He writes, and other letters and the papers confirm it, that several shocks of earthquakes took place last week in San Francisco, and reports come in from other parts of the state. I think that I have before told you that nearly or quite the whole of this state is subject to earthquakes. There will never be a high steeple built in the state, and in San Francisco the loftiest houses are but three stories high, the majority only two.

On December 23, 1862, Brewer experienced a moderate earthquake in San Francisco. The house where he was sleeping at 5:30 A.M. pitched about in an "uncomfortable way" and he felt slightly nauseated. "These mysterious quakings and throes of Mother Nature affected me as no other phenomenon of nature ever did," wrote the experienced traveler.

Brewer felt no earth movements the previous year at San Juan Bautista, where the San Andreas Fault was quite pronounced. The fault was in the form of a bluff, termed a scarp in geological parlance, that ran adjacent to the mission where Brewer attended mass. He noted the mission's thick adobe walls; of the surrounding landscape, he said only that it was "peculiar."

The fault dictated the site for Mission San Juan Bautista, the fifteenth of twenty-one missions founded by the Franciscans in California. It provided a commanding site above *El Camino Reál*, "the king's highway," a 650-mile-long trail from Baja through Alta California to San Francisco that followed the trace of the fault below the mission. From the fault issued a water supply for the mission in the form of two springs.

The mission was established by a corporal and five men in May 1797 on the orders of the successor to Father Junípero Serra, the founder of the California mission system. In July the first baptism of an Indian child took place. He was called Juan Bautista, in honor of St. John the Baptist, and his sponsor was the corporal. The first funeral was conducted two months later for the corporal's own infant son. From October 11, 1800, to the end of that month, there were as many as six earthquakes a day. The fathers slept outside in carts. Great cracks appeared in the ground and in the walls of the adobe buildings. There would be similar damage from major earthquakes in 1812 and 1906, and less damage and constant fear from smaller tremors.

The local population of Mutsun Indians were the first converts and slave laborers. Then the Franciscans, having exhausted that supply, went farther afield, sending expeditions into the Central Valley to save the souls of the Yokuts and capture, or recapture as the case may be, the Indians who built the structures and tilled the fields and sang in the choir. These raids were so successful that by the 1820s San Juan Bautista had one of the largest mission Indian populations in California.

The priests noted patterns in the behavior of the Indians: "They lie and cheat those who are not of their own color and class, but not their

The mission abuts the fault line and El Camino Reál

own people." There was a reluctance to learn Spanish, the language of the dominant race at the time: "It is not easy to find a means and method to teach the Indians to talk Spanish well, because it is extremely difficult to make them neglect their own idiom when the majority speak the native rather than the foreign tongue."

At San Juan Bautista, the Indians were confined in barracks-type structures. There was one room with one door and one window—sixteen square feet of space for each family. There were separate barracks for single men and women. New recruits were constantly needed. Father Felipe Arroyo de le Cuesta said in 1815, "The most common maladies are consumption, syphilis, and dysentery, from which many die. In a year the number of deaths exceeds that of births." Cuesta, who took the time to learn the Indians' languages, was depressed by his work, most of which involved running the mission. "There is hardly anything of religion in me, and I scarcely know what to do in these troubled times," he said.

San Juan Bautista was one of the richest missions in the chain. At the height of its prosperity in 1820, it could claim 69,530 sheep, 43,870 cattle, 6,230 horses, and 311 yoke of oxen. In the various storehouses, there were $75,000 worth of goods and $20,000 in gold specie. The massive church, twice the size of most of the missions, could hold one thousand worshipers within its three-foot-thick adobe walls. On the ceiling was painted a single, all-seeing eye.

By 1834, the number of recorded burials in the Indian cemetery, sandwiched between the church and the fault line, was 3,766. Originally, it was estimated that 2,700 Mutsun Indians lived within the Pajaro River drainage. (Eventually, more than 5,200 bodies were stacked one atop another in the half-acre plot. They were buried in canvas shrouds, while the padres and members of prominent families in the parish were entombed within the church in coffins.)

The smallpox epidemics of 1828, 1829, and 1838 were particularly devastating, killing off half the Indian population. Sometimes ten or more Indians a day were buried in the cemetery. The mission became a charnel house. The funeral records noted:

> The old graveyard close to the church is filled up with bodies to such an extent as to saturate the Mission with their smell. I blessed a new strip of land for a new burial ground on October 24, 1838.

Edwin Bryant, the Kentucky newspaperman who had seen the remains of the Donner party, visited the church in 1846. He wrote:

> During the twilight, I strayed accidentally through a half-opened gate into a cemetery, enclosed by a high wall in the rear of the church. The spectacle was ghastly enough. The exhumed skeletons of those who had been deposited here, lay thickly strewn around, showing but little respect for the sanctity of the grave or the rights of the dead, from the living. The cool, damp night-breeze sighed and moaned through the shrubbery and ruinous arches and corridors, planted and reared by those whose neglected bones were now exposed to the rude insults of man and beast. I could not but imagine that the voices of complaining spirits mingled with these dismal and mournful tones; and plucking a cluster of roses, the fragrance of which was delicious, I left the spot, to drive away the sadness and melancholy produced by the scene.

For a short time San Juan Bautista was the capital of Alta California and the axis on which pivoted a war of liberation that had its comic moments. Captain John Charles Frémont of the U.S. Corps of Topographical Engineers was the very incarnation of Manifest Destiny. He appeared on the scene in early 1846 with a force of sixty-eight heavily armed soldiers, scientists, mountain men, and Delaware Indians just as American immigrants were beginning to flock to the Mexican possession. Mapping was the cover for the impetuous Frémont, who was somewhat of a loose cannon. He sported a white felt hat, and strapped to his belt was a large bowie knife. Across his back was a rifle.

On the march down the fault line, Frémont's men shot a California condor and tethered the wounded bird to a cart, "but his savage nature would not permit of any approach," wrote the captain.

They sought trouble that was tinged with racism. They insulted the daughters of a relative of the Mexican general José Tiburco Castro, who ordered the "band of robbers" out of the country. Frémont was also secretly ordered to leave by the American consul in nearby Monterey. The captain chose to take calculated offense at the Mexicans, and he declined to depart.

The troop of Americans rode to the top of a 3,171-foot mountain, now known as Fremont Peak, towering above San Juan Bautista. There they built a rough fort from logs, unfurled the American flag, attached it to a thin sapling, and observed through a spyglass the preparations for an attack. Frémont wrote the consul: "I am making myself as strong as possible in the intention that if we are unjustly attacked we will fight to extremity and refuse quarter, trusting to our country to avenge our death."

Castro, too, beat his chest. Addressing his fellow citizens from his headquarters at San Juan Bautista, he said:

> Compatriots, the act of unfurling the American flag on the hills, the insults and threats offered to the authorities, are worthy of execration and hatred from Mexicans; prepare, then, to defend our independence in order that united we may repel with a strong hand the audacity of men who, receiving every mark of true hospitality in our country, repay with such ingratitude the favors obtained from our cordiality and benevolence.

When the flagpole fell of its own accord on the third day, Frémont seized that as an omen and descended unmolested from the mountain and rode northward. The mountain man and guide Joe Walker declared Frémont a coward and left the party. The Mexicans proclaimed victory. Needless to say, unlike the Alamo, the incident never became engraved upon this nation's historical consciousness.

Frémont and his men made their way to Tule and then Klamath Lakes, where they were intercepted by a messenger from Washington, D.C. They returned to participate in the Bear Flag Revolt and the surrender of the Mexican forces in Southern California, events that read like a Gilbert and Sullivan operetta. And thus did Alta California become first the territory and then the State of California.

It was to Fremont Peak that an elderly John Steinbeck retreated more than one hundred years later with his dog, Charley, after realizing that he couldn't go home again to Monterey and the Salinas Valley. Steinbeck mused how, as a small boy, he had found old cannonballs and rusted bayonets at the summit. As a young man with thoughts of death, Steinbeck

wanted to be buried on the peak "where without eyes I could see everything I knew and loved, for in those days there was no world beyond the
mountains." The author looked down upon where he had fished for trout,
where his mother had shot a bobcat, and upon "starvation ranch," the
family home. "This solitary stone peak overlooks the whole of my childhood and youth," he wrote in his farewell to the Salinas Valley.

When I gained the peak and found a camping spot for the night, there
was no view of the Salinas Valley, or any other valley, for that matter.
Dense fog choked the lowlands that evening and in the morning, leaving
only the mountaintops visible as floating islands. I could have been the
lone survivor of a vast catastrophe, gazing down upon the void where the
land had dropped beneath the boiling sea along the lines of earthquake
faults that perforated valley floors.

Gone were the places I knew and where I had lived or worked or visited along the San Andreas Fault: San Juan Bautista, San Jose, Silicon
Valley, San Carlos, San Francisco, San Rafael, and the Point Reyes Peninsula. When I finally descended, the headline in the newspaper was
STATE'S IMAGE SUFFERS IN POLL: CALIFORNIA DREAMIN' NEARLY A FORGOT-
TEN TUNE. Natural disasters were a major concern of the inhabitants contacted by the Field Poll.

Outside the front of the church, the seismograph display showing data
collected from Fremont Peak for the time that I had been there indicated no anomalies. But a posted fax from the U.S. Geological Survey
noted that in the four days following a 1992 earthquake along the San
Andreas Fault in Southern California, there had been three thousand
aftershocks.

This was the land of the serpentine barrens. When highly mineralized
rock is ground up into soil by the friction of the earth's movements and by
weathering, the resulting acidic soil inhibits, kills, or gives rise to its own
unusual plant growth.

Brewer was on his way to the nearby New Idria quicksilver mine in
the Diablo Range when he caught sight of the most spectacular serpentine barrens in the West. He wrote:

> The view from the summit is extensive and peculiar. It is to
> the north and east that the view is most remarkable—the

Panoche Plain, with the mountains beyond—chain after chain of mountains, most barren and desolate. No words can describe one chain, at the foot of which we had passed on our way—gray and dry rocks or soil, furrowed by ancient streams into innumerable canyons, now perfectly dry, without a tree, scarcely a shrub or other vegetation—*none,* absolutely, could be seen. It was a scene of unmixed desolation, more terrible for a stranger to be lost in than even the snows and glaciers of the Alps.

Serpentine rocks are found near earthquake faults. An ultramafic rock, serpentine originates in the upper mantle of the earth's crust and is thrust upward by tectonic action, resulting in soils rich in magnesium, iron, and silica. They could be barren, as Brewer noted, or contain one or a number of the two hundred fifteen known species, subspecies, and varieties of native flora that are restricted in whole or in part to such soils.

The geological use of the term *serpentine*—so called because of the rock's resemblance to the mottled skin of a snake—dates back to six-teenth-century Europe. Serpentine eventually became the state rock of California. Arthur Quinn, a professor of rhetoric at the University of California at Berkeley, described serpentine with greater license than any geologist would be allowed. He was discussing the enigma of the rocks found alongside the San Andreas Fault on the Point Reyes Peninsula when he wrote:

This greenish, slightly slimy rock is aptly named serpentine, the serpent rock. Moreover, it bears with it, like its rattles, the most arresting minerals of the whole Franciscan Formation, minerals for which there is no equivalent in the austere purity of Point Reyes: the lucent, aqueous crystals of the bluecrist; the fire-red powder of cinnabar; the complex green of jade, in appearance at once solid and liquid; and even the rare man-ganese ores, colored like the moods of a geologist confronted by the nightmare of the Franciscan, usually a blank black, occasionally a deeply shocked pink.

Serpentine landscapes were used to locate quicksilver, otherwise known as mercury. Also an indicator of a tectonic presence, mercury was used to process gold. Mining, grazing, geothermal power development, and outdoor recreation have taken their toll on these unusual environments. In the desert atmosphere of the New Idria barrens, off-road vehicles have torn up the landscape, perhaps also to the detriment of those who drove the trail bikes, dune buggies, and other all-terrain vehicles. There are asbestos fibers in the dust of serpentine soils.

There was another quicksilver mine to the north of San Juan Bautista. In his novel *Angle of Repose,* Wallace Stegner described the arrival of the easterner Susan Ward at the New Almaden Mine in the 1870s. Stegner, who was particularly attuned to the landscape of the West, wrote:

> But mostly what she felt in the moment of arrival was space, extension, bigness. Behind the house the mountain went up steeply to the ridge, along which now lay, as soft as a sleeping cat, a roll of fog or cloud. Below the house it fell just as steeply down spurs and canyons to tumbled hills as bright as a lion's hide. Below those was the valley's dust, a level obscurity, and rising out of it, miles away, was another long mountain as high as their own. Turning back the way she had come in, she saw those five parallel spurs, bare gold on top, darkly wooded in the gulches, receding in layers of blue haze. I know that mountain, old Loma Prieta.

Just before the start of the 1989 World Series in San Francisco, the earth shook underneath that mountain, shock waves radiated outward from the epicenter, and miles away bridges and freeways came tumbling down.

Following the miners, the settlers came, and none entered a more god-forsaken place than the Carrizo Plain. The desert steppe, just one mountain range over from the busy San Joaquin Valley and sixty miles northwest of Los Angeles, is retarded California: Because of its isolation and lack of water, everything took place later and on a much smaller scale. Time has begun to run backward on the plain, an unusual occurrence in California. Nowhere else on land is the San Andreas Fault a more visible or more powerful presence. The forty-five-mile-long plain is seared by the fault, much as the wrinkled hide of a cow is cauterized by a brand.

The fault line runs along the base of the hills

I was reluctant to enter the plain. I turned back the first time when the paved, one-lane road turned to unpaved mud. It was early spring in an above-normal-rainfall year following six years of drought. I returned a few weeks later on Easter Sunday and stopped at the gasoline station–restaurant-bar at Camp Dix, the southern entrance to this terra incognita where the California condor once soared.

Tom Pilcher, the owner of Reyes Station and Bar, warned me not to drive into the Carrizo Plain at night. Drug runners and the feds were there, he said, shooting it out, and I might get hit by a stray bullet. The others at the bar nodded agreement, although one warned me quietly that it was Pilcher's habit to blow smoke up people's asses.

The fault passed through Pilcher's property. He said it was hard to differentiate a quake from the large trucks passing by outside on Grocer Grade, or the motor on the big beer cooler kicking on and off. The walls of the bar were festooned with the relics of ranching, and Pilcher wanted me to hear the start of a poem dashed off in the bar by an old cowboy. He read from a much-folded piece of paper:

> *You can feel spirits long after they're dead.*
> *I was settin' in Reyes Station and talking to Tom,*
> *And I swear I heard voices long since gone.*
> *Old Poncho and Alice and Frank Pacheco,*
> *Backeroos who lived here long ago,*

And Tom he's a chip off the block.
You can tell he's related to Alice just from his talk.
He told me about Merle who stopped by and sang,
Just before Christmas, Haggard's his name.
He recorded Merle on a homemade tape and as he sang,
"I Stopped Loving Her Today,"
I looked over the relics that Alice left,
A couple old center-fired saddles and some Spanish bits.
As real as your breath.
And when Tom stepped outside to pump some gas,
I heard the jingle of spurs and some horses pass.

Pilcher insisted that I camp behind the bar, and since it was ten P.M. and I had driven far and had a few beers, I accepted.

I was up early and drove north on Soda Lake Road before sunrise. I took the fork to the right on the Elkhorn Trail and drove up a rutted dirt track that cut over the southern end of the Elkhorn Hills and dropped down into the Elkhorn Plain, a smaller and higher version of the parallel Carrizo Plain.

The fault-warped hills were alive with wildflowers. *Sunset* magazine, that paean to western living, had run an article on the wildflowers of the Carrizo Plain region; they were supposed to be exceptional this spring. Goldfields, which go from seed to flower to seed in five to six weeks, predominated. This relative of the sunflower family swept across the crests and flanks of hills like wildfire; pronghorn antelope had beaten paths through the thick, golden carpet.

There was a wondrous display of nature on these two plains and in the surrounding hills: vernal ponds, the last remnant of a great prehistoric lake called Soda Lake, the dramatic Elkhorn Scarp, abundant hawks, thousands of sandhill cranes, and more endangered species than anyplace else in California. The plains were home to nine threatened or endangered animal and bird species and three plant species. The Spanish had a good feel for the place. They named it the Estero Plain; and the mountain range to the east, the Temblor Range—*temblores* meaning "earthquake" in Spanish.

Prior to the coming of Europeans, both the coastal Chumash and the Yokut Indians from the San Joaquin Valley used the plain. A small year-round population gathered seeds from wildflowers and occasionally burned the plain to promote the growth of plant species rich in protein and oil. Others used the plain seasonally. Around the few springs was the carrizo grass which gave the plain its name.

The Chumash Indians recognized the power of the place. In the Chumash cosmology there were three worlds: The world below held dangerous creatures who entered the middle world after dark. The middle world, inhabited by humans, was held in place by two gigantic serpents; when they tired and moved, there were earthquakes. The world above was inhabited by such supernatural creatures as Sun. Sun was an old man who walked a path each day with a torch, which, if held lower, caused the blistering hot days of summer on the plain. A widower, Sun took people home each night. He cooked them over a fire, and with his two daughters ate them half-raw. They preferred blood to water.

Serpents were depicted on Painted Rock, a 120-foot-high sandstone rock formation that is a separate presence on the Carrizo Plain. The Indians adorned the rocks with red, black, white, and yellow symbols painted in diverse styles over the span of a thousand years. The rocks formed a natural amphitheater, and within this space the Indians held their religious ceremonies.

Two events stuck out in the relatively timeless expanse of history, as the Chumash conceived it: The Flood, and the physical displacement by strangers. When the Mexicans first moved sheep into the plain in the 1830s, and during the 1850s and later, when the Anglos brought more sheep and cattle, the amphitheater was used as a livestock pen.

In 1857, the earth lurched along the fault, and streams were displaced some three feet. The exposed fault has remained locked in place since that time. The barren, desert landscape reveals an array of tectonic features in textbook clarity. There are elongate grabens or troughs, sag ponds, scarps, linear scarplets and ridges, shutter ridges, medial ridges, folded ridges, tilted and rotated blocks, offset and deflected streambeds, and offset marine and river terraces. It is here, according to Robert E. Wallace of the U.S. Geological Survey, that "the surface expression of the fault is narrowest, clearest, and best defined."

It was also here, exactly at the intersection of an offset streambed and the fault, with the Elkhorn Hills on one side and a shutter ridge on the other, that I camped for two days. According to Wallace, there was a one in seven hundred chance that a great earthquake would occur here in any given year. He did not calculate the chance of being caught in a drug war.

When the federal government opened the plain for purchase shortly before the start of the Civil War, there was a land rush by San Francisco speculators. James M. McDonald, who owned a steamship line, bought fifty-seven thousand acres for as little as twenty-five cents an acre. In 1869, Chester R. Brumley was hired by McDonald to establish a permanent presence on the plain. Brumley and his family came from Brooklyn, New York. At the age of ninety in 1955, one of Brumley's daughters, Nellie, recalled for an interviewer how she was carried piggyback at the age of three across the Isthmus of Panama by a native. They took another steamship on the Pacific Ocean side of the isthmus to San Francisco. There her father got the job of overseeing the McDonald ranch.

From San Francisco the family took a rickety steamship to Avila, the port for San Luis Obispo County; and traveled by horse and cart to the desolate plain. They lived at first in a primitive lean-to and then in a two-story house. It was located high up on an alluvial fan near a spring. The view was toward the east and the serrated Elkhorn Scarp. El Saucito Ranch was the only permanent dwelling in the eight-hundred-square-mile plain for the next twenty years. It was twenty-five miles to their nearest neighbor, fifty miles to the mail stage stop, and a long two-day trip to town.

Nellie and her sister, Mary, were enthralled by nearby Painted Rock, where their father penned four thousand sheep each night. Myron Angel lived with the family while collecting material for a fanciful book on the rock. He dedicated the book to the sisters and wrote of them:

> These young ladies, fresh from the schools and society of the East, transplanted to the weird and wild waste of the West, were ecstatic in their enthusiasm for the strange vastness of their surroundings and overawed by the mysterious Painted Rock so near at hand, engaged at once as guardian angels of a sacred trust, measuring the rock in every dimension as engineers; studying and sketching the paintings as artists, and preserving them for reproduction in the form here given. Only through such strange circumstances as brought these ladies, "like caged birds let loose," said one of them, effulgent in the spirit of youthful romance, to these strange surroundings, could these pictures have been so preserved.

At the time, sheepherders had taken potshots at the rock art. Others picked away at the markings, taking chunks of rock home or selling them to collectors. Still others painted their names or initials over the weathered figures until they were almost entirely obliterated. Mary spent hours copying the symbols on a sketch pad with a lead pencil. Their brother explored the inner recesses of a cave blackened with the smoke of many campfires and found a nest of rattlesnakes.

One day a group of chanting Indians approached the house. Nellie hid inside while her mother went out to confront them. "*Agua*," they demanded. Her mother gave them dippers for the well water and a bucket of freshly baked cookies. Nellie said:

> They left just as abruptly as they had come. Stolid, unsmiling faces fading away like the climax of an old fashioned movie. Their loose tattered clothing flopping in the wind, their heads bobbing as they trudged beside their burros and finally disappearing into the dusty background.

A letter in the San Luis Obispo newspaper in 1884 from two travelers reported:

> Mr. C. R. Brumley has a pleasant residence in the southwest part of the plain, but with this exception there are no improvements but a few huts and corrals for sheepherders in this broad area of fertile soil and open country. Mr. Brumley has grown eucalyptus, peach, apple, cherry and other trees.

Chester Brumley died alone on the arid plain. His horse startled and his hand became entangled in the reins. He was dragged for miles over the fractured ground. The family left. Mary's drawings were given to the University of California. Nellie died at the age of ninety-six.

Human failure riddled the plain's history. As Marijean Eichel wrote in her history of the Carrizo Plain, "The fever for land was not always amenable to economic reasoning." The homesteaders came next. They lacked an adequate land base, there was little water, and the winters at the two-thousand-foot level of the plain were harsh. Drought around the turn of the century eventually eliminated most of the homesteaders.

Minerals, too, were a chimera. There was oil to the east and to the west of the valley, but nothing but dry holes were punched in the floor of the Carrizo Plain. Gold prospectors found nothing in the surrounding hills, and uranium prospectors had the same results. For a time sodium phosphate was mined from Soda Lake, but that enterprise ended during the Depression.

Tractors to sow and harvest, trucks to transport, and fences to keep out the cattle contributed to a wheat boom during the two world wars. There is still some dry farming on the north end of the plain where the rainfall is greater. During World War II, the air force used Soda Lake as a bombing and gunnery range, killing the operator of a county road grader by mistake.

In 1960, Richard Walker of Arizona, who claimed he was a former district attorney in that state, outlined his grandiose plans to the San Luis Obispo County Board of Supervisors for a twenty-five-thousand-acre development on "the history-soaked lands" of the plain to be called California Valley. Walker said he would spend $250,000 on advertising in the *Saturday Evening Post, Life,* and *Reader's Digest.* The supervisors bought the proposal. They paid no attention to the county engineer, who warned that there was not enough water on the plain to supply a bowl of baby goldfish.

A newspaper reporter wrote a story about the supervisors' hearing that was repeated in different forms and forums throughout the length and breadth of California. It went this way:

> From there on the meeting seemed to move to its inexorable conclusion as Walker, moving in the deliberate style of the court-trained attorney, began to whittle down his opposition, refuted contentious ranchers fearful of rising tax assessments, and skillfully painted a picture of snowbound easterners enraptured at the prospects of escaping their icy domain to bask in the golden sun of California Valley for $10 down, $10 per month.

The rubes bought it. Approximately 7,200 lots were sold on Walker's assurance that "development follows speculation," a play on the similarly spurious notion that "rain follows the plow." The speculation was extreme in this case. There were no sewers, no paved roads, no electricity, no phones—and no water. There did exist a hastily constructed airstrip, bar and restaurant, general store, gas station, motel, and a huge buffalo barbecue to entice buyers.

Three years later a reporter toured the development and wrote:

> A drive through the development shows the roads deeply rutted, wherever they have been used, washed out in places, patches of bog, the remnants of a few ranch houses, miles and miles of sagebrush, dotted by road signs and a little construction, all concentrated in one relatively small area.

Walker disappeared. The development changed hands a number of times. In 1977, another Arizona developer took it over and found sheep wandering through the motel's rooms. The pitch in that year of the energy crisis was raising the jojoba plant that miraculously would supply the nation's oil. The following year the come-on was a vacation and retirement community to which one could commute by small plane that would taxi to the front door. The developer said, "I think the difference this time is I'm basing it on the aviation community. I think the reason the others failed is the isolation."

The newest developer reopened the motel and gas station and remodeled the restaurant and bar. But the isolation remained. He said, "I hired this man for a recreation director. I gave him three thousand dollars to buy groceries for the store and he took off for Las Vegas—chartered an airplane, he wanted to get out of here so fast."

In the early 1980s it was solar power. Arco Solar Inc. built a solar plant that was supposedly the largest in the world at the time. The firm's plan was to fill five hundred acres with two thousand solar panels that would track the path of the sun. By the end of the decade Arco was gone, the victim of disappearing tax breaks for such developments and the economics of generating electricity.

Time began to run backward in the 1990s. The Nature Conservancy bought 82,000 acres, and the drive was on to create a huge nature preserve in the plain—"the Serengeti of California" was the hype. A 150,000-acre natural area was created around the lake.

My time on the fault line was peaceful. I departed without incident.

Some black birds thrived on civilization in California, while others became extinct. The raven is an example of the former; the California condor, the latter. I have seen ravens on a ridge high above Death Valley and on the concrete slabs of the Los Angeles riverbed. Fifteen years ago I saw my first and last California condor in the wild, just to the west of the Carrizo Plain in the Sierra Madre Mountains. I was hiking alone on a trail that followed a small stream in the San Rafael Wilderness when the giant bird flew from one tree to another. It perched there for a while, then departed. I had seen Andean condors in the mountains of Patagonia and, of course, many of the condor's smaller relative, the turkey buzzard, throughout California. All those large black birds glided with an effortless grace that belied their bulk.

A lot of time and more than $25 million has gone into the effort to save the California condor, perhaps time and money that could have been better spent elsewhere. That prehistoric creature with a ten-foot wingspan (the largest bird in North America—another extreme in the land of extremes) was doomed by time. We only hastened its departure.

I have often wondered what the fascination was with this carrion-eating vulture that is related to the European griffin. There was, of course, size and rarity and all that ferocious blackness topped by a bare neck and ruby-red eyes. The condor was a military symbol, as well as a meal ticket for ornithologists.

And now condors are being bred in captivity. Might that be our eventual fate, too? The condor pens at the Los Angeles Zoo are known as condorminiums. The giant birds are fed by aliens who manipulate condor puppets behind one-way-glass windows so as to minimize contact. They are studied for clues to their past existence.

There are pictographs of condors at Painted Rock. The Indians regarded condors as sources of power, and condors appeared in the dreams and visions of shamans. The birds were captured live, or killed and used in ceremonies. The Native Americans, like those who followed, contributed to their demise.

At one time the condor ranged from Baja California to British Columbia and as far east as upstate New York. They circled the California missions, competing with coyotes and grizzly bears for the carcasses of cattle that were slaughtered for their hides. In 1861 they were plentiful along the coast at Monterey, where Brewer noted:

Condor country adjacent to the fault

Buzzards strive for offal on the beach, crows and ravens "caw" from the trees, while hawks, eagles, owls, vultures, etc., abound. These last are enormous birds, like a condor, and nearly as large. We have seen some that would probably weigh fifty or sixty pounds, and I have frequently picked up their quills over two feet long—one thirty inches—and I have seen them thirty-two inches long. They are called condors by the Americans. A whale was stranded on the beach, and tracks of grizzlies were thick about it.

The pioneer California ornithologist James G. Cooper was tempted to kill a condor with his hammer, but, being weighted down with fossils and far from camp, he calculated that the additional twenty-five-pound load would be too heavy to carry. He wrote in 1890:

I can testify myself that from my first observations of it in California, in 1855, I have seen fewer every year when I have been in localities suitable for them. There can be little doubt that unless protected our great vulture is doomed to rapid extinction.

By the mid-1960s their range had been restricted to a region along the San Andreas Fault from Mount Pinos to San Juan Bautista, and the reasons for their disappearance had been well-documented. Besides the evolutionary process, they included: the depredations of Indians, white hunters, museum collectors, and egg collectors. Additional losses were caused by trapping, dwindling food supplies, loss of habitat, human intrusion, and collisions with manmade objects. A cornucopia of poisons were spread upon the agricultural lands of California, and the great birds died from poisoning by herbicides, pesticides, rodenticides, and lead shot.

About sixty condors remained in 1950, forty in 1965, and they dwindled throughout the 1970s. The plan of government agencies, some conservationists, and scientists was to capture condors, breed them in captivity, and then release their offspring. Over the years there was controversy, acrimony, consensus, and competition for government and private funds among the various parties.

The last condor was captured in the wild in 1987. The first two chicks were released in early 1992. A handful followed. One died from lapping up a mixture of oil and antifreeze used to keep rain gauges functioning during cold weather in a condor sanctuary, another after being electrocuted by a 17,000-volt power line, and a third from internal injuries sustained when either hit by a car or colliding with a utility pole. As of July 1993, five remained in the "wild," where each was fed forty pounds of uncontaminated, stillborn dairy calves a week.

Given the setbacks, perhaps California was not the best place to release condors. Perhaps releasing condors back into the very conditions that caused their deaths did not make sense. Perhaps in some far-off time the remnants of our race may face the prospect of being forced from the comfort of captivity into the hostile environment of their forebears, just to satisfy the curiosity of their captors.

Lloyd Kiff, the leader of the Condor Recovery Team, sought to justify the captive breeding program. He wrote:

> In a sense, the condor is a metaphor for California. Like the land itself, the condor has enjoyed periods of exceptional prosperity only to suffer grievous losses from human overpopulation, overdevelopment, and unwise agricultural practices. Like modern California, its salvation now rests upon technological solutions. The condor going back into the wild will be the offspring of environmental tinkering, but it will be no less a condor for such an origin.

For almost everyone in the San Francisco Bay Area, the abrupt shift of the earth's surface at 5:12 A.M. on April 18, 1906, was an exercise in sheer terror. There were a few exceptions.

Grove Karl Gilbert, a geologist, was greatly disappointed at having narrowly missed out on the 1872 Lone Pine and 1899 Alaska earthquakes. He wrote, "When, therefore, I awakened in Berkeley on the eighteenth of April last by a tumult of motions and noises, it was with unalloyed pleasure that I became aware that a vigorous earthquake was in progress." From his room in the faculty club at the University of California, Gilbert analyzed the direction of the movements.

The noted geologist learned of the San Francisco fire two hours later. "This information at once incited a tour of observation, and thus began, so far as I was personally concerned, the investigation of the earthquake." Gilbert was named to the State Earthquake Investigation Commission, as well as a commission formed by the U.S. Geological Survey; he spent the next few months in and around the Point Reyes Peninsula in Marin County, where the movement was most extreme.

As one of the leading scientific explorers of the West who worked for the geological survey during the golden age of American geology, Gilbert had an extensive familiarity with western terrain. He believed in the "rhythmic principle" of earthquakes—meaning that given a certain length of time, the earth would rupture from the accumulated strain. Gilbert scorned the policy of boosterism that sought to conceal the fact that California was prone to devastating earthquakes. San Francisco officials played down the 1906 earthquake and the number of deaths, hoping to gain the eastern capital that was needed to rebuild the city.

Gilbert wrote:

> It is feared that if the ground of California has a reputation for instability, the flow of immigration will be checked, capital will go elsewhere, and business activity will be impaired. Under the influence of this fear, a scientific report on the earthquake of 1868 was suppressed.

The terrain Gilbert selected to investigate was impressive for the massive groove that represented the fault. Brewer had spent two days on the shores of Tomales Bay in 1862 yet missed the significance of the landscape. (Others who came later, and were lulled by the tranquil setting, would also be deceived.) He wrote:

> The bay is a long narrow arm of the sea that runs up into the hills, surrounded by picturesque characteristic California scenery. The bay is pretty, and the number of waterfowl surpassed belief—gulls, ducks, pelicans, etc., in myriads.

The concept of a long, continuous fault emerged piecemeal. Around the turn of the century, Andrew C. Lawson of the University of California bestowed the name San Andreas Fault on a short segment running through the San Andreas Valley to the south of San Francisco. A Stanford University geologist, who was mapping a section of the fault in the Santa Cruz Mountains, named it the Portolá-Tomales Fault. Another geologist noted the Baulinas-Tomales Fault north of San Francisco. The rupture from the 1906 earthquake made it clear that there was one fault that ran nearly the whole length of the state.

I wonder now how they could have missed the significance of Tomales Bay, the most dramatic water-filled expression of the San Andreas Fault. Its long and narrow shape echoes the horizontal movement of the two massive plates underneath it, just as the plates have shaped the long, narrow Olema Valley to the south. The bay and valley form a thirty-mile trough that cuts the one-hundred-square-mile triangular Point Reyes Peninsula off from the mainland of California. The peninsula, termed an "island in time," has crept northwestward from a point off the Tehachapi Mountains in Southern California. It is a perfect example of suspect terrain.

The joining of peninsula to mainland is fragile. The fractured rock that underlies the bay and valley is a weak glue. To the west the rocks are granitic; to the east they are Franciscan sandstones, cherts, and limestones. From the erosion of rock comes soil, and vegetation follows soil type.

Two different lands—and cultures—are temporarily joined. The foliage is thick to the west and thin to the east of the fault. Shaded Inverness, the village on the peninsula edge of the bay, retains a New England preciousness; it was a summer town. Point Reyes Station, on the mainland edge of the fault, is exposed to sun and wind; it was a railroad and

Tomales Bay, with the peninsula to the right and the stakes marking the fault line

ranching center. Inverness was referred to as "aristocratic" at the time of the earthquake. Point Reyes Station was plebeian.

More recently, writer John McPhee stood on the shore of Tomales Bay. He was struck by the "complete dissimilarity of the two sides" and likened the mismatch to wearing a green wool sock on one foot (the cooler, wetter, west side) and a tan cotton sock on the other (the warmer, drier, east side). Scientists have also noted the differences: Sixty species of plants found on the peninsula are missing from the adjacent mainland. Alan J. Galloway, a geologist with the California Academy of Sciences, remarked upon the disparities in a technical publication:

> Point Reyes Peninsula is distinguished from the other coastal
> areas west of the San Andreas Fault largely by being a distinct,
> nearly sea-girt, tract of land separated from the mainland by
> the valley created by the fault zone, and by having a varied
> geology entirely different from that of the mainland. These
> factors have resulted in the peninsula having a distinct charac-
> ter of its own, geologically, botanically, and agriculturally.

I have lived on both sides of the bay for nearly twenty years and can feel the change when crossing the fault line on Sir Francis Drake Boule- vard—the transition is from light to shadow and from mainland to float- ing island, if the journey is made from east to west.

It was to this same road, where the land to seaward had shifted twenty feet to the northwest, that the geologist Gilbert came on his first day in the field. He noted, "The 60-foot space between the ends of the faulted road is a shear zone, occupied by shredded ground with many small horizontal faults."

The methodical Gilbert took a black-and-white photograph of the offset dirt road looking west at a horse and cart in the distance. He couldn't be certain that the road was in exact alignment at this point before the quake, but he thought it probable. Marshy ground underneath the roadway might have exaggerated the slippage, he thought. Nearby he found a barn that had been dragged 16 feet off its foundation, a fence was displaced 15.5 feet, a garden path 15 feet, and raspberry bushes 14.5 feet. Off the village of Inverness a pier was shifted "not less than 30 feet," Gilbert reported; but he wasn't sure whether that displacement represented the flow of mud or the movement of the earth.

There were other reports. An Olema man wrote his friend in San Francisco:

> The whole country from Tomales to Bolinas is completely wrecked. Buildings are down, bridges gone and the roads are impassable in many places. Those buildings which are not completely demolished are so badly damaged that they will have to come down.

There were other types of phenomena. Ships at sea off Point Reyes staggered, as if they had struck a rock or an iceberg. People heard a rumbling noise, a great roar, a sharp crack like thunder, or a rushing noise like the sound of wind followed by a soft slap on the face. The ground moved like swells at sea. Some aberrant animal behavior was reported to have taken place shortly before and during the earthquake: Horses whinnied or reared, their hooves vainly pawing the air. Cows in fields stampeded or were thrown to the ground. The hair on the backs of cats bristled, and they disappeared for days. Dogs ran for cover, their tails between their legs.

A fisherman on Tomales Bay saw the water suddenly retreat, leaving his boat stranded on the mudflats. It returned in the form of a "great wave, which looked a hundred feet high, but which was probably not more than ten." Others reported waves six or eight feet high. Sand boiled in the newly exposed mudflats, and ridges of mud were formed. The bottom shifted twenty-five feet to the southwest and filled what had been a deep channel. Piers cracked and bent.

At Point Reyes Station the 5:15 A.M. train for San Francisco was getting ready to depart. It lurched to the east and then toppled over to the west, the crew just managing to jump to the ground at the last moment. Brick chimneys, houses, and barns collapsed. Cows standing in barns and early-morning milkers were thrown to the ground. Gilbert wrote in two places in his report that a cow had fallen into the fault, an incident that was widely reported in the press. He recounted the story thusly:

> During the earthquake a cow fell into the fault-crack and the tail had disappeared, being eaten by dogs, but there was abundant testimony to substantiate the statement. As the fault-trace in that neighborhood showed no cracks large enough to receive a cow, it would appear that during the production of the fault there was a temporary parting of the walls.

Indeed, the testimony was understandably voluble, for the locals took the gullible scientist and press for a ride.

Gilbert was a more reliable observer of natural phenomena that could be verified firsthand by on-site inspection, such as the numerous landslides along Inverness Ridge: "There were many landslides of the dry type on hillsides, masses of earth and rock breaking away on steep slopes and tumbling to the bottom." The landslides blocked roads, as they would three-quarters of a century later during another natural disaster.

The patterns of seismicity vary along the San Andreas Fault. From north of San Juan Bautista to Point Arena—the northern segment of the fault that slipped in the 1906 earthquake—the fault is locked. This means that the land moves only during infrequent earthquakes of large magnitude. The two plates are also held tightly in place from Cholame, where the 1857 Fort Tejon earthquake originated, south through the Carrizo Plain to Big Bend. In the central section, from Cholame to San Juan Bautista, there is a steady, slight creep.

Within a twenty-mile stretch from Cholame to just north of Parkfield, six moderate earthquakes have occurred at fairly regular intervals since 1857. It was at Parkfield, the self-proclaimed "Earthquake Capital of the World," that scientists made the first formally endorsed earthquake prediction in the United States and located their earth-measuring devices in hopes of catching an earthquake in the act. And it was also here, in prototypical ranching and earthquake country, that actor James Dean met his fate.

Following 1857, there were earthquakes in 1881, 1901, 1922, 1934, and 1966. The last three quakes, for which there are seismological records, were virtually identical, all of them being of an approximate magnitude of 6 on the Richter scale. To the people on the ground, however, there were differences. Scientific and historical descriptions of the activity along the fault measured magnitude, but in different ways.

Anecdotal reports concerning the 1881 earthquake indicated an intensity similar to the more recent ones. Two weeks after the quake the following letter appeared in the Salinas newspaper from a Parkfield resident:

> On the 1st of this month we had seven shocks of earthquakes. The first were two very hard ones; they knocked down several chimneys, one adobe store room and one end of an adobe barn. I counted thirty quite large cracks in the ground running across the roads. It also opened several springs of water on Mr. P's ranch, one I noticed between the house and the road boiling up quite strong; just back of the house, it started sulphur springs and just where those sulphur springs are, the ground, about 20 paces square, is sunk about 4 feet.

The March 1901 earthquake was felt over an area of forty thousand square miles. Vertical motion of a foot and surface cracks were reported in the fault zone. C. W. Wilson, a resident of Cholame Valley, wrote his mother:

> Well, Ma, we have had a terrible shaking up down here. Last night at 20 minutes of 12 o'clock there was the heaviest shock

of earthquake I ever felt—my bed was jerked out in the middle of the floor, nearly all the goods in the store was on the floor in an instant and poor old Earth trembled and groaned like some person in great agony, and at intervals ever since has rumbled and shook. Daylight revealed a scene of destruction. All the chimneys in town were shaken down and the ground is seamed for miles, they tell me. Half the people around seem half scared to death. Lou Fisher was here this morning and said she felt 38 distinct shocks. And she like myself was not scared, only, as she said, it seemed queer to feel that what you had under you could not keep still.

In March 1922—twenty-one years after the last quake—the next temblor struck near Cholame. It was somewhat stronger and was felt over a one-hundred-thousand-square-mile area. This was how the scientists described the damage: "Pipelines crossing the Cholame Valley were ruptured by this shock, and considerable damage was done to existing structures in the Valley."

Buck Kester was two years old in 1901. He would experience the four earthquakes of this century. In 1922 he was working for Ben Carr, driving a ten-horse plowing team. Donnalee Thomason, a local historian, recalled what happened:

> The Carr house sits right on the San Andreas Fault. As Buck recalls, the shake arrived at around 3 A.M. Buck said he could hear dead limbs falling off the cottonwood trees, also the Carr children screaming from fright. He relates that a tramp had come the night before and had asked to sleep in the barn. The next morning the horses had stampeded and were gone from the barn—so was the tramp!
>
> Evelyn Fretwell Carr tells a story about Buck and the 1922 quake. It seems he was sleeping in a little room attached to the Carr house. The next day Buck made the remark, "I needed spurs to ride the bed."

While the first three earthquakes were about twenty-two years apart, the 1934 quake came ten years early. There was another distinguishing feature—the magnitudes recorded at the University of California at

Berkeley and the California Institute of Technology at Pasadena differed widely. Thomason was nine years old at the time and was participating in an end-of-the-year school program. She recalled:

> I remember being thrown back and forth against the walls of the narrow runway behind the stage. It seemed the hall was turning upside down there in the darkness for a few seconds. I could hear people screaming and trying to run to the exit, falling down of course. The program came to a halt for the second time that evening. People stood around this time, discussing what they should do. It was decided that the show must go on!

The epicenter of the 1966 earthquake was close to the northern limit of faulting of the 1857 quake; and once again just to the west of the Diablo Range, a series of fault-folded hills dominating the east side of the valley. By this time numerous strong-motion measuring devices had been implanted in the area. The quake, felt over a twenty-thousand-square-mile area, was "similar in many respects" to the 1934 disturbance, the scientists noted. It was not the same to Thomason, who was now forty-one years old. She wrote:

> Between 8 and 9 P.M. on a hot Monday evening in June the heavy foreshock arrived, and there was no rumble. The very first thing my ears recorded was similar to a great drawing in of a breath, or a suction sound may be a better description. Then a blast of hot air hit my back as the shock wave rushed through.
>
> I tried to hang onto the door frame. I looked toward the ceiling in the kitchen. To this day I do not see how walls could have buckled that much and then snapped back into place. Dust was flying everywhere.
>
> What I didn't see was the door coming straight at me from the right; it hit me hard enough to knock me down on one knee, but I kept trying to see and record in my mind what was happening. Looking at the china cabinet, I was shocked to see its latched doors burst open as the wall buckled. The colored glassware shot straight out of the cabinet and stood in mid-air.

> After a few seconds all the pieces crashed and shattered on the kitchen floor. What a noise all this made, what with both fireplaces going down and boards popping and cracking, and items falling and breaking all over the house. I glanced at the floor and felt immediate motion sickness. At this time the electricity went off and everything was dark, but the noise continued.

Elsewhere, oak trees fell, and tombstones toppled over in the Parkfield cemetery.

Scientists took note of the regularity of earthquakes and in 1985 proposed a specific recurrence model for quakes in this region. They predicted that the next one would occur before 1993. One of those scientists was Allan G. Lindh of the U.S. Geological Survey's Menlo Park, California, office. Lindh, in the tradition of geologists like Gilbert, sported a full beard. He was chief seismologist in 1993, by which time the predicted earthquake had not yet occurred. The odds of a major earthquake taking place within the lifetime of a Californian, he said, were somewhere between one in two and one in ten.

What were the odds of James Dean being killed in the San Andreas Fault Zone? Dean's chaotic life came to an end where all that seismic activity was tearing the earth apart. It was a California life, a California death, and a California life after death.

Dean was on the cusp of making it big in Hollywood in September 1955 at the age of twenty-four. He had just finished filming *Giant* with Elizabeth Taylor and Rock Hudson, *Rebel Without a Cause* was ready for release, and *East of Eden*, an adaptation of a Steinbeck novel, was already a success. With the money beginning to flow in, his life seemed to be headed in a number of directions at once. Money managers were hired and life insurance policies were considered by that side of him that sought stability. The side that sought speed and danger had just purchased a silver Porsche 550 Spyder, a serious racing car. Internally, Dean was struggling with his sexual orientation.

There was to be a road race at Salinas, where Dean had been on location for *East of Eden*, and he was up early on Friday, September 30. He watched the Porsche mechanic, Rolf Wütherich, check the new car in

the Hollywood garage of Competition Motors. With two friends following in a station wagon towing a car trailer, Dean and the mechanic set off
in the Porsche. The actor wore his regular outfit of faded blue jeans,
white T-shirt, red nylon jacket, and dark lenses clipped onto his regular
eyeglasses.

He sang, smoked cigarette after cigarette, joked, and drove very
fast—so fast that shortly after crossing the San Andreas Fault where it
bisects Highway 99 (now Interstate 5) north of Los Angeles, he got a
ticket from a California Highway Patrol officer on the long, steep grade
known as the Grapevine that ends in the San Joaquin Valley.

Dean turned west in the late afternoon of that hot fall day on Highway
46, a straight shot across the flat valley floor to the distant Diablo Range.
Wütherich later recalled, "It felt like driving on an endless ruler." At the
remote intersection called Blackwells Corner they stopped at a store.
There Dean bought some apples and bumped into an acquaintance,
Lance Reventlow, Barbara Hutton's son, who was also on his way to the
road race. They talked about cars and admired Dean's new purchase.

Dean headed west again into the setting sun, the distinctive whine of
the four-cam, air-cooled Porsche engine reaching a climax of revolutions
before he shifted. The car quickly climbed the 1,100-foot grade to Polo-
nio Pass and then dropped into the Cholame Valley at speeds later estimated at between eighty and eighty-five miles per hour.

It was nearly six P.M. The day had begun at the garage at eight A.M.
The blinding sun, heat, speed, fatigue, and the torpor of straight roads
were all operating on Dean. For the California Polytechnic College student named Donald Turnupseed who was driving east in a two-tone 1950
Ford toward the Central Valley and a weekend at home, there was a left
turn to be made onto Highway 41 and a serious miscalculation in the
speed of the oncoming silver bullet. "It was," according to Car and
Driver magazine, "the most famous highway accident in history."

It was also the violent convergence of two very different young men of
approximately the same age from two very different Californias. Given
the solidity of the steel Ford and the fragility of the aluminum-frame
Porsche, the student suffered minor injuries and a stigma that would last
a lifetime. Dean's neck was snapped—instant death at high speed, the
coroner ruled. A weekly San Luis Obispo County newspaper lumped two
fatal accidents together and gave Dean's a one-paragraph description
under the headline COUNTY HIGHWAY DEATH TOLL REACHES 28.

Where Dean died, the vibrant green grasses and wildflowers of spring had become the dull yellow pastiche of early fall. The arid scene, softened by the light of the sun setting over the Cholame Hills to the west, was dotted like a pointillist painting by stands of valley oaks and cattle. Underneath the landscape of earthquake country, the two plates rubbed against each other in ways incomprehensible to humans. One of Dean's favorite quotes was "What is essential is invisible to the naked eye."

A myth was born in a state that supported similar myths, whether their last names be Valentino or Monroe; there was life after death. The student, who eventually settled in the Fresno area, refused to talk about the accident. The mechanic, who recovered from serious injuries, returned to his native Germany, where he remained obsessed by his part in the drama. He was killed in an automobile accident in 1981. Dean's crumpled car was put on tour by the Greater Los Angeles Safety Council and was stolen, never to be seen again. Perhaps the Porsche was cut into pieces for souvenirs, or sits in the garage of some devotee, or was stripped for parts and shipped to Japan as scrap metal.

It was a wealthy Japanese fan who built the elegantly simple memorial to Dean in the nearby hamlet of Cholame, just to the west of the fault. A procession of 1950s vintage cars wends its way from Los Angeles to Cholame on September 30 each year. On any given day, numerous motorists on the coast-to-valley road stop at the gleaming, stainless steel–and–aluminum monument surrounding an ailanthus tree, known as the tree of heaven. The day that I was there, during a light spring rainstorm, a young couple in a silver Porsche stopped. He took a picture of her reading the inscription that states:

> James Dean is all the more with us today because his life was so fleeting. In Japan, we say that his death came as suddenly as it does to cherry blossoms. The petals of early spring always fall at the height of their ephemeral brilliance.
> Death in youth is life that glows eternal.

In the nearby restaurant, which constitutes one-half of downtown Cholame (the other half being the post office), a framed newspaper article on the wall told of lawsuits filed by Dean's heirs against those who sought to appropriate the actor's image without permission.

My first encounter with the San Andreas Fault occurred when I lived in Belmont on the San Francisco Peninsula. I worked for the *San Carlos Enquirer,* the weekly on which I got my first newspaper job shortly after arriving in California in 1960. Newcomers were flooding the peninsula at that time. It was my job to put together a "get acquainted" edition, which meant taking photographs of merchants, from which an artist sketched a likeness displayed in the advertisement that I had sold. Strangers would then recognize one another.

The fault was over one ridge to the west, and I drove that way on summer evenings, discovering California. Six months there—followed by two years in Turlock—and I returned to the Bay Area to live in Sausalito while working for the *San Rafael Independent-Journal* in Marin County. Weekend excursions north of the Golden Gate took me to the Point Reyes Peninsula, where I returned to live in 1977 after time spent in Los Angeles and Sacramento.

I lived in a four-level pole house atop Inverness Ridge; I wrote books and articles for *Audubon* magazine in my top-floor office. The poles were

The monument and the Porsche

sunk in the granite of the ridge top. The house was a perfect seismograph, registering every quake much like a tuning fork.

I was acutely aware of the presence of natural disasters in California, having witnessed my share as a journalist throughout the state. Fires, floods, and earthquakes were very much on my mind in the first home that I owned. My next-door neighbor and I consulted on what to rescue from each other's dwelling in case one of us was absent during such a disaster. In cities they worried about crime, and organized neighborhood watches.

Most of the peninsula, in effect my backyard, was included in the Point Reyes National Seashore and thus was a unit of the national park system. Its beauty had been officially recognized and preserved. That and the excellent libraries in nearby Berkeley were why I had moved there.

The end of the 1970s and the early 1980s was a dark time in the San Francisco Bay Area. We were visited by a plague of violent deaths: More than nine hundred people, mostly from the Bay Area, docilely and knowingly swallowed a poisoned drink and died gasping in Jonestown, Guyana. A Bay Area congressman who had flown there to investigate the People's Temple was assassinated. Back in San Francisco a supervisor ran amok and killed another supervisor and the mayor. And on the shores of Tomales Bay, Synanon, which had made its reputation curing alcoholics and drug addicts, was stockpiling weapons, intimidating outsiders, and turning its members into zombies.

Marin County, a larger peninsula to which the Point Reyes Peninsula is tenuously attached, was known for its affluence, its affectations, and San Quentin Prison. The phrase that was customarily applied to the oddities of behavior displayed in the county (which had gained a national reputation for such mannerisms, mainly through Cyra McFadden's book *The Serial*) was "only in Marin." It was in this balmy place that people like myself had sought refuge from the harsher realities of life, namely those found on the urban streets of America.

Yet it was here, in what we thought a sanctuary, that blood began to be viciously shed on the dirt trails that wove through the suspect terrain. It was murder in the cathedral come to California. We had been blessed, and now we were cursed. A paradise turned into a nightmare, especially for women hikers who enjoyed the solitude of nature.

There were three murders on the slopes of bucolic Mount Tamalpais within fifteen months. In August 1979, the body of a middle-aged woman was found in the position of a supplicant on the slopes of the mountain that dominated the eastern and southern portions of the county. She was shot in the back of the head, execution style, by a large-caliber weapon. A hooded man had been seen nearby. In March 1980, a twenty-three-year-old woman was hiking with her Labrador retriever on the mountain. She was hacked to death by a man yielding a large boning knife, while her dog barked. A witness described the attacker as a lean man with a beaklike nose. In October a third woman, described as blond and quite attractive in the California way, drove to a point near the top of the mountain and got out of her car to admire the sunset. She seemed reflective to those who last saw her. The woman was reported missing, and her body was discovered a few days later. She had been raped and then shot in the head.

A detective who participated in the searches said, "I found myself always looking up because I envisioned him, whoever he was, sitting high in the rocks, like some sort of vulture." A park ranger would later liken the killer to a spider, who suddenly dropped from above to fatally sting its prey.

Fear ruled what appeared to be a gentle land. The killer was endowed with cunning and extrasensory talents by law enforcement officers, who were neither accustomed to getting out of their patrol vehicles nor dealing with violence in the wilderness. As it turned out, the killer was also an alien in this environment. In retrospect he was quite careless: He dropped his prison prescription glasses, got rid of the knife carelessly, and was treated in a hospital for cuts, after which he was questioned by police. But it would take some time, and more deaths, before he was linked to the killings.

The Marin County Sheriff's Department, within whose jurisdiction the murders took place, thought they might be connected. Serial killers were not unknown in California, although they had never before crossed the Golden Gate Bridge, other than as prisoners destined for San Quentin. California, as it so often did in other categories, led the nation in serial killers—a fact, like earthquakes, that was not trumpeted.

The Encyclopedia of Serial Killers, whose British authors said it was outdated as it was being printed in 1992, lists one hundred and four such murderers in the United States and thirty-four in England, the second highest total in the world. California, with 12 percent of the nation's population, accounted for 27 percent of this country's serial killers, with Southern California the location for the vast majority of these crimes.

California serial killers have, at times, held the U.S. record for number of victims. One such record holder, with twenty-five deaths to his credit, was displaced by two male killers. They worked in tandem and accounted for thirty-two male victims, who died after being sadistically tortured and raped.

The encyclopedia, written by two crime experts, is gory reading, particularly those entries that deal with California crimes. These killers go by names such as Juan Corona and Charles Manson, and nicknames including the Candyman, the Hillside Strangler, the Night Stalker, the Freeway Killer, the Sunset Strip Killers, Zodiac, and—as the Marin murderer came to be known—the Trailside Killer.

Then, suddenly, there was no doubt that a serial killer was lurking in the forest near my home. On a late November day in 1980, the friends of two women reported them missing. When last seen, they were hiking on Sky Trail in Point Reyes National Seashore. A search party set off the next morning. A high-school teacher with wilderness skills stumbled upon the two badly decomposed bodies of a couple who had been missing for six weeks. A short distance away were the two nude bodies of the women, startling white against the backdrop of dark green vegetation. The four were killed by the same gun that had ended the life of the blond woman on Mount Tamalpais, just to the east.

The high-school teacher, who frequently brought his classes to the national seashore to walk the earthquake trail and forested paths, said, "Point Reyes is a cathedral; and to me it was a strange, bizarre crime. It was almost like someone coming in and shooting this couple that had gone to church." A park ranger had been killed by a deer poacher in 1973, but longtime residents could not recall any other murder, and certainly no multiple murders, on the peninsula.

The National Park Service, acting with unaccustomed speed, posted Day-Glo orange signs that read:

WARNING FOR HIKERS:

Because of the tragic discovery of four bodies in the park on Nov. 19, 1980, *please do not hike alone*. Everyone should hike/camp with at least one other person and they should stay together as much as possible. Women should be especially cautious, and under no circumstances travel alone.

Fear and anger gripped our small community and contorted our lives. Women accused men of being violent creatures; men postured, admitted violent behavior, or were silent. A tourist who passed through at the time accused us in a letter to the editor of having it too good, of living in the midst of too much beauty. Thus was guilt added to the mix. Innocence had already been lost.

We armed ourselves and suspected—and in some cases chased—any strange male. I was a single man at the time and could feel people stiffen when I approached, particularly when I hiked alone—a practice that I continued. I was not going to let that bastard take my park from me.

There actually were two malevolent spirits loose in the national seashore; there had been a series of burglaries of houses that bordered the park, including mine. Burglaries had been unknown, and we usually left our doors unlocked.

The thief stole into our homes, sometimes while we slept, and helped himself to camping equipment, food, and shiny baubles. We knew who he was—a local boy who had taken a wrong turn. He had been convicted of a burglary and sent to a halfway house, where he didn't tarry. He lived in the woods like an Indian. Occasionally someone would find one of his abandoned campsites or catch a glimpse of him.

I came home one night to find the toaster and a brandy bottle out on the kitchen countertop. He had used a heavy silver candlestick to light the scene of his brandy-and-English-muffin feast. Then he walked out of the house with camping equipment, warm clothing, a double-barreled shotgun, and jewelry that had belonged to my grandfather. He slept in the woods that night on the other side of the dirt road. A neighbor came across him in the morning in my sleeping bag, with the shotgun by his side. By the time the sheriff's deputy arrived, he was gone.

I felt violated and angry. I borrowed my son's dog and combed the top of the granite ridgetop looking for a burglar or a killer. For all I knew, they were one and the same. Others were out there, too. A neighbor heard me and fired his rifle in one direction, hoping to flush me toward him. I took a chance and yelled, "John! For Christ's sake, it's me." John emerged sheepishly from behind a tree. A friend was also out hunting with his dog, and we managed to track each other down in the relict forest. Fortunately, no shots were exchanged that time.

There was some humor involved in middle-aged men turned manhunters. I was a writer, my neighbor was a lawyer, and my friend was an artist. It was that kind of community. In retrospect, we were ridiculous, at the very least, and potentially quite dangerous to one another. I now see the hunt as a form of vigilante justice, an activity that was not unknown in California.

The burglar was captured in early December in San Francisco, where someone thought he fit the description of the killer. The killings had moved south along the fault line to Santa Cruz County. The killer was said to be a brilliant tactician. When he was finally captured in May 1981, he turned out to be a very ordinary looking man named David J. Carpenter, who stuttered and wore glasses. Hikers returned to the trails of Marin County. Carpenter resides on Death Row in San Quentin, where he has a view of Mount Tamalpais. From that angle the mountain resembles a reclining woman.

Besides serial killers, California was known for its notorious prisons: namely Alcatraz, San Quentin, and Pelican Bay. Alcatraz was closed and San Quentin, while still the site for all executions, no longer housed the state's most dangerous prisoners. That distinction was now reserved for Pelican Bay, which sounded like a summer-home colony. The name, however, disguised a harsh reality. Author Mike Davis called it "an Antarctica of solitude moored to the picturesque Redwood Coast."

As I write, crime is the main concern of the current governor and legislature. Carpenter had a prior prison record, as did the burglar. The solution being considered involves tougher laws, which will mean a larger prison population and more prisons. But that is not a new remedy in California. Pelican Bay was part of a previous administration's similar quick fix—a solution resulting in as much money being spent from the general fund for prisons as is allocated for higher education. From 1980 to 1990, when the state's population was increasing by 24 percent, the prison population increased by nearly 300 percent. Violent crimes, such as murder and rape, accounted for 40 percent of all offenses.

In the pantheon of prisons, fog-shrouded Pelican Bay stood out, much as Alcatraz did in its day. When the governor traveled to the forested clearing just below the Oregon border in 1990 to dedicate the new state-of-the-art prison, he said that it would serve as a model for the rest of the nation. The high-tech prison was designed to deprive the state's most dangerous prisoners of any stimulation. It was modeled on the concept of "the hole" or solitary confinement, as isolation techniques were once called. With the aid of computer technology, electronic devices, and modern building materials, true isolation for longer periods of time could be achieved.

There are no signs pointing the way to Pelican Bay, where there are neither pelicans nor a bay north of Crescent City on Lake Earle Drive. Nor are there many visitors, as most of the prisoners are from the Los Angeles area, and the prison, located at the far end of the state, is not accessible by public transportation. William Brewer visited the area in 1863. He wrote:

> There are some lakes in here, beautiful sheets of water. I went out to one—with grassy swamps around it and rushes and reeds growing up in the shallow margin. Dense forests surrounded it.

The dense stands of redwoods in what was known as the Malarkey Forest were logged in the 1950s, leaving a second-growth forest to sprout in their place. Depressed Del Norte County welcomed the prison. Two hundred forty acres were cleared of all vegetation and the prison facilities were constructed on one hundred acres, leaving a wide buffer strip partly shielded from the road by a thin stand of trees. The blue guard towers, which resembled inverted flashlights, were set at regular intervals around a perimeter consisting of two parallel chain-link fences topped by razor wire. Inside were the slate-gray concrete cell blocks. No one was visible inside the perimeter the day I paused there.

Pelican Bay's sterile tone was Orwellian. Gone were the quaint fortress effect and warm earth tones of San Quentin. Pelican Bay could be a bomb factory, for all a stranger would know. It was vulnerable, however: The prison was built to withstand most earthquakes, but there is the possibility of the ground turning to jelly in a large quake—a potentially devastating process called liquefaction. Killers would then be set loose upon the vibrating land. The Del Norte and Grogan Faults, which are hooked into the San Andreas Fault, are located offshore.

After the prison was opened, all was not as secure as planned inside the supermax, as it was called. The prison had three categories of inmates: those in isolation, called the Security Housing Unit (SHU); maximum custody general population; and minimum custody. An information sheet for Pelican Bay stated, "The SHU is a one-of-a-kind design for inmates who are management cases, habitual criminals, prison-gang members, and violence-oriented maximum custody inmates."

It didn't take long for the clever prisoners to learn how to pop open the pneumatically operated doors. They were replaced with mechanically locking doors at a cost of $6 million. Inmates fashioned weapons out of wood, plastic, or aluminum foil honed into sharp objects and attacked other prisoners. Guards killed inmates. There were escapes from the minimum security portion of the prison, and murders, suicides, and racial fights.

Compared to California's population and the general prison population, there were extremely high percentages of Latinos and African Americans in the SHU unit: The two races accounted for 80 percent of its prisoners. As Mike Davis pointed out in his book *City of Quartz,* the specter of a brown and black criminal underclass had replaced the red menace in California.

Stories about the brutality of the place, both inflicted by guards and by the lack of stimulation, began to emerge and filter outward. An American Civil Liberties Union official said, "Many of us feel that Pelican Bay is the most frightening supermax built to date." A suit was filed and a trial held in Federal District Court in San Francisco. "When you're in Pelican Bay you're a caged animal," said a psychiatrist on behalf of the inmates. State prison authorities replied that wild animals needed to be caged.

I visited San Quentin when I was a reporter for the county newspaper and interviewed prisoners for feature stories. Minor prison breaks were covered by telephone. Fortunately, there were no major disturbances or executions during my watch. When I returned to Marin County, I regularly passed the prison on my way back and forth from Berkeley, where I taught at the university. I marveled at how easy it was to ignore the presence of the prison when living in this sylvan place. It was also easy to disregard the violent side of nature.

Some local residents who informally grouped themselves under the name of the Tomales Bay Explorers Club gathered on the Great Beach, a twelve-mile stretch of sand along the outer coast of Point Reyes Peninsula, on New Year's Day of 1982. They sat around a bonfire, eating and drinking. A large, dark cloud slowly floated toward the peninsula. They took note, became alarmed, and then, quickly gathering up children and belongings, fled in a long, straggling line to their cars, parked one mile distant.

It seemed like an eternity in time. Wind and hail tore at their backs. One of them, Michael Sykes, later wrote:

> *And the lightning crackled overhead, around our fragile souls,*
> *blooming on the earth not ten feet away—*
> *a glowing sphere of dense magnetic streams*
> *that threw an instant cloud of fiery, white light*
> *across our hearts and made the hair*
> *on the backs of our necks stand on end.*
>
> *In this manner we made our way home,*
> *one fractured ankle among the lot and no great harm*
> *done, save what we soon surmised:*
> *This was the forefront of a steady frontal assault,*
> *gathered in the basin of eternal rains,*
> *which beat against the rooftops of our lives*
> *until they soon collapsed, within seventy-two hours,*
> *from the constant drone. Entire hills*
> *resigned their contents to the sea,*
> *valleys rushed like wind into the bay,*
> *houses caught the sweep of quick debris*
> *and gathered crisis in their arms for all to hold.*

A warm, tropical air mass loaded with moisture, the product of a periodic warming trend generated by a condition known as *El Niño*, had collided with a cold Arctic front, a more regular winter occurrence. It rained hard for three days; then it rained harder (twelve inches in twenty-four hours) and harder still (four inches in one hour). Inverness Ridge was clogged with vegetative matter accumulated over years of fire suppression and watershed protection. It badly needed cleansing.

The peninsula became bloated with water. Between 11 A.M. and noon on January 4, debris flows burst almost simultaneously from the mountainsides at nearly uniform heights. The land swept down the twisting canyons. Heavy masses of debris, consisting of mud and ferns and pines and boulders and whatnot, crushed any structures that lay in their paths.

The single road leading from the peninsula was blocked by landslides. The community was isolated for days. There were no deaths, nor had there been any in 1906, when the land had also moved.

After the storm, I walked through a canyon with hydrologist Luna Leopold. The accumulated debris had been swept away, revealing the gleaming bedrock of granite into which a streambed had been incised over a period of some 2,000 years. Later, scientists would date wood fragments in these canyons and determine that such massive flooding had occurred as far back as 46,500 years ago and may have been repeated at intervals since then, varying from 30 to 2,000 years.

To know that the violent rearrangement of the landscape was part of a natural cycle was, for me, a tremendous relief. I felt a sense of renewal, symbolized by the appearance of the granite to which I periodically returned and caressed. The weight of the clotted growth was lifted. The dark times that had left a stain on the land were washed away.

The cycle continued. A dozen years later the vegetation had returned, and the granite was no longer visible. There are many new people living here who do not know of Synanon, the murders, the burglar, the storm, or the fault. They see this place only as it appears to be at the present moment. (In January and March 1995 the torrential rains and floods, spawned by the same conditions, returned to Northern California.)

Shortly after I was married in 1988, the house on the ridge top burned to the ground, and my wife and I lost all our belongings in the fire. No material trace of our prior lives existed. We were rootless in California. We left the fog and mountainous terrain on the west side of the fault and sought the sun and flat ground on the east side.

We were again reminded of our tenuous existence on October 17, 1989, when the land moved violently along the fault line just before the third game of the World Series was scheduled to start in San Francisco. It was the first earthquake that was broadcast live on television. The immediate aftermath was also well documented. Tom Brokaw sprayed his hair into place before sitting down to anchor the news, with the wreckage serving as the background for the NBC broadcast.

Sixty-two people were killed, thousands injured, and there was more than $6 billion in damage, making the Loma Prieta earthquake one of the costliest natural disasters in the nation's history at the time. (A hurricane in Florida and the 1994 Los Angeles earthquake would supersede it.) For the San Francisco Bay Area, the Loma Prieta quake was the largest since 1906.

On the road, now paved, where the movement had been greatest in 1906, there were just a few cracks. There was little damage this time along the shoreline of Tomales Bay. The pictures on the walls of our rented house were knocked askew. The next morning, as previously planned, I flew to New York, where television cameras at the airport panned over the arriving passengers as if we were aliens from a strange land.

After the earthquake, we built a home on the edge of a bluff with a marvelous view back across the fault zone toward the dark-green ridge where we had once lived. Our new house stood on forty-two concrete piers sunk twenty feet into the ground; we carried earthquake insurance. Scientists calculated in 1990 that there was a 2 percent chance of a major earthquake within the next thirty years along this segment of the San Andreas Fault System. For the Bay Area as a whole, the thirty-year probability was 67 percent.

Yet the majority of California residents were relatively unconcerned. Researchers, working under a National Science Foundation grant, surveyed Northern and Southern Californians before and after the Loma Prieta earthquake. They concluded:

> Some California pundits have said that it takes an earthquake to induce California residents to pay attention to the earthquake hazard. In 1989 California experienced just such an earthquake. Yet a majority of our survey respondents continue to eschew investment in earthquake insurance or in any of the many measures they could adopt to reduce their vulnerability to earthquake-related damage.

After surviving these catastrophes, and others that preceded them in Southern California—toward which I now turned my camper—I came to the conclusion that there were only three options: depart California, take up religion, or learn to live and enjoy one day at a time. For various reasons, I ruled out the first two alternatives. I am working on the last. We are criticized by outsiders for living this way, but I see it as environmental adaptation.

VII

The Profligate Province

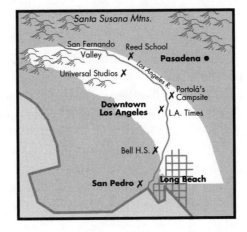

The first Europeans to cross the Los Angeles Plain marched to the cadence of muffled drums from within the unstable crust of the earth. The Portolá expedition of 1769 experienced its first earthquake at the Santa Ana River, designated the *Rio de los Temblores* by the Spanish. "It lasted about half as long as an Ave Mariá, and about ten minutes later it was repeated, though not violently," wrote the expedition's diarist, Fray Juan Crespi.

The Spaniards and mixed bloods and Indians from Mexico continued across the plain. They walked or were mounted on horses and were accompanied by a long pack train. They drove spare horses and mules that left extensive droppings behind them. The scouts, the main force, and the rear guard creaked and sweated in their leather armor; their metal parts jangled. They carried lances, swords, shields, and guns and alternately cursed and prayed in the hot summer sun. The expedition stretched for miles. It was a formidable apparition to the native population.

They passed through a beautiful and fertile country that was periodically scoured by earthquakes, fires, and floods. These natural disasters set

315

the erratic rhythm of the violent history that was to follow. They rode across "a broad and spacious plain of fine black earth with much grass, although we found it burned." Crespi's diary was full of such contradictions.

He continued: There were "many green marshes, their banks covered with willows and grapes, blackberries, and innumerable Castilian rose-bushes loaded with roses." They ate the watercress growing in the streams and noted that "a goodly portion of the valley" could be easily irrigated. "In the afternoon we felt another earthquake."

On July 31, they fought their way through thick brush and low woods. Leather chaps protected the legs of the riders, while the horses and Indian slaves were scratched and gouged by the underbrush. They felt another earthquake. The next day they decided to rest and celebrate the Jubilee of Our Lady of Los Angeles of Porciúncula. Mass was said and they took communion.

Crespi then noted: "At ten in the morning the earth trembled. The shock was repeated with violence at one in the afternoon, and one hour afterwards we experienced another." Game, such as antelope, was plentiful on the alluvial plain that shook like jelly.

They continued their *entrada* the next day, Wednesday, August 2, and encountered "a very spacious valley, well grown with cottonwoods and alders, among which ran a beautiful river from the north-northwest, and then, doubling the point of a steep hill, it went on afterwards to the south." They halted for the day on the bank of the river, which they designated the Porciúncula, the name of a small Italian town where the Church of Our Lady of the Angels was located. It was there that St. Francis of Assisi, the patron saint of travelers, prayed when the Jubilee was granted to him.

Nothing was to last long in this region. The sonorous name was replaced by "Los Angeles River." It was the first of many rivers eventually tapped to succor a settlement, then a city, and finally a region that became another reincarnation of the Califia myth, or so the land agents claimed in their honeyed phrases. Water would make the golden-hued inhabitants fabulously wealthy, the modern fable went. The rivers were then killed so that the people could live beside the dead streams.

The drums of violence continued to resonate from deep within the earth. "Here we felt three consecutive earthquakes in the afternoon and night," related Crespi. The explorers noted evidence of "great floods" where logs had been stranded high on the banks of the river.

The view toward downtown Los Angeles from the spot where the Portolá expedition camped

Indians visited the encampment, and the Spanish traded gifts with "the heathen." The next day they forded the river, and the Indians "began to howl like wolves" when the explorers approached their settlement. They spurned the Indians' gifts of seeds and marched on with some trepidation. "This afternoon we felt new earthquakes, the continuation of which astonishes us," wrote Crespi. They thought there were volcanoes nearby because of the earthquakes, the presence of the Santa Monica Mountains, and the "boiling and bubbling" pitch in the La Brea tar pits, where prehistoric animals and humans were entombed within the thick asphalt.

They crossed the mountains via Sepulveda Canyon on August 5 and found "a very pleasant and spacious valley," now known as the San Fernando Valley and part of the sprawling City of Los Angeles. The valley was thick with trees and reeds, through which ran a small stream—the upriver extension of the Porciúncula. Three days later they passed out of the valley and left the beat of drums behind them.

Portolá's party continued north and discovered San Francisco Bay. They plodded back across the Los Angeles Plain in January 1770. Crespi observed:

> We crossed the plain in a southeasterly direction, arrived at
> the river, and forded it, observing on its sands rubbish, fallen

trees, and pools on either side, for a few days previously there had been a great flood which had caused it to leave its bed.

The Los Angeles River is now encased in concrete. Its sole source of water, when there are no winter storms, is treated and untreated toxic wastes and sewage. This travesty of a river flows sluggishly for little more than fifty miles through the most densely populated portion of Southern California. The region contains nearly 60 percent of the state's population, who live on 7 percent of its land area. *California del Sur,* the Cow Counties, subtropical California, Southern California, the Southland, or La-La Land, as this portion of the state has been variously called, extends from just north of Santa Barbara to just south of San Diego, and from the Pacific Ocean to the crest of the encircling mountains. Southern California *is* California, only more so.

It is a province of geographical extremes and great disparities of opportunities for its inhabitants; a region of quick alternations in topography and the movement of different peoples across its surface; a prodigious place where *biggest* and *greatest* are common adjectives; an area walled off by mountains and separate within a state that is isolated from the remainder of the nation. All characteristics of California are exaggerated here.

The Profligate Province resembles the blade of a scythe, which is the ancient symbol of death or the rebirth that follows the harvest—destruction or creation, take your choice. Actually, the shape of Southern California resembles more the blade of a brush hook, a local adaptation of the classical scythe. A scythe wouldn't work here—what is needed is a thicker, stubbier blade to clear the thick brush known as chaparral, the dominant fire-prone vegetation of the area.

The cutting edge of the tool is where the tapered coastal plain meets the ocean in a wide arc that takes in some ninety degrees of curvature from the east-west trending coastline at Point Conception in the north to the north-south slant of the coast at the Mexican border. The thick portion of the blade, what gives it heft, is the chain of mountains along the back edge that seals the region off from the remainder of California and the nation.

In a clockwise direction, reaching eleven-thousand-foot heights and with names betraying the early Hispanic presence, are the mountains:

Santa Ynez, San Gabriel, San Bernardino, San Jacinto, Santa Rosa, and Laguna. The Santa Monica and Santa Ana Mountains are intermediate rises, ripples in the blade that form interior valleys. Imperfectly forged and riddled by numerous faults, the heavy blade appears to be descending on the heads of the seven Channel Islands.

The volatile landscape is being constantly rearranged. From the mountain heights that can be locked in snow and ice during the winter while palm trees sway gently in seventy-degree weather along the coast, a few streams, such as the Los Angeles, San Gabriel, and Santa Ana Rivers, have carved deep canyons into the steep hillsides during their short, mad descents. As the mountains rose during periodic jolts from earthquakes deep within the earth, so have intense winter storms worn them down from above. ("Intense" can mean 11.5 inches of rain in an hour and twenty minutes.)

This land-rearranging process has been going on only recently, in terms of geologic time. Some twenty-five million years ago movement along the San Andreas Fault began carrying Los Angeles north from mainland Mexico to its present position, just across the artificial border imposed in 1848. Some fifteen million years ago uplifting began; the province began to emerge, dripping from the sea. The rise accelerated in the last two million years as the barrier mountains thrust upward.

The ascent was continuing in 1861 when William Brewer, newly arrived in California, hiked into the Santa Susana Mountains at the head of the San Fernando Valley and found fossil evidence of ancient seabed origins. He wrote:

> The ridge was strewn with them, as thick as any seabeach I have ever seen, and in as good preservation—oyster shells by the cartload, clam shells, in fact many species. I cannot describe my feelings as I stood on that ridge, that shore of an ancient ocean. How lonely and desolate! Who shall tell how many centuries, how many decades of centuries, have elapsed since these rocks resounded to the roar of breakers, and these animals sported in their foam? I picked up a bone, cemented in the rock with the shells. A feeling of awe came over me.

Meanwhile, periodic floods dumped massive amounts of alluvium on the coastal plain that is deeper in places—some 14,000 to 20,000 feet thick—than the mountains are high. When there is an earthquake, this loose sediment shakes. Destruction and death follow on the now densely populated plain.

While the formation of the province in its present form is relatively recent, there are rocks at the headwaters of the Los Angeles River dating back 1.7 billion years. At one time, long before the Portolá expedition, there was a period of volcanism.

Violence and transience in the landscape predated the arrival of the first Europeans, who found oak and willow trees, shallow ponds and marshes, grizzly bears and coyotes in the basin. The oak trees are now mainly preserved in the names of such communities as Sherman Oaks and Thousand Oaks; the water was drained to make way for farmland and homes; the last grizzly was seen eighty years ago; occasionally a coyote will attack a baby or small child within the city limits when no other food is available.

Kem Nunn, a third-generation Californian and native of Pomona, accurately portrayed the frenzied pace of change in that inland city, and elsewhere in Southern California, in his novel *Pomona Queen*. He wrote:

> The haze thickened with each passing year. And then there was the coming of the San Bernardino freeway, the end of the Korean War, the loss of jobs at the local defense plant. Suddenly many of the jerry-built tract homes were standing empty behind weed-choked yards and dying elms. Lenders had foreclosed. Buyers were needed. At which point there came the Watts riots, the Great Society housing programs. Black families intent upon escaping the ravages of south central Los Angeles could suddenly move to a tract home in Pomona for no money down. The white folk headed north. It was, in retrospect, the last act in a veritable shit storm of change begun just prior to the Second World War.

I left Southern California in 1975, after having lived and worked there eleven years for the *Los Angeles Times*. I returned during the summer of 1993 for a seven-week stay. During that time I visited old haunts as widely dispersed as Mt. San Gorgonio in the mountains, Santa Catalina Island off the coast, Santa Barbara to the north, and Ensenada to the south. I crossed and recrossed Los Angeles in every direction, using my aging van and public transportation. I was searching for history in a place where, during the most recent race war, a shopkeeper said of his burned-out store, "That's history"—meaning, here one minute and gone and forgotten the next. This dismissive phrase originated in Southern California.

I was also looking for memories of a younger self, a person who had felt the rhythms of the place through such work-related experiences as the 1964 Santa Barbara fire, the 1965 race war and its aftermath elsewhere, the first violent confrontation between police and anti–Vietnam War protesters at Century City in 1967, later antiwar and university demonstrations, the assassination of Robert Kennedy in 1968, the Vietnam War that was physically launched from California's shoreline, the 1971 San Fernando earthquake and Malibu fire, and other fires and floods too numerous to mention. Strange, I thought, that I recalled only the violent events.

My personal life was fraught with the same type of dangers. The yearly fire-flood cycle swept the hillside where our small family first lived, leaving it charred and the sunken living room of the rented home a mud hole when the rains came. Then there was the trauma of divorce. A woman I loved, who had given me shelter after the divorce, lost her home in the 1971 Malibu fire. (She sold her charred lot below the Serra Retreat. The area was rebuilt with expensive homes, and it was again ravaged by fire in the fall of 1993 and by mud slides in early 1994.)

When the land shook, my apartment on Point Fermin teetered on a slide-prone cliff above the ocean. My aged parents in New Jersey frequently telephoned and asked, in voices filled with puzzlement and fear, "Are you all right?" I became a true Californian during those years.

On my return, the most striking change was really no change at all. Rather, it was the latest manifestation of the ongoing cycle of people of different tribes and ethnicities rapidly and violently displacing other people. When I left Los Angeles, my tribe—non-Hispanic whites, in census jargon—was numerically dominant. Now it was just about even with Hispanics, and, when others were factored in, it was a minority. The Anglos were about to fall behind the Latinos, who had been a minority for many years after losing their majority status in the last century when conquered by the whites. The Hispanics had displaced the Gabrielino Indians, who in turn had forced out an earlier culture. So it had gone throughout California.

While representing only a little more than one-third of Los Angeles County's population, whites retained political power and economic hegemony in the early 1990s; that foretold trouble in future years. The Hispanics (40 percent), Asians (10 percent), and African Americans (10 percent) who made up most of the population wanted their rightful shares, too, but they were far from being a united force.

There were wider implications. The County of Los Angeles, whose nine million inhabitants represented 30 percent of the state's population, was the indicator county for California, where whites would soon become a minority. Other states faced the same prospect, but California was the most ethnically diverse and the most fractious. In Los Angeles County alone, there resided the largest Mexican, Salvadoran, Guatemalan, Armenian, Korean, and Filipino communities outside their native countries. Some eighty languages other than English were spoken by schoolchildren. The rich were getting richer, and the poor were getting poorer. There were communities with median family incomes above $100,000 and those with median incomes below $20,000.

There had been numerous, drastic shifts in Los Angeles's human content before the last decade. A population expert wrote, "Demographic changes have always played a key role in its development. Streams of immigrants have fueled its soaring economy." They included, in their times: Native Americans, Spaniards, Mexicans, Chinese, Anglos, Japanese, Mexicans, African Americans, more Mexicans, more Asians, and Central Americans.

The evidence for the most recent displacement was graphically presented in a chart derived from 1980 and 1990 census data and prepared by the Marketing Research Department of the *Los Angeles Times*. The dislocation between 1975 and 1993—the period of my absence—was actually greater, but the census data over that ten-year time span still told the story and set the demographic tone for my journey down the Los Angeles River. The captions for the series of multicolored maps representing changes in the areas of residence of four races over ten years noted in militaristic terms:

The county's multi-ethnic population was on the move throughout the past decade.

White: Shrinking by just over 8 percent, the White population staged a broad retreat across Los Angeles County, declining in all but 19 of 163 communities measured by the census.

Hispanic: The Hispanic population grew throughout the county, gaining population in 148 of 163 communities.

Asian: From the Palos Verdes Peninsula to the Wilshire district to broad stretches of the San Gabriel Valley, the population of Asians and Pacific Islanders, once thinly distributed across the county, is increasing its visible presence.

Black: While the county's Black population grew by only 1 percent during the 1980s, much of the Black population redistributed itself, abandoning the central area and fanning out into suburban communities where Blacks traditionally have not lived in great numbers.

The effect of the splotches of color on the maps, displayed on the wall of the *Times*'s newsroom (along with detailed instructions about what to do in case of an earthquake), and the impression I gained from my *entrada* across the county, were of various enclave civilizations whose tribal members barely mixed. In the offices, classrooms, corridors, and cafeterias of public agencies and public schools, where affirmative action was most rigorously practiced, people of different races worked together, but they gossiped, ate, drank coffee, played, and lived separately.

Novelist Charles Bukowski watched his Hollywood neighborhood change from lower-class white to lower-class brown during that decade. He wrote:

> My tax consultant had suggested I purchase a house, and so for me it wasn't really a matter of "white flight." Although, who knows? I had noticed that each time I had moved in Los Angeles over the years, each move had always been to the North and to the West.

The summer of 1993 was a fascinating but wearying time to be in Los Angeles, once the chamber of commerce prototype of the upbeat city. Los Angeles was suspended in dark and violent times; it was at the nadir of its brief existence. The Depression had been bad, but at least Los Angeles had not suffered alone. Now it felt set upon, like some Old Testament city assaulted by pestilences and plagues. Yet to come were the devastating wildfires of October, the earthquake of the next January, and the O. J. Simpson murder case. Things would change—extreme change was the dominant characteristic of the region. But the violent shifts in the landscape and the culture were becoming more frenzied.

During my extended visit, the jobless rate was 10 percent, compared to a national rate of 7 percent. Between 43,000 and 77,000 homeless were on the streets on any given night. There were no championship sports teams. My time there was bracketed by the guilty verdicts of the two policemen in the Rodney King beating case at one end, on the other by their sentencing and the beginning of the trial of the blacks accused of beating the white truck driver during the 1992 racial conflict. All of these events, including the shopkeeper's dismissal of history, were caught by cameras in this most filmic of all cities.

They were capturing images on film almost everywhere I went that summer: in Royce Hall on the campus of the University of California at Los Angeles, in a parking lot at the Santa Monica beach, at the La Brea tar pits, on the urine-stained streets of downtown Los Angeles, and among the palm trees at Rosedale Cemetery. It was the summer of a new mayor, one whose sparsely attended inauguration on the south steps of City Hall was beamed by live television to late-morning viewers in their homes. The next day the inauguration was hailed thusly in a front-page, banner headline in the *Los Angeles Times*: RIORDAN SEES RENEWAL FOR L.A. No matter that the editors knew the office was weak and the man was politically untested.

They were grabbing at straws that summer, the summer when the most widely promoted feature films, ones starring Clint Eastwood and Arnold Schwarzenegger, contained scenes of great violence that took place in the city—the city in the county of the same name that had four times as many registered assault weapons and owners of such deadly firearms as the next highest county in the state, that being neighboring Orange County. But nobody really knew how many of the more than seventy types of such weapons covered by the California Assault Weapon Control Act were really out there.

It was also the summer of the latest immigration uproar, with first a reported "armada" of seven ships carrying Chinese immigrants headed toward Southern California mysteriously shrinking to three vessels deflected to Mexico. Once again, immigration became a highly emotional issue. California's elected officials, both Republicans and Democrats, joined the frenzy. They advocated calling out the National Guard to stem the flow of illegal immigrants, imposing a head tax on all persons crossing the border to finance increased patrols, and cutting off all public assistance to illegal aliens. In Orange County, where there is a sizable Hispanic and Vietnamese population, the grand jury called for a three-year halt of all immigration to the United States, citing the increase in diseases, deaths, drugs, crimes, and the escalating costs of welfare programs and education. Hardly anything undesirable was left off the list.

Along with anger and fear, hate once again rippled across the land. Skinheads or neo-Nazis, really Anglo kids not too long out of high school, plotted to decimate African Americans, Jews, Latinos, and Asians. The *Los Angeles Times* concluded, "What is certain is that ethnically diverse, economically depressed Southern California has become a flash point for the white supremacy movement—a loose-knit collection of tightly wound individuals and ragtag organizations who generally believe that non-Christians, homosexuals and people of color do not belong in the United States." But the region had historically contained such groups, from vigilantes to Birchers.

The locals took umbrage at what they called "L.A. bashing" by the eastern media. *Time* magazine ran a cover story, whose headline IS THE CITY OF ANGELS GOING TO HELL? was splashed across a painting depicting the burning of Los Angeles. The cover contained elements of the painting "The Burning of Los Angeles," described in Nathanael West's novel *The Day of the Locust*; televised images of wildfires, earthquakes, and racial conflagrations; and the opening scene from the futuristic film *Blade Runner*, which depicted a seamy, multiracial Los Angeles. *Newsweek* waded in with a critical article. An uncomplimentary documentary appeared on CBS television. California native and former L.A. resident Joan Didion, who had fled east a few years earlier, chronicled the "sad, bad times" of Southern California in *The New Yorker* magazine. And the self-righteous *New York Times* climaxed the attack with a three-part series.

The solution that was offered to stem the deluge was to create a "total PR package" similar to the "I Love New York" campaign that seemed to have lifted that city from the budget-crime doldrums of the 1970s. "Depict the positive aspects of L.A." was the message on a public-radio talk show and also the one given to a meeting of downtown businessmen a few days earlier. They were told by a Pittsburgh public relations executive: "Every adjective used to scorn L.A. has an opposite term of praise. Instead of chaos, think diversity. Instead of ferment, think energy. Instead of recession, think transition." That the advice came from Pittsburgh may have been the unkindest cut of all.

In an unusual aside, the reporter who wrote the story for the *Los Angeles Times* added his personal encouragement. "Now that's a little more like it," he wrote. I wondered if the writer and the editors who passed on the story, or who may have inserted the sentence themselves, knew that catchphrases had been used for nearly one hundred fifty years to paper over the realities of this place.

I was on the last leg of my journey. I drove south to the headwaters of the Los Angeles River and then descended the length of the river by car, foot, and bicycle. There were detours off the concrete riverbed, but only to places the river had once covered before being encased in a concrete straitjacket.

The river had once been a wild beast when swollen with floodwaters. It had roared and writhed and cut its way across most of the Los Angeles plain. The river swung violently and precipitously in wide arcs, ranging from the San Gabriel River in the east to Ballona Creek in the west, where it emptied into the ocean at what is now known as Marina Del Rey.

To most Angelinos, if they were aware of the Los Angeles River at all, any reference to it seemed like a joke; a smile or chuckle or jest followed. If it was a joke, it was a bad joke to laugh at—Los Angeles *is* the river and the river *is* Los Angeles.

I followed three rivers from their headwaters to their endings during my circumnavigation of the state. The Owens River, the first, was the remnant of an ancient river gone dry because of the changing climate and increasing use by humans. The Tuolumne River was still a fairly vital river, as far as California rivers go. The Los Angeles River was simply surreal. It had the qualities of a California nightmare and was by far the strangest watercourse of all.

The Los Angeles River assumes that name at the junction of Calabasas and Bell Creeks at the western end of the San Fernando Valley. It flows southeast for twenty miles along the south side of the interior valley and then jogs south at Griffith Park, continuing in that direction for thirty miles across the Los Angeles plain to Long Beach Harbor. The river and its major tributaries—Tujunga Wash, Arroyo Seco, and Rio Honda—drain an area somewhat less than nine hundred square miles.

Quite short as rivers go—fifty miles long, not counting the length of its tributaries—the gentle curves and straightaways are shaped either as a box or a trapezoid and lined with concrete or grouted rock, except for the last 2.6 miles, which are lined with rock alone. The algae-flecked discharge during the dry months is mostly confined to the low-flow channel incised in the middle of the concrete riverbed.

The river is relatively devoid of native wildlife. A recent study by the Natural History Museum of Los Angeles County determined that, in terms of mammals, the river was "inhabited mainly by rats, house mice, feral cats, dogs, and human transients." The last documented sighting of a grizzly in Los Angeles County was a female bear that was shot in Tujunga Canyon in 1916. The seven species of native fish are gone, too. Steelhead trout that once migrated up the river were last caught near Glendale in 1940. Such exotic species as goldfish and mosquito fish flourish, while the only native species that manages to hold its own is the arroyo chub. Santa Ana suckers are on the verge of disappearing.

Nature has taken some strange twists in this artificial environment. Sunlight on the concrete river bottom—over which a thin sheen of nutrient-rich wastewater passes—attracts algal growth, which in turn lures the invertebrates that shorebirds feed upon. The result along the last few miles of the river is the largest concentration of shorebirds on the Los Angeles County coastline, which is also a comment on the lack of wetlands.

The only remaining salt marsh within the historic course of the river is the highly degraded Ballona wetlands, once the mouth of the river but now mostly a huge marina and high-density residential complex. The only dunes are those at the western end of Los Angeles International Airport, just south of the diminished marsh.

I selected Bell Creek as the headwaters of the Los Angeles River because it seemed to be the tributary that penetrated farthest inland, and there was a constant flow of water from the crest of the Simi Hills during the dry summer months. In the few miles before the creek reached the valley floor and became encased in concrete, it actually seemed like a real stream.

The morning fog lay like a cool shroud over the steep ridges and rounded sandstone boulders of the Simi Hills. Through these hills many a bandit, posse, cowboy, and Indian had ridden at breakneck speed when Hollywood churned out westerns in this dry terrain. My father and I had watched Hopalong Cassidy gallop through these very same hills back in New Jersey. The scratchy black-and-white images were obscured by the occasional snow and frame-skip on our first television set. In 1950 the production of potboiler westerns was replaced by the testing of weapons for the cold war in those east-west–oriented hills inhabited by rattlesnakes and coyotes and aerospace technicians.

The Simi Hills, varying in height from 900 feet to the summit of Simi Peak at 2,403 feet, are a seismically active region bisected by numerous active faults. The friable crust is randomly fractured, and a series of intermittent creeks flows past the coast live oaks, valley oaks, poison oak, sycamores, and cottonwoods along the canyon floors. The dense chaparral is periodically swept by wildfires.

Everyone was not welcome in the exclusive area known as West Hills at the far end of the San Fernando Valley. The sign on Woolsey Canyon Road read PRIVATE ROAD. NOT FOR PUBLIC USE. USE RESTRICTED TO PROPERTY OWNERS AND INVITEES ONLY. ROCKETDYNE. The new homes built by such contractors as Chateau Homes Inc. were monstrous; steel gates were readied to be lowered and locked into place at a moment's notice. The road ended at the guardhouse for the Rocketdyne Division of Rockwell International, an aerospace firm. My escort met me there.

On that misty morning, the twenty-six hundred acres of the Santa Susana Field Laboratory were a study of a once-vibrant industry reduced to skeletal remains because of shifting national priorities. A gray patina covered the landscape, as if a sheet had been placed over these artifacts of a dying civilization: the rusting machinery, abandoned buildings, and towering superstructures that once held rocket engines belching flames and gases and steam in floodlit splendor at night. Fifty thousand gallons of water per minute had cooled the flame deflectors under the rocket engines. The shock waves shattered windows in nearby Bell Canyon.

Those engines were being tested to lift humans into space and speed nuclear payloads to the Soviet Union. The engine for the Sidewinder missile, developed at China Lake Naval Weapons Center in the desert, was also fired up at what was referred to in Rocketdyne literature as the "largest rocket engine test facility in the Western Hemisphere."

The remote site was selected in 1947. The first crew consisted of four engineers and thirteen mechanics. The engines for the Atlas, Thor, and Jupiter missiles were tested on what was called "The Hill." The legend of buried gold under a rock shaped like the head of a giant lizard was one topic of conversation on the ride up the long, steep road. There were other oddities. One project engineer recalled, "Some drivers swore that even though it was uphill all the way, there was a portion where they felt their car accelerating—as though being attracted by some strange and powerful gravitational or magnetic force."

Four months after the first Russian Sputnik, a Rocketdyne engine in a Redstone missile propelled America's first satellite into orbit. Rocketdyne engines powered the first manned flights and the unmanned satellites that were flung into the far distances of space. They put men on the moon and powered the space shuttles, and a leak in one of the solid rocket boosters caused the Challenger tragedy in 1986. Scientists also experimented with solar and nuclear-powered electrical generating facilities at the field laboratory. At the height of activity there were six thousand workers at the sprawling facility.

This verse was sung to the tune of "Clementine":

> When there's thunder on the mountains,
> Every evening just at nine,
> And your walls begin to tremble,
> It's not God,
> It's Rocketdyne.

In the last few years the workforce shrank to six hundred, and the neglected site became mute testimony to the impermanence of yet another industry that the region had come to depend upon. Few people were visible. Deteriorating signs indicated where the safety showers had been located, to register here before entering the test area, that this piece of machinery was designated inactive. Power lines crossed and recrossed one another; some were askew. A string of lights dangled from a huge steel superstructure. This way to the Department of Energy (formerly the Atomic Energy Commission) facility. NASA was located over there. Both, like their host Rocketdyne, had seen better days.

Their spoor lay about, like the coyote scat in the nearby hills. The area was rife with such toxic wastes as radioactive materials, kerosene-type fuels, heavy metals, asbestos, and chemically laced water used to flush rocket engines after tests. Rockwell had paid over a quarter-million dollars in fines after the state filed suit charging twenty-seven violations of state and federal hazardous waste regulations.

Eight hundred thousand to one million gallons of water were pumped daily from one hundred forty wells. The water was treated at seven groundwater treatment stations and held in storage in reservoirs and tanks. Some of that water was used for the few tests still conducted at the site. The remainder was released downstream, and this used water was the start of the Los Angeles River. Rocketdyne had acceded to the request of downstream Bell Canyon property owners to release water at a steady rate so that the creek appeared to be a natural, year-round stream.

Two unnamed streambeds descended from the Rocketdyne property to a point just inside the Bell Canyon subdivision where they joined to form Bell Creek. I picked the westernmost stream to follow, as Rocketdyne was releasing no water into the other stream that day. The treated water was piped to the Area 2 Reservoir, where the space-shuttle engines were once tested in the nearby Coca Area.

A great blue heron perched on a large sandstone rock. The still water reflected the fog-shrouded images of a propellant-loading facility to the left and a giant test-stand bristling with cranes in the background. The water flowed over the lip of the pond and into a gunite-lined streambed that quickly gave way to a natural watercourse.

Up until 1968, Bell Canyon was a working cattle ranch. Then on one weekend in February 1969, nine hundred lots went on sale, and all were sold within two days. There were backup buyers for most lots in case the original buyer failed to come up with the financing. The minimum cost of a lot was $10,000. Bell Canyon was promoted as being "a guard-gated community." The other attraction, according to the developer, was, "It's quiet."

While blacks who could afford the move fled to such outlying communities as Pomona following the Watts riot of 1965, affluent whites drifted out of the more urban areas of Los Angeles to such places as the San Fernando Valley. The gated subdivision did not flourish at first.

Financing was difficult to obtain for home construction because of legal disputes with the developer, and the impetus to move to the eastern edge of Ventura County was not yet great.

Then along with changing demographics, rising crime, sinking public schools, increasing affluence for some, and the general feeling of a lack of "security" in the 1980s, construction boomed in Bell Canyon. The minimum lot price rose to $75,000. Home prices rose to between $500,000 and $1.5 million. The barricade and the quiet made the difference.

But all was not perfect in this setting where 550 families lived on 1,700 acres of land. On the twentieth anniversary of the subdivision ("What is twenty years old, still growing rapidly, and has the vitality of an energetic youth? Yes, it's Bell Canyon!") the newsletter of the homeowners' association noted:

> Bell Canyon is virtually a crime-free community, but every few years we seem to have a small rash of vandalism, which has almost always been traced to teenagers or young adults living in the Canyon. Recently, in the last three months, we again have observed some vandalism and even some burglaries. The Sheriff feels that, once again, these are coming from

Area 2 Reservoir and the test facilities

within our Canyon and appear to have descriptions of the
youths involved. Even gate-guarded communities are not
immune to crime, so please do your part by reporting any sus-
picious looking characters to the guard gate.

The association assured homeowners that measures were being taken
to keep out unwanted intruders: Tire spikes were installed at the exit "to
stop unauthorized visitors from entering the canyon." The guards behind
the smoked-glass windows of the red, tile-roofed guardhouse had a new
computerized Rolodex system. Guest lists must be brought to the guardhouse in advance of a party. A real estate or sales agent would escort
prospective buyers from and back to the gate when there was an open
house. Ty Hall was one of the guards; he played semi-pro football for the
San Fernando Freelancers, homeowners were told.

During the recession that stretched into the early 1990s, the architectural committee of the Bell Canyon Association issued strict guidelines
for the numerous For Sale signs that sprang up on the lawns of the mock
châteaus, villas, Spanish adobes, and ranch homes. The dark brown signs
with white lettering must be thirty-four inches wide, twenty-four inches
high, with the words *for sale* inscribed in three-inch-high capital letters.
The name of the real estate agent was to be in two-and-a-half-inch upper-
case letters with his or her phone number in two-inch-high numerals.
There were additional caveats.

On my first attempt to enter the gated community in my dented van,
I was not admitted. The guard gave me a name and number to call. My
request was submitted to the board of directors and was approved. I
entered on a clear, hot afternoon following my morning visit to the
upstream Rocketdyne property.

All was quite orderly. The For Sale signs were uniform. There was an
equestrian center, tennis courts, common areas, a small park and playground, and quiet along the canyon bottom and the ridges into which
homesites had been blasted in the sandstone rock. There were newly
planted California fan palms, glistening sod lawns, and leaded glass windows. The children returned at 4:30 P.M. in buses and vans from day
camps, which explained why the local playground was deserted.

I parked on Buckskin Road. The names of streets recalled the fake
West of the old westerns filmed in these hills: Bronco Lane, Roundup
Road, Hitching Post Lane, and for the weapons minded, Flintlock Lane.

I walked up an old Chumash Indian trail to the paradisiacal waterfall on Bell Creek, whose setting was marred only by black and white graffiti on the beige sandstone. "Rev" was one signature. "Rev" for revolution? I wondered.

Dragonflies and butterflies flitted about in the late-afternoon shafts of warm sunlight. I sought the shade. There were flowers, deep grasses, willows, oaks, and the sound of what easily could have passed for pure mountain water running, gurgling, falling, and splashing into a foam-flecked pool. It was the only place of near-natural beauty that I was to encounter along the whole length of the river, and only the rich who made the effort to walk there could enjoy it.

At the mouth of Bell Canyon, just inside the Los Angeles County line, the creek picked up its concrete escort and entered the flatlands of the San Fernando Valley. Reyner Banham, an architectural historian and connoisseur of landscapes, wrote, "In terms of some of the most basic and unlovely but vital drives of the urban psychology of Los Angeles, the flat plains are indeed the heartlands of the city's Id." He added in *Los Angeles: The Architecture of Four Ecologies* that "the history of Los Angeles is a story of the unscrupulous and profitable subdivision of land."

The palm tree, planted in rows along Sherman Way and elsewhere in the valley and the lower plain, was the principal image used to sell this land. It became the preeminent icon of Southern California—the scene-setter in movies, novels, and advertisements.

Palm trees were not native to the flat plains. They were first imported in large numbers during the real estate boom of the 1880s when Bible-toting midwesterners arrived in overwhelming numbers. They sought a biblical connection and tropical verdancy in a semi-desert; the stately palm tree gave them both. Social status was determined by the sweep of green lawn from the road to the house, and the number of palm trees.

The only palm native to the state was the California fan palm (*Washingtonia filifera*), and that tree was found only in the desert. An imported palm was planted at the site of the San Diego Mission when the Spaniards arrived in 1769. The tree may have been planted by Junípero Serra, the Franciscan missionary who founded a chain of missions in Baja and Alta California. When the palm died in 1957, it was the oldest known

Palm trees and the river

palm tree in the state. Workmen cut it down and found lead bullets, circa 1850, embedded in the trunk. It may have been used as a backdrop for target practice or executions, or both. That particular type of palm came from Moroccan date groves.

"Palmania" reigned in cycles, alternating with the more eastern look of shade trees. An 1897 nursery catalog listed eighty-seven varieties of palms. They came from Japan, China, Mexico, the Canary Islands, South and Central America, Africa, and the Mediterranean region. Palms were not the only immigrant trees: from Japan came the podocarpus, from China the camphor, from the Himalayas the deodar, from Peru the pepper, from Brazil the jacaranda, from Abyssinia the banana, from Australia the eucalyptus, and from South Africa the silver tree. "These foreigners have transferred the landscape of southern California into a vast botanical garden," wrote Victoria Padilla in *Southern California Gardens.*

But palm trees fit best; they imparted the right feeling. Developers planted them in monotonous rows. "It should be pointed out that many of these poorly-planted palms were set out by real estate dealers who were only concerned with their immediate appeal to prospective lot buyers," wrote a researcher. On a practical level, there were also certain advantages: The hollow cylinder of the trunk bent easily in the fierce Santa Ana winds that periodically swept down the mountain passes from the desert. Palm trees weathered extreme temperatures, thrived in a dry climate, and were resistant to smog. Vandals could not harm them, as their conductor vessels were buried inside the tree.

There were certain disadvantages: Palms were a fire hazard. A palm tree pyromaniac terrorized the San Fernando Valley a few years ago, setting fire to hundreds of trees. Palm trees were ignited near the Civic Center during the 1992 racial eruption. Palms could be infested with scorpions, rats, and other vermin. A tree trimmer suffocated to death when the heavy, fan-shaped skirt of a palm slipped and buried him.

But palm trees multiplied, and their image drew immigrants to Southern California. In his 1939 novel *Ask the Dust* John Fante, a writer who came from Colorado, described the view from his Bunker Hill apartment:

> Through that window I saw my first palm tree, not six feet away, and sure enough I thought of Palm Sunday and Egypt and Cleopatra but the palm was blackish at its branches, stained by carbon monoxide coming out of the Third Street Tunnel, its crusted trunk choked with dust and sand that blew in from the Mojave and Santa Ana deserts.

To H. J. Whitley, who wanted to be known as "The Great Developer," palm trees were a necessary adornment. He had them planted in 1911 along the twenty-two miles of Sherman Way that traversed the length of the San Fernando Valley. The boulevard's broad sweep was patterned after the *Paseo de la Reforma* in Mexico City and served as a sales enticement, much like the fountain-punctuated, artificial lakes of modern desert developments.

The San Fernando Valley, that national model of suburban decorum and unease through which the Los Angeles River runs, was transformed by Whitley and his four powerful associates from grazing and wheat lands into the beginnings of communities eventually known as Canoga Park, Woodland Hills, Reseda, Tarzana, Encino, Van Nuys, and Sherman Oaks. The change the valley underwent reflected every major landscape and cultural phase in the history of Southern California.

The oval valley, which tilts toward the southeast, contains two hundred forty square miles of gently sloping land. When Portolá camped on the valley floor, he encountered the Fernandeno Indians, a branch of the Gabrielino. The Spanish first used the valley as a route between points of greater interest. During the fourth crossing of the valley, made in the year of the outbreak of the American Revolution, the explorer Juan Bautista de Anza noted "a plague of fleas." The first Spanish settler was induced by the Franciscans to give up his claim to a well-watered spot so that a mission could be established there in 1797.

The padres recognized two classes of people living in the valley: themselves and their look-alikes being the people of reason, and the Indians being people of instinct and appetite. By the 1850s, most of the Indians had been wiped out by diseases, ranging from smallpox to alcoholism. Well over two thousand Indians were buried in the small mission cemetery between 1798 and 1852.

In 1812 an earthquake damaged the San Fernando Mission. Earthquakes had punctuated the valley's history, the Fort Tejon quake of 1857 being one of the severest. "Trees were uprooted. People rushed out of homes. The Los Angeles River leaped from its bed," wrote a local historian. Otherwise, there was little damage because there was little development at that time. More earthquakes followed in 1860, along with a devastating wildfire, a drought, and then rains that melted adobe homes and flooded the valley floor.

Cattle, sheep, and wheat—a great deal of each in their respective times—came and went in the San Fernando Valley, leaving the landscape a barren plain and foreshadowing the rapid coming and going of other such one-crop economies as the aerospace industry. The valley was still virtually deserted after the Civil War. Historian W. W. Robinson wrote:

> During the late 1860s and early 1870s a visitor riding horseback through the high, wild mustard would have seen no evi-

dence of human activity in the whole wide San Fernando Valley except the Mission, perhaps Lopez Station, a few other
near-by adobes, and cattle grazing in fields.

A new industry, one that could flourish on the depleted land, was
invented around 1870. "Subdivision was heard in the land—meaning to
break a large ranch into smaller parcels. Prairie schooners brought in
many people. Newspapers spoke of 'cutting up ranchos.' Prices began rising," wrote Robinson, describing the economic premise that Angelinos
would rely on in future years. Water would make that assumption a reality. But at the turn of the century, a court decision prohibited the valley
from withholding Los Angeles River water from the downstream city. All
that potentially rich land would turn to dust unless something was done.

Whitley was a doer. He moved to Los Angeles at the age of thirty-four
after laying out town sites in North Dakota and the Oklahoma Territory.
A Mason, a Republican, and a Presbyterian, the stout boomer fit right in
with his peers in Los Angeles. Soon he was subdividing the Hollywood
Hills. He built a turreted showplace at Hollywood Boulevard and Wilcox
Avenue, the grounds extensively landscaped with palm trees. (Within
twenty years, the mansion would be torn down and replaced by apartments and homes. Progress would do the same to Whitley, who has been
all but forgotten.)

Whitley moved on to greener pastures in the San Fernando Valley
where he built a three-story, Spanish-style home surrounded by porticoes
and palm trees. His wife, Margaret Virginia, was a formidably dressed
matron who was frequently mentioned in the society columns of the local
newspapers that Whitley owned. Known as "H.J." to friends and associates, these papers trumpeted Whitley as "The Great Developer" and
"the Father of Hollywood." The latter phrase followed Whitley to his
grave via a *New York Times* obituary.

Whitley, who would become associated with the owners of the *Los
Angeles Times* in a huge San Fernando Valley land deal, knew the value
of the right kind of publicity. He once told an audience, "In our development work, we have always had the newspapers with us, and they have
done much."

In later years, an embittered Mrs. Whitley wrote an unpublished
autobiography in which she sought to give her dead husband what she
thought was his rightful place in the history of the valley. She wrote:

It was he who made possible the transformation of a 47,000 acre tract on the south side of the Valley into productive ranches, the building up of all these towns which were the result of his planning and financing, that of a master town and community. Pointing to a map he said:

"These shall be towns, and in a few months time these blue-penciled specks shall be towns busy and alert with electric light, telephones, post offices, churches, school-houses, boulevards, banks etc."

And so it was.

There were four other partners, and they were the true movers and shakers: Harrison Gray Otis, the owner and editor of the *Times;* Harry Chandler, business manager of the newspaper, the son-in-law of Otis, and heir to his post upon Otis's retirement in 1914; Otto F. Brant, head of the Title Insurance and Trust Company and the self-proclaimed inventor of a real estate sales procedure known as escrow; and Moses Hazeltine Sherman, a member of the city water board and owner of streetcar systems. Sherman and Otis both went by the honorific title of general. Their names and those of their close relatives would come to adorn valley streets and towns. These men were not strangers to owning land. Their holdings in groups or as individuals extended from the Central Valley to below the Mexican border.

Catherine Mulholland, a granddaughter of William Mulholland, the engineer responsible for bringing water south from the Owens Valley to the San Fernando Valley, wrote of these men in her book, *The Owensmouth Baby*:

> They spoke of vision, of progress, and of upbuilding Southern California. They adored the game of Profit and Loss, none more so than the man who was to be their field marshal in the new Valley campaign: Hobart Johnstone Whitley, The Great Developer.

The key to profit in Southern California was water; at that time, water to supplement the depleted Los Angeles River. The plan was to transport water by aqueduct from the distant Owens Valley, which drained the eastern flank of the Sierra Nevada. Men of this cut were up to that gigantic task; they played a decisive role in accomplishing it, to both lasting praise for furthering the growth of the city and eternal condemnation for stealing the water and profiting from land sales.

The outline of this story was depicted in the 1974 movie *Chinatown*, the best of many cinematic portrayals of the region. Los Angeles River water was manipulated to increase the value of land that was surreptitiously bought in the San Fernando Valley. Noah Cross (John Huston), an evil version of a Whitley-Mulholland-Otis-Chandler blend of characters, lectured Jake Gittes (Jack Nicholson), a private eye attempting to solve a murder. Cross repeatedly mispronounced Gittes's name.

> You see, Mr. Gits, either you bring the water to L.A., or you bring L.A. to the water.
> How are you going to do that?
> By incorporating the valley into the city. Simple as that.
> How much are you worth?
> I have no idea.
> Why are you doing it?
> The future, Mr. Gits, the future. Now where's the girl?

Whitley was one of five men constituting the Board of Control of the Los Angeles Suburban Homes Company. But Whitley, the managing partner, was the gofer. Mrs. Whitley wrote sarcastically of "that wonderful 'Board of Control' composed of such great and energetic men who would use him, but never be used." She attributed the following words to her husband: "I had been warned of their selfishness, but I knew what I could do."

For eight years prior to the consummation of the deal, Whitley, in his wife's words, combed the valley seeking out "those poor people who have been almost buried alive on their little patches of land leading through to the prize." The ranchers' land was mortgaged to the hilt, they were unable to make a living, and they were "willing to sell and get out of their misery." She added, "Of course, they knew nothing of his plan."

In 1910, the syndicate purchased 47,500 acres for $2.5 million, which worked out to a little over $53 an acre. Tract 1000, as it was designated on the subdivision map, was the largest single development in Los Angeles up to that time. One estimate placed the total take some seventeen years later, when the last piece of syndicate property was sold, at $98 million, which put the per-acre price at $2,063.

Through the intervening years, the board acted quite frugally, which may be why such persons were successful businessmen. Whitley reported that the board was asked to sponsor the celebration of the aqueduct's completion because "we were the most largely interested, and probably chumps enough to pay the whole bill." The board refused to donate any money, but did put on a huge bash when lots went on sale at Owensmouth (later renamed Canoga Park because Owens River water actually arrived some twenty miles distant). There was a "monster free Spanish barbecue," auto races, free entertainment, and cut-rate streetcar fares.

Besides having a hand in furnishing the water and in selling land, the board's most lasting contribution to the valley was to separate the races. There were the following notations on the subdivision map: "The Board of Control reserves the right to refuse to sell any of these lands to any person whom in their best judgment will be undesirable." Concerning the nicer sections of Owensmouth, deed restrictions stated that "no part of said premises shall ever, at any time, be sold, conveyed, leased or rented to any person of African, Chinese or Japanese descent." While not singled out by name, the previous occupants of the land—those of Mexican descent—were confined to Cholo Town.

White midwesterners flocked to the valley. One recalled:

> You see, from the Midwest, my mother had never seen a Mexican before and the train was half full of Mexicans, and my mother had heard that the children had lice, so she wouldn't let us near them. And they tell me they're the cleanest people—always washing. She told my aunt how worried she was, so my aunt thought it would be a good idea as a preventative to wash our hair in kerosene. It's a wonder we didn't catch on fire.

The same restrictions applied elsewhere in the valley, which became the whitest part of Los Angeles. Segregated housing and schools bridged the gap between the genocidal violence of the vigilantes in the mid–nineteenth century and the open racial warfare in the latter half of the twentieth century.

The Great Developer eventually left the valley and purchased forty-eight thousand acres near Paso Robles to the north. The oil bug, the next one-crop economy, bit him. "Perhaps we may strike it rich," he wrote his wife. But the crash of 1929 ruined Whitley. After his death two years later, Mrs. Whitley wrote, "He was a genius—who succeeded—but failed when he became too grasping. 'Just one more million,' he said."

I crossed the Sepulveda Flood Control Basin, once part of the syndicate's holdings, on my bicycle. It was an overcast Sunday morning, but the fog would eventually lift. The mountain bike gave me increased mobility and greater contact with the landscape, but I was also more exposed to the vicissitudes of the cultures that I passed through.

My only protection was a plastic safety helmet and a microcassette tape recorder, which I used for taking notes. I also found it handy, when someone made a threatening move in my direction, to raise the tape recorder to my lips. It appeared that I was in communication with higher authorities, and they backed off. There were certain times when it was advisable not to use the recorder for note taking: for instance, when I was among many illegal aliens who might mistake me for *La Migra.*

I entered the vast, uncluttered space of the basin along the trape-zoidal channel from the northwest and followed the river through the basin to its entry into the residential core of the valley. Trees and plants have been allowed to grow within the channel in order to give it a more natural look. I passed baseball and soccer fields, a golf course and a walking-jogging trail, fields where sod was being grown for lawns, a sewer treatment plant and its adjacent Japanese garden where wedding cere-monies were performed, a wildlife preserve, and a small lake. The wildlife preserve and recreational lake, where people were canoeing, were filled with treated wastewater that was also dumped into the river, thus augmenting the dry-season flow that originated in the Simi Hills.

I paused at a model-airplane landing strip where one version of the aerospace industry was thriving. Helicopters were hovering, and all types of fixed-wing planes were taking off, landing, and circling at an altitude not to exceed two hundred feet, according to the posted regula-tions. Above the models a real plane circled over the basin. Below, a con-fused bird flew through the flight path, veering right and left to avoid the models.

A Los Angeles Police Department sergeant drove up. He walked slowly over to a group of aviators who were sitting at a picnic table. Con-versation died. He asked if anyone would like four Dodgers tickets.

There were no immediate takers. No one seemed quite sure how to interpret the offer. The conversation slowly resumed, with the uniformed sergeant taking the lead among the white males. The talk had to do with the 1965 and 1992 racial conflicts. I listened to the sergeant with the expert marksman's badge on his chest and white hash marks on his sleeve, and to the others who formed a chorus.

> *Sergeant:* They were smoking them pretty good left and right. You didn't hear a lot about this stuff, but I understand what went down in those days. But when they declared martial law, they shot looters then.
>
> *Others:* That's the way it should be handled.
>
> *Sergeant:* Well, I know it should be. We didn't shoot them this time. Let me tell you this. We didn't shoot them. God knows we wouldn't do that, you know.
>
> *(General laughter)*
>
> *Others:* What fingers did you crack?
>
> *(More laughter)*
>
> *Sergeant:* The National Guard came in and they had their Jeeps and machine guns on the back end of them.
>
> *Others:* No kidding.
>
> *Sergeant:* Boy, I tell you, that National Guard. We had snipers coming in on us, you know, shooting at us for a good half hour. The National Guard backed up their Jeep and sat there with their machine gun and tore that goddamn building apart.
>
> *(Laughter)*
>
> I was just reading a teletype. Newton division had a shooting. Two CRASH officers, these are guys who go after gangs, go rolling in I forget what street it was [the 1700 block of 42nd Street]; but there was about thirty gang members out in the street with guns. They split in all directions, like they do. One little asshole is walking away, spins, starts firing a gun at the officers. Then another asshole spins and fires one round, a shotgun round. It hits the side of the police car. That car gets one round off at the guy. The second car goes whizzing through seeing the action and they pump ten rounds at some-

body. Ah, there's probably a hundred rounds being fired back and forth between the four officers who pulled in and got cover. Fired at the assholes. No one was hit.

Others: Geez.

Sergeant: The two police cars finally rolled out. They called for assistance, and they surrounded the area and got twelve of the bad guys. Recovered two guns. One gun is an air gun. (*He laughs.*) Yeah, you bring a knife to a gun fight.

(*General laughter*)

[A Newton Street Division sergeant was quoted in that morning's *Times* as stating, "We're getting shot at more and more often now. They have guns and we have guns and it's bound to happen." Twelve suspects were detained after the shoot-out in South Central Los Angeles, but only five were booked on charges ranging from attempted murder to weapons violations.]

Sergeant: As fast as we put them away, there are ten more on the streets.

You know, we had the Black Panther raid. We shot at least four thousand rounds at that building and they shot probably a third more back at us. And we finally dropped a satchel bomb through the roof on that one. That was a minor shoot-out. It looked good on television.

Others: Did you see the Black Panther shoot-out?

Sergeant: We filmed the whole thing. We have a tape. I've seen some segments. My favorite segment is some newsman asshole was standing there and he gets a round. "Oh, geez, this is real," he says.

(*General laughter*)

[One of the others identifies himself as a retired LAPD officer.]

Retired officer: This country is going down the toilet.

Sergeant: You should see, they are hiring guys now they wouldn't touch, they wouldn't look at before.

There followed some talk about the relative merits of different types of handguns. They asked the sergeant what he carried. He patted a

Beretta stainless with a sixteen-round clip. It turned out that the sister of one of the others had the same gun. The sergeant again asked if anyone wanted the Dodgers tickets. This time there was a taker.

The river passed out of the basin just beyond Sepulveda Dam. On a previous day, I had climbed down into the concrete interior of the dam, completed by the U.S. Army Corps of Engineers in 1941. It was designed to form a 16,700-acre-foot reservoir, should floodwaters ever need to be detained in order to lessen the danger downstream. If that happened, all of the streets and recreational facilities would be inundated. I walked through the cool passageway underneath the dam that rose fifty feet above the riverbed. A sign read:

> Sepulveda Dam crest gate operation. For normal operation set valves and mechanism as indicated below. Note: in the event of a flood the operation of the crest gates is entirely automatic with valves and mechanism set as indicated. No further adjustment is necessary.

My escort from the Corps of Engineers explained that one gate, called a political gate, was immovable. Should the choice lie between the safety of the dam and flooding downstream, and should a politician try to interfere with the dam's operation, the corps' answer would be that it had no further control over downstream releases.

Downstream from the dam the staple-shaped riverbed began a series of sinuous curves, an attempt to imitate a natural riverbed. The concrete trough passed comfortable single-family homes, apartments, and condominiums on tree-shaded streets. The black plastic and cardboard shelters of the nearby homeless were hidden in the bushes fronting on the dirt maintenance road that paralleled the concrete river. Here, not far from twenty-four-hour supermarkets, were the prairie schooners of the homeless—the stainless-steel shopping carts that began to litter the river in earnest. They were either stranded high up and molded to bridge abutments by powerful flood flows or they lay mangled or intact in the shallow water. The abandoned shopping carts, found all the way to the end of the river, were another L.A. icon.

I had some prior acquaintance with this part of the river. My mother-in-law lived in a riverfront condominium in Studio City. I have flown into town for brief holiday visits over the years. I have jogged alone or walked her dog along the flood-control channel. The vegetation that partially hid the concrete and the lack of visible structures across Valleyheart Drive from her condominium provided a sense of isolation in the midst of a city.

There were disadvantages. The isolation made it easier for burglars to gain entrance and steal cars from the gated garage. My mother-in-law and her elderly friends feared the night. The alluvial soil underneath the condo magnified the shock of the 1994 earthquake, and the three-story structure was heavily damaged. (Her son, driving from West Hollywood to Studio City in the early morning hours of that cataclysmic day, said the clarity of the stars and the jangling burglar alarms were the most noticeable sights and sounds. The electricity was out; the valley lay blackened, except for scattered fires that cast no glow upon the sky.)

I was entering movie country. At Jerry's Famous Deli on Ventura Boulevard, which backed up against the river, I placed my order at the counter and then went to the men's room. A distinguished-looking, elderly gentleman was shooting up over the washbasin. He turned to me and said, "Don't ever get diabetes. It's a pain in the ass."

I said I hoped that I would not get it, but that my mother had been diabetic, and I had heard somewhere that it was hereditary.

"No. No one in my family had it," he said. "It is stress related. I had a big job in the movies for a number of years."

I zipped up and left. The pager went off on the man sitting next to me at the counter; he plucked a cellular phone from his pocket and dialed a number. Not to be outdone, I took my tape recorder from my pocket and spoke into it. People all over L.A. that summer were talking into or listening to cordless devices. A few days later in a different neighborhood in Southeast Los Angeles I was waiting to speak to the police chief. A man sitting across from me walked outside and dialed a number into his cellular phone. I wondered if he was calling his bail bondsman or his stockbroker.

The farther east I went the more media-related the landscape became. A few blocks away at CBS Studio City, at the junction of the Los Angeles River and the Tujunga Wash, Shakespeare was being badly bent to fit the L.A. scene: A production of *Romeo and Juliet* ranged across the sets for *Seinfeld, thirtysomething, Hill Street Blues,* and *Falcon Crest.* It was complete with every imaginable Los Angeles icon (with the exception of shopping carts): Uzis, 9mm pistols, gangs, motorcycles, convertibles, palm trees, drive-by shootings, stabbings, steamy sex, drugs, Beverly Hills land developers, and a cast that included African Americans, Latinos, Asians, and Anglos.

At one point Romeo whipped out a cellular phone. He wooed Juliet from the roof of his white Datsun 240Z. Juliet was perched on the balcony of a Victorian home, above a sign for a security service that promised an armed response. The backdrop for the Queen Mab scene was the Los Angeles River.

The director's note in the program stated: "In our own city, random violence has now developed to such an extreme that we are all packing guns. Paranoia is everywhere." The director, who described himself as being half Armenian and half Jewish, added, "I wanted to set this play, written over four hundred years ago, in the environment in which we now live, and try to bring Shakespeare's genius directly into our lives."

The director was not alone in his interpretation of Shakespeare: The previous year *A Midsummer Night's Dream* was set in post-riot Los Angeles. A cutout of city hall dominated a graffiti-splashed landscape, while the forest was represented by the last four letters of the HOLLY-WOOD sign.

Following Whitley's departure, the inevitable growth slithered up the Los Angeles River corridor. The East Valley absorbed the impact first. With the arrival of Owens Valley water, irrigated fields climbed to a high of seventy thousand acres in 1926. At one time 90 percent of the nation's baby lima beans were raised around Van Nuys. Then the irrigated acreage plummeted as people and industries replaced agriculture. Universal, Warners, Disney, and Republic studios located in the East Valley, as did Lockheed Aircraft, which built twenty thousand planes during World War II.

In 1948, $130 million in new construction set yet another record for the "Valley of Destiny," as a local newspaper called it. The single-family home was dominant. Chevrolet opened an assembly plant. Of four hundred thousand valley residents at the half-century mark, five thousand were nonwhite. The population doubled by 1960; four years later it reached one million. Hurricane-force winds, hail, snow, wildfires, an earthquake, and flash floods punctuated the event. "The great year of 1964," as it was called, was the heyday of the Valley Girl. One half of valley households owned two or more cars.

The 1970s opened with a devastating earthquake that killed sixty-four people and caused $550 million in damages. The economy slowed down; there were layoffs in the aerospace industry, a bitter school desegregation fight, crueler crimes, and a noticeable shift in the population. Latinos were moving into the valley. Of all the valley schools the impact was being experienced first at Walter Reed Middle School, the closest to central Los Angeles.

Reed was an all-white, predominantly Jewish school in the 1960s. The swift change that was so typical of the region began in the 1970s when Reed became the first valley school to experience mandatory busing and the influx of Hispanics. Between 1980 and 1990 in North Hollywood, from which the junior high school drew most of its pupils, the Latino population grew and the white population decreased. At Reed in the early 1990s, the whites were a minority, with 35 percent of the students. Hispanics were a majority, with 52 percent. Blacks accounted for 8 percent and Asians were 5 percent. Reed, located at the junction of the Tujunga Wash and the Los Angeles River, was a microcosm of the region.

A school, I thought, was one of two institutions that was a fairly accurate mirror of the community, the other being a newspaper. The difference was that a school, in this case two schools that I visited within the Los Angeles Unified School District—Reed Middle School, and Bell High School farther downstream—contained the seeds of the future, while the newspaper, in this case the *Los Angeles Times,* represented the past. The dynasty that had guided the *Times,* and the city, for more than one hundred years was no more. Both the schools and the newspaper were indices of that perennial California affliction, racial conflict.

Segregated schools evolved from segregated housing, and the San Fernando Valley led in both. By the 1920s, the Los Angeles school board had institutionalized a policy of segregation. Following a shoot-out between police and Mexicans in 1921, the white residents of Owensmouth requested "immediate steps to segregate Mexican pupils in the grammar school." Another building was acquired and called an annex or branch of the school. Mexican students were sent to it. Two years after white parents in San Fernando petitioned the school board for separate schools, white students were allowed to transfer elsewhere. The result was the San Fernando school was attended entirely by Mexicans, a strategy that allowed the board to say there was no official segregation policy.

Bowing to community pressure elsewhere, the board redrew school boundaries in South Central Los Angeles, creating predominantly black high schools, one of which offered the only course in Los Angeles schools on how to be a maid. When the Los Angeles district took over Bell schools in southeastern Los Angeles in 1926, parents from that community, then predominantly white, objected to having their children attend school with Mexicans and Negroes. The board obligingly drew the boundary at Alameda Street, stating that it would be too dangerous for children to cross that busy thoroughfare. Anglos lived to the east of Alameda, where the school was located, and minorities to the west. For a time, Bell High School remained predominantly white.

One day in 1961 a thirteen-year-old black girl wore a short dress to school. Her mother was too sick to lengthen the dress that had shrunk in the wash. She was paddled. The girl whirled and struck the vice principal, who had administered the two whacks. She was charged with assault and battery and sent to Juvenile Hall. The American Civil Liberties Union became curious about the case: What was the policy on corporal punishment? Was it administered prejudicially and with racial vehemence? These questions led to the broader inquiry as to whether the schools were segregated. Representatives of the ACLU met with the school superintendent, who stated, "We're not going to bus them." The words were prophetic: More than thirty years later, after court rulings that first favored busing advocates and then the antibusing forces, who were strongest in the valley, there was no mandatory busing; schools like Bell High School became 96 percent Hispanic.

The summer of 1993 marked a period of unprecedented crisis for Los Angeles schools. In 1970, Anglo and minority students were almost evenly divided in the district. Since then, white and black enrollments have declined, while Latino students have dramatically increased from 22 percent to 65 percent. The most recent figures showed blacks at 15 percent; whites at 13 percent; and Asians, Filipinos, and Pacific Islanders at 7 percent. What was termed "combined minority student enrollment" by school authorities was placed at 87 percent. The remainder were whites.

For the first time the school board was dominated by minorities, and there was a black superintendent. The district faced drastic budget cuts. There was a move in Sacramento to break up the district, the second largest in the nation. Burglaries and vandalism in schools cost $6.5 million yearly. Schools were frequently sprayed with graffiti, requiring full-time maintenance employees to eradicate the spray paint.

Bullets also zipped about. There were racial brawls and two killings, one at a Valley high school where two students faced down each other in the corridor, both thinking the other had a gun in his pocket. One did. Three hundred metal detectors were handed out to school administrators. A policy of automatic expulsion for bringing a gun or its look-alike to school was adopted by the board. Expulsions doubled over those for the previous year. Eight- to seventeen-year-old students were caught with a variety of weapons: .25-caliber, 9mm semiautomatics, .357-caliber magnums, sawed-off shotguns, and accurate replicas of real guns.

I was extremely nervous about entering the schools, but once inside I never felt threatened. I came away with feelings of great hope and deep despair.

At Reed, a brightly colored water pistol lay on the principal's desk amid the clutter of telephone messages, papers, and confiscated baseball cards. Draped over the back of his chair was a white jacket with green lettering that stated MR. TASH, WALTER REED SCHOOL. The mid-June day had started at 7:30 A.M. with a faculty photo in front of the main building, a 1939 Works Progress Administration attempt at combining institutional architecture with a Spanish motif in what had once been a walnut orchard. One-third of the faculty did not show for the photo. Now two angry mothers were sitting in Larry Tash's office.

The school attendance boundaries stretched from the bougainvillea-clad homes on the heights below Mulholland Drive to the barrio along Lankershim Boulevard. These two parents were definitely from the heights, and they were incensed that their ninth-grade children had not been accepted into a high-achiever high school program. The "highly gifted magnet centers" accepted only students with a minimum of a 145

IQ score, if there was room. There was also a racial quota of 60 percent nonwhites, the district's nod toward integration. These magnet schools were the best chance for acceptance from Los Angeles's high schools into a quality college or university.

Most of the Reed faculty and administrators prided themselves on their high-achiever program, created in 1971 by Paul Mertens, who was retiring this June. Geneticists, astronomers, musicians, lawyers, and Harvard and Stanford University graduates—with IQs between 145 and 200—who had passed through this program would be on hand at the retirement ceremony to honor the teacher. Tash promoted the program heavily in order to attract gifted students and to retain white students, who might otherwise go to private schools. Twenty-seven parents from the Westwood area surrounding UCLA paid two hundred dollars a month to have their children bused to Reed. If the students could not transfer to an equivalent program in high school, Reed would be much less desirable.

Some teachers, especially those who taught in the bilingual program, thought too much attention and too many resources were being devoted to the Individualized Honors Program. The rate of transfer from the bilingual program into the regular curriculum, let alone the honors course, was so minimal that it amounted to segregation of Hispanic students within the school. Of the eighteen hundred students at Reed, one-third had language difficulties. Few from the Hispanic majority were in the honors program.

The two parents fumed. Tash had to do something, they said. One of the mothers mused, "I'm French-Canadian, will that help?" She wondered if she had to return to the expensive private school that had accepted her daughter and beg on her knees for her admittance. The family had turned down the acceptance, thinking that she would get into the free magnet school.

The principal explained that more students than usual from other schools had applied to the North Hollywood High School magnet program, and that was why a large number of Reed students had not been accepted. Meanwhile, he left messages with the secretaries of downtown school administrators, who did not want to come to the phone to deal with the problem.

He warned them through the secretaries: "I'm trying to head off a major headache for the district. These are big, powerful people."

The messages were not returned; the two mothers eventually departed, threatening to take their case to the school-board member who represented the district. At a meeting held that night at the school, it became apparent that qualified Asian and African American students had also not been accepted. To an Asian mother and an Anglo father who had switched their child's school-designated ethnicity to Asian, thinking he would have a better chance of acceptance, the fact that they had company was a relief.

This problem, like others, came and went all day for Tash, a rotund, balding, middle-aged man who wore ties adorned with all types of odd objects in hopes the students would relate to him. And they did: "Hey, Mr. Tash, are you going to hear our report on overpopulation?" He promised to return as he dashed through the hall, walkie-talkie in hand, to make a midmorning check. "Usually I get out of my office earlier than this," he explained.

"This is Reed Seven," Tash said into his radio, directing a maintenance employee to remove the "love" graffiti from the telephone booth at the school's main entrance. He made his way across the asphalt school yard, expertly dodging the places made sticky by spilled food and drink. Tash called most students by name. He reminded some not to be tardy, picked up a forgotten backpack, and asked others to clean up the area where they were eating snacks. A few stopped to admire his Snoopy tie.

In Mertens's classroom on the second floor, one student went up to another and quietly said, "Hey, I got in," referring to the magnet school of his choice. The daughter of the French-Canadian mother entered the classroom late. She had a hoarse voice and did not seem well.

Mertens used no textbooks for the United States history class. Instead, he relied on outlines, oral and written presentations, and the reading of some sixteen books. The students, divided into groups, worked on half-hour presentations of their concepts of a utopian society. Their societies had different goals: happiness, equality, freedom, and goals that had not yet been clearly defined. They had questions of one another:

> Do we have to consider overpopulation?
> How do we get paid if there is no money?
> Do we have professional sports, regular sports, or no sports?
> Should every job be equal and important to society?

Mertens, in a white, short-sleeved shirt and tie, moved from group to group. He interjected a comment or two and answered questions here and there. The interchanges were lively. The students, mostly whites and Asians, were serious and affluent enough to be able to fax assignments to the school if they were sick.

Back in Tash's office, there was a new flurry of telephone messages. Besides the threatened parents' revolt, he had to deal with the procedure for getting a substitute to replace a teacher who had to go to court on a school-related suit, replace the sound system in the auditorium that had blown out in the middle of the school play the past weekend, hire a new financial manager, advise a student who was being bullied by others, consult with a vice principal on teachers who had not held student body elections in their homerooms, discipline a student who called a teacher fat, tell a woman seeking work that speaking Arabic was no advantage at Reed, get the paperwork going on a student who had punched a teacher, and deal with a parent who wanted the time of graduation changed.

At one point Tash, holding the telephone in his hand, said, "I forgot what number I called." He laughed and redialed.

An eighth-grade girl knocked and asked in a quiet voice, "May I talk to you about something personal?" Tash gulped. The last such query had resulted in a long, horrible tale of sexual abuse at home. This time it was simpler. The girl was pregnant. She was going to have the baby. The boy and her family supported her. Tash said there was a special school she could attend while her pregnancy showed. (The birthrate among teenage girls in Los Angeles County was 40 percent above the national average. Latinas, such as this girl, accounted for 70 percent of all new teenage mothers, although they made up only one-half the county's population in that age bracket. Some teachers called them "breeders.")

Shortly before noon the principal dashed downtown with a report that was due that day at the Special Projects Office. He walked through the well-funded office admiring the new equipment and the abundance of people who did not look very busy. On the way back over smog-choked Cahuenga Pass on the Hollywood Freeway, Tash related how in fourth grade he wrote a story about a math professor and knew he had written about himself.

After graduating from San Fernando High School and UCLA, Tash taught math. He then began moving up the administrative ladder in South Central Los Angeles, where a lot of attention was paid to personal safety. A sniper killed two children at one junior high school where Tash worked. Tash made it back to the Valley where he was raised and took over the principal's job at Reed in 1989. He was due to be transferred soon. Tash was paid $62,000 a year and had not made it home before ten P.M. for the last four nights.

Back at Reed it was lunchtime, and the glucose intake was evident in the noise level and raw energy being expended in the homeroom basketball championships. This was also the week of the National Basketball Association play-offs, and every player thought he was a Michael Jordan. There were a lot of high-fives and daps.

Someone came over and got Tash, who was watching the game. A girl had collapsed and was having trouble breathing the smoggy air. The principal comforted her. He radioed for the school nurse and a wheelchair and instructed the office to call her parents, who had a pager number.

While he waited with his hand on her shoulder, he handled a half-dozen other problems. He told a boy to take off his baseball cap; hats were not permitted. If someone wore a cap, it was a clue he was an outsider who had sneaked onto the grounds. Ten minutes later the nurse arrived.

After lunch Tash worked with a vice principal on hiring new teachers for the fall. He wanted to exceed his quota of minorities—28 percent of the faculty—with the first new hires so he would later have the freedom to hire who he wanted. Young teachers were best because they cost less; young and bilingual was nearly perfect.

Then the flasher problem intruded. For the last two days two girls on their way to school had been confronted by a man who jumped out of the bushes at Riverside Drive and Colfax Avenue and exposed himself. It happened again this morning to one of the girls and her mother. Tash could not get the school police to respond to his phone calls. District policy dictated that he go through them first before contacting the Los Angeles city police, who, Tash thought, probably knew who the flasher was. The school officer appeared when Tash was in a closed-door conference. He didn't wait. Tash finally got permission to contact the city police, who showed up too late the next morning to catch the flasher.

There were other crime-related problems: A mentally disturbed girl had reported an attempted rape on her way to school, but Tash and the police believed she had imagined it. Two boys went down by the Los Angeles River and got drunk. A Cadillac repeatedly cruised past the school, and Tash thought it might contain a drug dealer.

There was a marked letdown after lunch. The last class of the day for Pat McHarg, a veteran teacher, was a United States history class composed of regular students. McHarg used a textbook last updated in 1973, despite the fact that the State Board of Education had recently gone through the torturous process of approving a more politically correct textbook for just such a course. The school had no money to buy new books, there were not enough old books to go around for everyone, and students were not allowed to take the textbooks home.

McHarg asked questions by rote that might appear on the final exam. "What was the Progressive Period about?" "Who was the first president to use the power of the federal government for change?" "Who discovered the cure for yellow fever?" (Walter Reed, after whom the school was named.) "Name the five Great Lakes."

"The Colorado River," someone answered. The class was getting restless. They talked to one another.

"By the way, we are not done yet," said McHarg.

The conversations continued on divergent planes from the topic at hand.

"I can't very well go on. I think I lost it for the day."

Mayhem broke out.

McHarg next tried silence. There were shouts of "shut up" and "quiet" that only contributed to the din. Facetious answers to his questions were flung into the air. McHarg stood silent in front of the class. No one paid attention to him anymore.

Finally, he said, "You are on your own. I can't possibly listen. You know, it is incredible, I can't hear you all at once. I will wait until tomorrow when you have one day to go to your final."

There was momentary silence and then a fresh outbreak.

McHarg retreated to his desk and busied himself, head down. Various soft objects flew through the air, some in my direction. There was a steady din punctuated by a few futile "shut ups." Then, mercifully for all concerned, the bell rang and school was over for the day.

Tash was on the front steps of the main building with his radio supervising departures. He contacted a staff person behind the school.

Reed Seven to Reed One.
Reed One.
David, is it pretty calm back there?

Yeah.

Okay, I'm going in.

A Reed student in 1947 wrote a story titled "Our School." It began:

As the last lingering student leaves the grounds, a heavy quiet
descends on the gracious building. The great structure seems
to heave a sad sigh as it prepares for another lonely night.

I was allowed past the guard and onto the grounds of the Lakeside Country Club in Toluca Lake, Bob Hope's old haunt. The river formed the serpentine boundary to the south, and across the concrete arterial rose the
superstructure of Universal City. The sounds of birds and sprinklers were a
constant. There was the occasional click of a well-hit golf ball. It was midmorning; the summer fog was lifting, the pine trees were coming into
clearer focus, and the freeway traffic had settled into its steady midday roar.
The fresh smell of new-cut grass mingled with the aroma of oil-soaked air.

While I waited for the club member with whom I had an appointment, other members were paged. I imagined that those telephone calls
contained the seeds of multimillion-dollar deals that would one day
entertain me, for I was within the placid center of the hurricane that is
the entertainment industry. Latinos carefully raked the sand traps, while
on the wall of the reception area were the color photographs of the white
males who had served as club presidents.

Late that afternoon I stood on the nearby Lankershim Boulevard bridge
that crossed the Los Angeles River. A trail of Styrofoam cups and other
plastic eating implements flowed into the river from the storm drain under
the black-sheathed buildings that were the working studio portion of Universal City. A lone duck waddled across the stained, concrete riverbed. My
companion on the bridge was a homeless man who held a cardboard sign—
stating WILL WORK—like a buffer against the oncoming commuter traffic. I
drove up the hill to one of Los Angeles's many diversions from reality.

It cost five dollars to park and enter what was termed "the essence of
Los Angeles" by its developer, MCA Inc. I had been searching for just
such a summation. The essence of Los Angeles, according to the archi-
tect of CityWalk, was conglomeration. "The theme of Los Angeles is that
there is no theme. The fact that Los Angeles has no theme, as a theme, is

a brilliant idea," he said. But he was wrong. The theme was violence, but that was not allowed here.

CityWalk, the newest attraction on the Universal hill, was a safe, crime-free, upscale, pedestrian-oriented distillation of Los Angeles. In other words, it was not Los Angeles. But people loved it. There were thousands upon thousands of them there that night, walking beneath the stately palm trees.

The milling, sauntering crowds traversed the two blocks, reversed themselves and walked back, or dropped out along the way at the many shops and restaurants. There were lines for everything. MTV-type images of scenes from Universal movies, other attractions on the hill, and the fast-forward construction of CityWalk itself flashed across a huge television screen suspended like a cyclopean eye in the night sky. With mouths slack, people gazed upward at parts of films, films of how films were filmed, a film of the place where films were filmed. They were on the set for Los Angeles. They were in films.

Some walked into Out-Takes, described thusly in a press release:

> A full-service fashion and portrait photography studio with a most unique twist. Through computer-generated composite

The homeless and the river

The movie industry and the river

photography, subjects can be placed in one of hundreds of movie or television scenes, or added to one of thousands of stock photographs—wonderful for gifts or promotional items.

It was a city of shops and restaurants, of flashing lights, of neon signs, of no homeless and no racial tensions, with no visible recession and a place to walk, talk, gawk, and buy. There were stores that specialized in giant objects, items from other worlds, and contemporary clothes. The latter were sold in a store named The Wave. Outside, mechanically generated waves broke upon a bit of sand and rock at one end of a concrete trough.

At the other end of the malled city was a giant sandbox containing chairs and tables with umbrellas that shaded the night sky. A rock band played. There was a notice stating that at such-and-such a time a beer commercial would be filmed there; those present would be scripted into the scene that promoted lite and dry. In the center of CityWalk, children ran gleefully through water jets that were both a playground and a decorative fountain. A street performer coaxed an exotic beat from giant plastic water bottles.

Private security police, uniformed sheriff's deputies, and three men with PROBATION DEPARTMENT stenciled on the back of their windbreakers wove through the crowds. It was, said an MCA executive of the hugely successful commercial venture, "a controlled environment. I call it 'safely chaotic.'" As a place to walk at night, CityWalk had replaced Hollywood

Boulevard and Westwood, around which there were no fences and within which there was danger.

Movies were, or had been, the chief attraction at all three places. The eighteen-theater Cineplex Odeon at Universal City was the largest-grossing theater complex in the nation. Two years ago, at the opening of a movie about South Central Los Angeles, youths fired shots in one of the theaters, wounding five. No place in the city was free from violence.

This summer the Toronto-based theater chain that owned the cineplex took no chances. It did not screen a movie set in the same locale by the same director on the film's opening weekend, when a riot or shooting was thought to be most likely to occur. A spokesman for the Canadian-owned theater complex said, "Our film programmers in Los Angeles are ensuring that the theater is programmed with an upscale demographic to make sure that CityWalk's environment is kept safe with a family atmosphere."

Interestingly, that was the same demographic the *Los Angeles Times* pursued. The newspaper's promotional ads appeared on the many screens before feature presentations at the cineplex.

I left the San Fernando Valley and entered the city proper. That portion of the river that ran through downtown Los Angeles—the locus of the power yielded by the *Times* for over one hundred years—was the most unreal and the most revealing portion of the journey.

The campsite that Crespi called "this delightful place among the trees on the river" two hundred twenty-four years ago was now a city park whose triangular shape was bounded on the west by the concrete river and whose sides were hemmed in by the convergence of North Spring Street and Broadway in the Lincoln Heights section of Los Angeles. There was a sandbox with no sand and a dry wading pool cordoned off by a fence. About a dozen men, whom I took to be illegal aliens, lay on the grass by the fence. They were shaded from the sun by palm trees.

A police helicopter passed overhead. A Sheriff's Department black-and-white jail bus climbed the incline of the Spring Street bridge and disappeared on the downslope, belching diesel fumes. It was headed toward the many jails, prisons, and detention centers in downtown Los Angeles, which housed the most concentrated population of prisoners in the nation.

There was a small plaque set in a concrete base in the park. It noted that "the beginning of the great city of Los Angeles" had taken place on this spot. The concrete base was covered with graffiti, as was the concrete river. Graffiti was layered on top of graffiti until no individual mark could be distinguished. Bridge abutments, the trapezoidal banks, and the concrete arches of freeways far above the river were the favorite targets of the taggers, who sought momentary fame from the placement of their signatures in highly visible places.

A short distance from the river and within its ancient floodplain a tagger had left his mark at the second-floor level of the gleaming white wall of the *Times*'s new $224 million printing plant. A man in a cherry picker extending from the truck of a "graffiti abatement" firm was erasing the mark. The *Times* had previously editorialized, "The worst is that in the wake of graffiti is left a sense of things spiraling downward."

For the downtown portion of the journey, I sought an escort from the Corps of Engineers. There was no bicycle trail along either side of the river, and railroad yards on both sides blocked off access by vehicle. I wanted to travel the actual riverbed, and I knew the area was dangerous.

David G. Weaver, a project engineer, drove a Chevrolet Celebrity with U.S. Government plates down the incline underneath the Sixth Street bridge and out onto the flat riverbed. The river was covered with a thin sheen of water, the overflow from the low-flow channel in the center. The large car, much like a slow boat, plowed steadily through the water that was clotted with drifting algae.

Like a pilot on the Mississippi River, Weaver dodged the debris. The riverbed was at its filthiest here. I compiled a list of unwanted objects: Styrofoam cups, plastic water jugs, take-out shrimp containers, Snapple bottles, Taco Bell soft-drink cups, tires, shopping carts, egg cartons, and small pieces of a smashed mirror. A foul-looking liquid issued from a storm drain; the milk-colored substance had eaten away some of the concrete.

Tucked under the downtown bridges and into the openings of storm drains was the litter of the homeless: torn mattresses, broken furniture, cardboard, plastic drop cloths, plywood shelters, and spiderlike webs of rope attached to bridge abutments that held belongings out over the trapezoidal banks and maximized living spaces. Weaver said, "There is no way to know who is living here or how many have drowned in the flash floods. It is impossible to keep people out. We put razor wire up one day, and the next day it is taken down and sold to other security-conscious people."

A couple scurried into a storm drain near the Sixth Street bridge. I aimed my camera. The woman shouted something, began to walk toward us, and made the time-out signal. The man ducked into the drain. Weaver sped away. He said, "Damn good chance we could be two dead people. I think: We stopped, he's on the lam, he comes out of there with a gun aimed at you and wants your car, wants your money." I thought she wanted payment for the photograph, or had something to sell. I wasn't about to argue, however; I wasn't the expert.

The homeless were all about. They rose specterlike out of the early morning mist along the putrid river, drifting inland to confront office workers hurrying to their jobs. They were omnipresent on the stinking streets, stationed in parks, on sidewalks, at freeway on-ramps, outside of businesses like the *Times,* restaurants, and the Music Center on Bunker Hill. The homeless were not new to the warm weather of Los Angeles. They had been called bums, beggars, hoboes, tramps, stiffs, bindle stiffs, vagrants, and other names in the past. I remembered when I could afford to give them a quarter. No longer. There were too many. The future seemed to belong to them.

There was an alternative future for the river proposed by Weaver and other development-minded people. He thought platforms should be

Cruising the river

built over the river; on these platforms would be such commercial ventures as shops, restaurants, apartments, and a monorail. Weaver and others saw the river's future in terms of a transportation corridor. They wanted to turn the bed of the river into a freeway, a high-speed railroad, or the conduit for a huge freshwater reservoir to be constructed in Long Beach Harbor. And, of course, there was the other extreme: There were a few who wanted the river returned to its natural state.

The talk remained just that in the early 1990s. The river was a joke, a graffiti-covered eyesore, the conduit for sewer flows, a deadly force during flash floods, an illegal kayak or canoe run, the home of the dispossessed, a place to train bus drivers and law enforcement officers in their vehicular specialties, and the set for movie chases.

The floods, like earthquakes and wildfires, scoured the lower plain periodically and contributed to the erratic rhythm of its history. Between 1811 and 1959, by which time most of the river had been encased in concrete, there were between twenty-one and twenty-six major floods, depending on who was counting. In a 1915 report, somewhat suspect because it was aimed at promoting flood control, a board of engineers told the county supervisors, "History shows a mean interval of three and a half years between damaging floods in this county."

In its time the river alternately sustained and destroyed the Indian village of Yangna; the Spanish settlement of *El Pueblo de la Reina de los Angeles,* founded in 1781; the Mexican pueblo of the 1830s and 1840s whose name was shortened to Los Angeles; the wild Anglo-Mexican frontier town of the 1850s and 1860s; and the city that sprang from the town which, up until 1913, was solely dependent on the river's surface and underground flows for a water supply.

The roiling walls of water that traveled at great speeds and were accompanied by the sound of a strong wind could appear without warning; it took only a violent storm in the far-off mountains to loosen the furies. On a moonlit night in the spring of 1825, Don José del Carmen Lugo, a prosperous rancher, heard a "dreadful noise." He hurried outside to investigate.

> I went to the bank, and I discovered that it was a sea of water
> which was overflowing vegetable gardens, fences, trees, and

whatever was before it. The water was running with great violence, making enormous waves.

He ran to warn the sleeping inhabitants, who fled to higher ground on
the nearby bluffs. Adobe homes, cattle, and vineyards were destroyed. It
was after the flood, Lugo said, that the first wild mustard was noticed in
the region. The Hispanics, who had occupied the site for only the previous forty-four years, noted another first: The river, which had previously
emptied into Santa Monica Bay to the southwest, had shifted its course
due south to San Pedro Bay.

The southern end of the plain was inundated. Juan Sepulveda
recalled, "It was so great a flood that it went against the opposite bank
and on this side toward Alameda Street. It was quite a sea. From the
ranch of Mr. Bixby [near Long Beach] to the house on the [Dominguez]
hills belonging to Juan Dominguez it was all a sea of water." The new
channel partially drained the marshes of South Central Los Angeles. The
forests began to disappear, a process that was speeded up later in the century by land clearing, firewood cutting, cattle grazing, and pumping from
wells, which lowered the water table.

Time passed. There were floods in 1832 and 1842. William Brewer
arrived in San Pedro by oceangoing steamer in December 1860. He
transferred to a smaller steamer and rode six miles up the Los Angeles
River, where he took the stage to town. It rained hard in December, and
then rained harder in January. Brewer and his party were camped just to
the east of Los Angeles. He wrote:

> It has now rained about seventy hours without cessation—for
> forty hours of that time, over twenty consecutive, it has
> rained like the hardest thundershowers at home [New
> Haven, Connecticut]. No signs of clearing up yet—fire out
> by the rains, provisions getting rather scarce—one meal per
> day now.

The floods of that year were called the Noachin Deluge. Brewer continued:

> Four men were drowned near here in the recent rains, and
> much damage done. In town vast damage was done—*adobe*

houses hurt—the *adobe* cathedral which has stood over half a century is nearly ruined, some of its walls fallen.

On Christmas Eve of 1861 it began to rain again. It continued to rain for thirty days, with two slight interruptions. Of one break the *Los Angeles Star* said, "A Phenomenon—on Tuesday last the sun made its appearance." On another day the *Star* noted:

> The embankment lately made by the city, for the water works, was swept away—melted before the force of the water. The Arroyo Seco poured an immense volume of water down its rugged course, which, emptying into the river, fretting and boiling, drove the water beyond all control.

A huge lake extended from Los Angeles to San Pedro. It was impossible to get to the port for one month, unless a horse was made to swim across the dips in the higher ground to the west. "We thought the ocean had come up here," said one old-timer.

A terrible drought followed the floods. J. J. Guinn, an early-days Los Angeles historian, wrote:

> The loss of cattle was fearful. The plains were strewn with their carcasses. In marshy places and around the cienegas, where there was a vestige of green, the ground was covered with their skeletons, and the traveler for years afterward was often startled by coming suddenly on a veritable Golgotha—a place of skulls—the long horns standing out in defiant attitude, as if protecting the fleshless bones. It is said that 30,000 head of cattle died on the Stearns Ranchos alone.

When John Guess drove to Wilmington in his wagon, he had to tie a handkerchief over his nose. "You could walk for miles on the dead cattle," said Guess. "The whole slough and river down below Bixby Hill was full of them." Jotham Bixby said that the floods of 1866 brought the willows back. Bixby bought the twenty-seven-thousand-acre ranch between Long Beach and Wilmington for twenty thousand dollars in the middle of the drought. Two shearings of wool from his sheep paid the debt. He and others later reminisced about the terrain. Bixby recalled:

It was a mass of jungle in those days. The valley was practically impassable except for two or three trails, and it was covered by a dense growth of willows, weeds, briars, some blackberry vines, and marshes grown up with tules. In those days there was bear in the valley as well as other wild animals.

Bixby and his clan looked forward to the floods because they brought rich new soil that fertilized the land. Floods also meant hard work, and they brought conflict. George Bixby said:

> In our case the water began washing away our land and we wanted to protect our place and get rid of the water. Many times I have gone up the river, taking a few Indians with me to move a few boulders about and get the water going the way we wanted it to. By placing a double row of boulders across the stream it would raise the water level enough to throw the water over the other way. Of course, it brought opposition from people on the other side who did not want the water either. It finally became serious; and when a man wanted his work to stand, it was necessary to stay with it armed with a double-barreled shotgun.

Duck and geese were hunted along the shores of Nigger Slough, which expanded to five thousand acres during the wet season, almost reaching Redondo Beach. When it rained, the locals strapped boards on their feet so as not to become mired in the thick mud. They swam across the river, pushing a board in front of them. People crossed the plain in small boats during floods. With the coming of drier weather, Nigger Slough contracted and thousands of carp were trapped in shallow water. The stink was terrible, and they were set on fire.

In its placid moments, the Los Angeles River and the surrounding region could be a "veritable paradise on earth." That was what Ludwig Louis Salvator, archduke of Austria, observed when he rode along the river in 1876. The broad riverbed, through which only a trickle ran, passed "through country that is still virgin, uninhabited, and where the silence of Nature is unbroken."

The archduke, who spoke a dozen languages and had a passion for travel and science, noticed the graceful willows, the thick oaks, and the luscious orchards of olive, peach, pear, and apple trees. However, not all was peaceful. The author of some forty books, Salvator wrote in *Los Angeles in the Sunny Seventies*: "Earthquakes occur fairly frequently, usually in August; this is probably why many rich families prefer to live in wooden houses."

There were problems in this earthly paradise. The Hispanics, who served Salvator a juicy watermelon, "spoke wistfully of life in the past and of the days when there was great freedom." Of the Indians, he wrote, "At one time long-lived, they are now becoming strongly addicted to brandy. The women drink as well as the men." The Chinese were under siege: "A group has been formed that has declared, both by word and deed, a veritable war against this peaceful Mongolian invasion."

The history of the human occupation of the lands inundated by the river mirrored the capricious violence of the region's natural history. The Indians were first: They were plied with liquor, sold as slaves, plied with liquor again, and sold again and again and again until they died. Horace Bell, who was a witness to the process, wrote, "Those thousands of honest, useful people were absolutely destroyed in this way." He added, referring to the 1850s and 1860s, "Indians did the labor and the white man spent the money in those happy days." Both Hispanics and Anglos participated in the practice of slavery that continued in Los Angeles after the Civil War.

In 1869, the editor of the Los Angeles *Semi-Weekly News* criticized the inefficient practice of using Indian slaves, which discouraged the importation of more worthy immigrant labor and retarded the economy. He said the Indians built no houses, owned no lands, paid no taxes, and "encourage no branch of industry." The editor added:

> The habits of the Indians are such that decay and extermination has long since marked them for their certain victims. They have filled our jails, have contributed largely to the filling of our state prison, and are fast filling our graveyards, where they must either be buried at public expense or be permitted to rot in the streets and highways.

Bell was a lawyer, newspaper publisher, and vigilante ranger who did not mince words and had an independent mind, being one of only two men from Los Angeles to fight on the Union side during the Civil War. He put the blame on the system, whose weekly cycle he described thusly:

> On Sunday the streets would be crowded from morn till night with Indians, males and females of all ages, from the girl of ten or twelve, to the old man and woman of seventy or eighty.
>
> By four o'clock on Sunday afternoon Los Angeles Street from Commercial to Nigger Alley, Aliso Street from Los Angeles to Alameda, and Nigger Alley, would be crowded with a mass of drunken Indians, yelling and fighting. Men and women, boys and girls, tooth and toe nail, sometimes, and frequently with knives, but always in a manner that would strike the beholder with awe and horror.
>
> About sundown the pompous marshal, with his Indian special deputies, who had been kept in jail all day to keep them sober, would drive and drag the herd to a big corral in the rear of the Downey Block, where they would sleep away their intoxication, and in the morning they would be exposed for sale, as slaves for the week.
>
> Los Angeles had its slave mart as well as New Orleans and Constantinople—only the slave at Los Angeles was sold fifty-two times a year as long as he lived, which did not generally exceed one, two, or three years under the new dispensation. They would be sold for a week, and bought up by the vineyard men and others at prices ranging from one to three dollars, one-third of which was to be paid to the peon at the end of the week, which debt, due for well-performed labor, would invariably be paid in *aguardiente*, and the Indian would be happy until the following Monday morning, having passed through another Saturday night and Sunday's saturnalia of debauchery and bestiality.

The frontier town of Los Angeles during these years was disheveled, lawless, and in a constant state of siege. There were rumors of revolts or invasions by the recently displaced Mexicans, who, it was said, would throw the gringos out. Besides losing their country to superior firepower and Manifest Destiny, the Hispanics, although a numerical majority, were relegated to minority status—meaning, in this case, that they were hunted down on the mean, unpaved streets of Los Angeles and in the arid backcountry of Southern California. "The Americans of the Angel City were in the habit of amusing themselves by hanging some luckless Mexican," said Bell. If cheap and sometimes poisoned liquor were the weapons of choice employed against the Indians, the gallows was the hastily constructed device that claimed many a Latino life.

It is amazing that moviemakers and the writers of western pulp fiction, many of whom have made Southern California their home, have ignored Los Angeles's role as the foremost western locale for barbarity and summary justice. For a time, murders averaged one a day. A common question at breakfast was, "Well, how many were killed last night?" The celebrated Tombstone, Arizona, was a kindergarten in comparison to Los Angeles's graduate school in frontier violence. Los Angeles far outdid San Francisco, whose vigilantes have received much more attention.

Records were spotty because few vigilantes wanted to draw attention to their illegal activities. Since the self-appointed guardians, or at least their leaders, were from the "better" classes, they also tended to control recordkeeping. There may have been eight killings by vigilance committees in San Francisco in the mid–nineteenth century, while there were between thirty-two and thirty-five similar lynchings in Los Angeles. Additionally, in Los Angeles there were thirty-eight ad hoc lynchings, meaning a more impulsive act than those arranged by the various committees, usually after some type of impromptu hearing. There were also forty legal hangings. (These figures do not include those Chinese lynched in the 1871 race riot. Nor do they include the many unsolved murders of the day that featured the knife and the gun.)

Most of the lynching victims were Hispanic. In his study of Spanish-speaking Californians from 1846 to 1890, Leonard Pitt wrote, "Vigilante justice had a distinctiveness in Los Angeles, however, in that every important lynch-law episode and most minor ones involved the Spanish-speaking." He added that Anglos of the time described such activities as "the war of the races."

Richard Griswold del Castillo, an associate professor at San Diego State University, referred to this time as "the bloody chronicle of racial warfare in Los Angeles." The lynching of Hispanics continued into the 1890s in such outlying communities as San Bernardino and Santa Ana. Castillo also wrote of the unequal treatment under the law, when it was applied. An Anglo accused of murdering a Mexican was sentenced to one year in jail. He served only seventy days. A Hispanic was given ninety days for disorderly conduct under terms of the "Greaser Law," a state law that provided for the jailing of Latinos on vague charges.

Horace Bell wrote that "police statistics showed a greater number of murders in California than in all the United States besides, and a greater number in Los Angeles than in all the rest of California." Bell sometimes employed exaggeration as a technique; but violent death was certainly no stranger in Los Angeles.

There was another witness. A sickly Presbyterian minister, James Woods, arrived in Los Angeles in 1854 and spent a tormented year railing in his private journal against the idolatry and popism of the Catholics and horse racing on the Sabbath. Rather than the City of Angels, Woods referred to it as the "City of Demons." Frequent offhand references to violent acts were scattered throughout his journal, such as: "Two men were shot last night one mortally the other slightly. Both low mexicans. Rum the cause Murders are almost of daily occurrence. This is an awful place."

Another observer who was on the scene, William Brewer, foreshadowed the extremes to which Los Angeles citizens would go to arm themselves in the late twentieth century. He wrote on December 7, 1860:

> This southern California is still unsettled. We all continually wear arms—each wears both bowie knife and pistol (navy revolver), while we have always for game or otherwise, a Sharp's rifle, Sharp's carbine, and two double-barrel shotguns. Fifty to sixty murders per year have been common here in Los Angeles, and some think it odd that there has been no violent death during the two weeks that we have been here.

Writing of vigilante actions in the *Southern California Quarterly,* the publication of the Historical Society of Southern California, Robert W. Blew stated in 1972:

> Los Angeles is a city of violence. Murders, riots, slaughter on the highways, and crimes of passion seem to be daily occurrences in her history; furthermore, during the years 1835 to 1874, the first and last recorded lynchings in the city, the citizens resorted with distressing frequency to vigilante actions.

Not only does Los Angeles have a long record of individual (and mass) murders, but also her history includes the unenviable record of at least three major race riots.

What might be viewed as the first major race riot, or just the continuation of the existing pattern of racial warfare, erupted on the night of October 24, 1871, when the bullet, the knife, and the noose were employed against the Chinese.

Los Angeles had the distinction on that October evening of hosting the first large-scale mob attack on the Chinese, a practice that soon spread throughout the state (witness Truckee, Eureka, and Turlock, among other places) and nation, much as the white-black urban conflict radiated outward from Southern California nearly one hundred years later. A local historian referred to the attack on the Chinese as "probably the biggest mass lynching in the history of California, if not the whole country." Writing in the publication of the Historical Society of Southern California, Paul M. De Falla termed it a "pogrom."

The fury was intense. In four hours the mob and police—using non-repeating guns and laundry lines for nooses—managed to shoot, stab, and hang more victims than were killed on the first nights of the riots in South Central Los Angeles in either 1965 or 1992.

As with the subsequent riots, no pundits predicted the violent outburst. The Chinese were hardly a threat. A few had migrated south after completion of the transcontinental railroad in 1869. They accounted for 172 inhabitants, or 3 percent of the population of Los Angeles in 1870. (The violent outburst instantly decimated 10 percent of this segment of the population.) The Chinese had replaced the Indians as domestic servants and laborers. They were believed to have hoarded gold.

The Chinese lived in a ghetto, hemmed in by others. Hispanics posted signs stating that the Chinese were not wanted in their adjacent neighborhood. The Celestials were stuffed into small apartments off Nigger Alley, as it was known colloquially. *Calle de los Negros,* or Negro Alley, as it was designated in print, was Los Angeles's sin street. The alley, a short, narrow dirt street leading off the central plaza, was flooded sometimes when the river overflowed its banks. The alley's name was later changed, and the extension of Los Angeles Street to Alameda Street obliterated it in 1888.

The raw materials for racial conflict were present. Alcohol, the drug of preference at the time, was omnipresent. In a city of some 6,000 residents, many of whom were women or children, and 2,400 registered voters, some of whom may have been nonexistent, 12,000 drinks were served on election day in the month preceding the riot. The breakdown was: 7,000 beers, 3,000 whiskeys, and 2,000 glasses of wine. It was a primary, not a general election. A drink could be purchased in 110 out of 285 Los Angeles business establishments. There was no drunk ordinance.

The newspapers were virulently anti-Chinese and condoned summary justice. Crime was rampant. The police were corrupt. They openly served as enforcers for the two main Chinese companies, popularly known as tongs.

The conflict was ignited by two tongs fighting over a woman who was attempting to escape bondage. It was heralded by the following announcement in the newspaper:

MELICAN MARRIAGE CEREMONY

The Whites' marriage ceremony is becoming quite fashionable among the followers of Confucius in this city. Such a contract was consummated between a "John" and a "Maly" last evening by Justice Trafford.

It is supposed that these marriages are contracted to evade paying the purchase money to the Company whom the woman may have been previously bought by.

A few days later, two police officers drinking in a bar heard shots coming from the alley. They found a strongman from one of the tongs lying in the dirt; blood spurted from a hole in his neck. Seeing the officers, the Chinese scattered. Most of them made for the apartments in the Coronel Block on the west side of the alley. A leader of one of the companies came to a doorway and loosed some shots that superficially wounded a few bystanders. He ducked back inside. A policeman charged into the room, only to emerge without his gun and a wound in the shoulder. He blew his whistle.

A former saloonkeeper–turned-rancher and another policeman fired round after round indiscriminately into the apartments. No shots were returned, and the officer left to get more ammunition. The rancher walked toward one of the doors and was greeted by a fusillade of shots. He staggered backward with a chest wound, muttering, "I am killed," and died one hour later.

The attitude of city officials was to see no evil. The city marshal, who doubled as the chief of police, arrived at the scene. He gave the order to shoot to kill, and then left. The county sheriff arrived, and soon thereafter he, too, decamped. The mayor arrived and departed.

Learning that the rancher had died, the gathering mob became enraged. The rumor spread quickly throughout the city that the Chinese were "killing whites wholesale." Orators of the moment harangued the crowd in Spanish and English. Water from a fire hose was played on the structures in a futile attempt to flood out the Chinese, but the mob soon tired of this play.

The unruly crowd—whose size was variously estimated at five hundred, one thousand, or three thousand people, the latter figure representing one-half the city's population and, therefore, obviously an exaggeration—seized a man being conducted to jail by a police officer named Emil Harris. A rope was borrowed from the owners of a bookstore, and they strung him up on a corral gate, just across the street from the only Protestant church in Los Angeles. The scene was vividly described by a merchant, Harris Newmark:

> Emil Harris then rescued the Mongolian; but a detachment of the crowd, yelling "Hang him! Hang him!" overpowered Harris at Temple and Spring streets, and dragged the trembling wretch up Temple to New High street, where the familiar framework of the corral gates suggested its use as a gallows. With the first suspension, the rope broke; but the second attempt to hang the prisoner was successful.

Afterward there was a great deal of fudging about who made up the mob. The actual composition was of less interest than who it was said to have contained. The persons who tried to disperse the mob and shelter the Chinese, it was reported, were "the better element in the population." The mob itself was composed of "people of all nationalities" who were "the scum and dregs of the city," meaning those of Irish and Mexican descent. The civic whitewash continued nine years later when the poet-historian A. J. Wilson wrote that "American 'hoodlum' and Mexican 'greaser,' Irish 'tramp' and French 'communist'—all joined to murder and dispatch the foe." Most likely the mob was a cross section of Los Angeles society at the time.

Whoever composed the mob, or at least part of the mob, had political clout and received protection. The mob included, said historian De Falla, "a great many people of consequence in Los Angeles." The lists of names compiled by both the coroner and a grand jury were never made public and were conveniently lost. Of the proceedings before the coroner's jury, the *Los Angeles Daily News* noted, "But a peculiar ignorance of names was manifested by the majority of witnesses."

There was no doubt, however, about the mob's frenzy and rapaciousness during the riot. The crowd, including a constable, fired at two Chinese women, wounding one. They chopped holes in roofs and poured gunfire into crowded apartments where women and children huddled. When men fled into the street, they were immediately cut down by a barrage of gunfire.

Other wounded Chinese were hauled from apartments, kicked, stabbed, and then hanged. They tried to smoke out the Chinese with fire but extinguished the flames when calmer persons pointed out that the fire might spread to engulf the whole city. A young Chinese doctor was shot in the mouth, robbed, and then hanged. Apartments were looted and the pockets and mouths of the dead were searched in the relentless quest for gold. Even a fourteen-year-old boy was hanged. Horace Bell noted in his memoirs that Los Angeles had "sown the wind and reaped the whirlwind."

The most commonly cited number of Chinese deaths was nineteen. Of the fifteen who were hanged, some were shot or stabbed, or shot and stabbed before they were dropped from three makeshift gallows. The white rancher made a total of twenty deaths. There was no mention of the wounded or injured, who must have been numerous.

Based on an indictment known to be faulty, eleven scapegoats were tried and nine were convicted of manslaughter. After serving a little more than one year of their sentences in San Quentin, the State Supreme Court ruled that the indictment was flawed; they were released. (There would be echoes of imperfect justice years later in the Rodney King beating case.)

Through attrition and because of a large influx of white midwesterners, the few remaining Indians, Hispanics, and Chinese were quickly subsumed into the new culture. Los Angeles had evolved into a white city by the 1880s in a process that historian Carey McWilliams labeled "genocide." Extreme change was a constant in Southern California. McWilliams wrote:

> From 1850 to 1870 Los Angeles was "the toughest town in the nation," but it became the most priggish community in America after 1900. A glacial dullness engulfed the region. Every consideration was subordinated to the paramount concern of attracting church-going Middle Westerners to Southern California.

In California, power, like the landscape, continually shifted; longevity in power was the exception. For the next hundred years one family, the Otis-Chandlers, who owned the *Los Angeles Times,* would have more to say about that holy California trinity of water, real estate, and race than any other private entity in the state. In California, such a length of time was considered an eternity.

The ingredients for great wealth and power in California were simple: Water was needed to make the real estate valuable. The races were manipulated in order to provide the cheap labor that yielded great profits, yet they could not be allowed to devalue the real estate or compete with the rulers. The Otis-Chandlers and the *Times* were Los Angeles to the extent that the family headed the oligarchy that ruled the region, and the newspaper set the agenda for a longer period of time and with more consistency than any other institution.

Of the late nineteenth century, McWilliams, for a period a *Times* employee, wrote:

> For here was a unique situation: a small group of men, under the leadership of Otis [Harrison Gray Otis, the first *Times* publisher], had not merely "grown up with" a community, in the usual American pattern, they had conjured that community into existence. Having taken over at a moment of great crisis, when the older residents had suffered a failure of nerve, they felt that not only had they "saved" Los Angeles, but that it belonged to them as a matter of right.

That right was never overtly asserted in public and only hinted at in private documents. Otis wrote in an historical sketch of the city and newspaper that circulated among stockholders in 1908:

> To a great extent the growth of Los Angeles has made possible the growth of THE TIMES. On the other hand it may be truthfully said, without boasting, that the wonderful growth of Los Angeles is to some extent due to THE TIMES. Thus are the fortunes of the leading journal of the Great Southwest and those of the City of the Angels so closely inter-related as to be practically inseparable.

Otis's heir and son-in-law, Harry Chandler, made that essential link clear in 1931 when he wrote in the company's in-house publication, *Among Ourselves*: "We—and that means all of us from President to printer's devil—are of, for and devoted to Los Angeles, as is the newspaper of which we are part and parcel—the only such one left to this city, more's the pity." He was referring to the recent acquisition of the *Los Angeles Evening Express* by the outsider William Randolph Hearst and the absentee ownership of the other three newspapers.

David Halberstam, an easterner, cited the congruence of family, newspaper, and region nearly fifty years later. He wrote of General Otis; Harry Chandler; Harry's son, Norman Chandler; and Norman's son, Otis Chandler, in *The Powers That Be*:

> It is a dynasty, one of the few remaining in American society, surviving the dual contemporary onslaught of modern inheritance taxes and normally thinning genes. Its power and reach and role in Southern California are beyond the comprehension of Easterners, no Easterner can understand what it has meant to California *to be a Chandler*, for no single family dominates any major region of this country as the Chandlers have dominated California, it would take in the East a combination of the Rockefellers *and* the Schulzbergers to match their power and influence.

The Chandlers were not the only long-lived and powerful family in California. Newspaper dynasties were a California oddity, sort of like copper-mine owners in Arizona and bankers in New York, only more dominant within their various areas of influence. The state was so geographically fractured that each had their own fiefdom. Two political-science professors at San Jose State University, referring to the transience of the population, wrote, "Nowhere in the country are the media more important than in California." Few residents had a sense of history, except for what passed for such in that day's newspaper or television newscast.

Other newspaper dynasties of varying length were scattered throughout the state: in the Central Valley there were the McClatchys of the three *Bee* newspapers; in Los Angeles and San Francisco, the Hearsts of the *Examiner* papers; in San Francisco, the de Youngs and the Thieriots of the *Chronicle;* across the bay in Oakland, the Knowlands of the *Tribune;* the Ridders, in San Jose and later Long Beach; and the Copleys, in San Diego.

Except for the McClatchys, all were staunch Republicans. There probably was no grand collusion; rather, it was a matter of similar ways of viewing matters and minions conferring among themselves and then, perhaps, talking to their publishers. James Bassett, once a key aide to Richard Nixon and later editor of the *Times*'s editorial pages, interviewed William F. Knowland, the publisher of the *Tribune* and a former Republican senator, in 1973 for a history of the *Times* that was to be published in book form. The question was framed in a manner that begged denial:

> Bassett: Bill, this might be a good time to ask about the continuing, or perhaps it's died away now, myth that used to come up of the so-called *Tribune, Chronicle, Times* axis, or however you want to put it. The implication being that the publishers or their political representatives would get together and plan for the Republican Party.
>
> Knowland: No, that is not a statement of fact. It is true that the three of them on many occasions supported statewide candidates who were the same, but this was not the result of any getting together and planning to do so, though that term was used by the opposition at times to try to, I suppose, strike at the press, or the "establishment," so called. But there was never such an "axis" as such in any formal or informal sense.

Halberstam picked up on the idea of "The Axis." It was created by Harry Chandler, he wrote, "so that those three conservative papers, acting as one, could totally dominate California politics" through mid-century. But all was not complete harmony among the kingmakers. They were bitterly divided on water, whenever its proposed use might cross circulation boundaries; but they agreed on the need to develop the real estate within their respective spheres of influence with water, whether it came from the Owens Valley or Yosemite National Park. The need for cheap labor was also a given, as was its inherent dangers. The dynasties were the haves; the minorities, the have-nots.

The greatest of them all, the Chandler dynasty, was no longer at the head of the *Times* in the early 1990s. The genes had finally given out, and both the once tremendously profitable newspaper and the region were in decline. But the beginning of the city's and the newspaper's growth, on the eve of the boom of the 1880s, was a time of great optimism. The history of both was characterized by the continuing conflict between the haves and have-nots.

Beginning in the 1880s and for the next sixty years, exclusionary immigration acts of Congress, anti–alien landowning and other restrictive laws passed by the state legislature, school segregation, prejudicial housing covenants, and the rantings and ravings of the press replaced physical violence as the means of manipulating minorities. Harrison Gray Otis, a Civil War veteran who thought himself a military man running a military operation, led the parade in his pugnacious manner.

He and his newspaper were yoked to the river and the city from the start. The Potter flatbed press was powered by a waterwheel that was fed Los Angeles River water via a ditch. The press broke down on occasion when a carp or other fish became stuck in the mechanism.

Shortly before Otis took over as editor in 1882, the paper had excoriated the Chinese at the height of one of the frenzies of racial hatred that periodically swept the state. Chinese washhouses, it was held, were responsible for the appearance of "syphilitic sores" among "moral people." As evidence, the newspaper cited "the well-known fact that so impure has become the Chinese nation as a whole that the custom of handshaking is unknown among them."

The newspaper tipped its hand as to the real reason for the anti-Chinese crusade in a December 1881 article headlined THE HEATHEN CHINEE. The story purported to detail the lifestyle of the Chinese in Los

Angeles. It managed to pillory two minorities in one sentence: "The average Indian possesses more luxuries and can be no more unsavory" than
the average Chinese. The article then went on to pinpoint the main
grievance against the Chinese: "The strictest domestic economy is practiced and herein lies the secret of their ability to under-bid white labor
and thrive on a pittance."

This was not Otis's position once he took over the newspaper: He
favored the use of Chinese and African Americans as cheap labor. An
1885 editorial stated:

> Among all the nations, tongues and tribes of the world the
> now unfettered colored man is entitled to a place—to an
> open field and a fair race—and even California has a need of
> him, for the State presents a broad field for labor, in which
> there are still too few workers. Give the black citizen a
> chance, say we.

Under the headline of DROLL STORIES, the newspaper printed racial
jokes copied from other sources. This one came from the *Cincinnati
Times-Star*. A southerner was given a bottle of bad whiskey and passed it
on to "one of his faithful negro retainers." He asked his servant if he liked
the whiskey. " 'Suited me zactly kunnel,' replied the colored man. 'That
whiskey was shorely jus right foh me. Any better, you'd a drunk it, en any
wusser, nobody could drink it.' "

When the presence of Native Americans threatened the whites in
California and in neighboring states, an Indian became "a dirty, lazy,
treacherous animal, gorging himself whenever he has the chance and
murdering in an objectless manner for the very lust of blood." Do-
gooders in the East were "theoretical philanthropists, who, seated in
their comfortable studies, view the Indian from a distance of three thousand miles as a picturesque abstraction—a chivalrous being of noble
instincts, whose rights are continually being trampled upon by brutal
white settlers." After the Indians were subjugated, there were a number
of editorials entitled "Lo, the Poor Indian."

Jews, who tended to be advertisers, were treated kindly; the *Times*
took off after "Jew baiters" and anti-Semites. There was no mention of
Mexicans until 1928 in a chronology of articles gathered under the title
"Racial Attitudes of the Times (1881–   )." The chronology was com-

piled in the 1970s by *Times* historians. It was as if the Mexicans were an invisible race up to the 1930s. But their time would come.

The "Japs" and the "Yellow Peril" were quite visible at the time of the Russo-Japanese War. "Race ties bind Russia and America together," declared a *Times* editorial in 1904. The paper predicted in 1905, "The day will dawn when the Jap will be cursed from Orient to Occident by those who now lavish friendship upon him." The war ended, and Otis, ever the admirer of cheap, productive labor, now referred to the Japanese as "the Brown Hustler" who worked instead of loafed. But there was a limit to Otis's tolerance of the Japanese. He admired docility, stating:

> For thirty years we have been preventing the Chinese from coming to our shores and have not allowed them to have the privilege of citizenship. Now it is proposed to allow the Japanese to have this right which we have not allowed the Chinese. I am opposed to that very strongly, as personally I am of the opinion that the Chinese is a great deal better than the Japanese.

Of the marriage of a white clergyman's daughter to the family's Japanese cook, Otis preached tolerance: "With the young people themselves it is all a matter of taste. If Miss Emery preferred the almond eye and bronze complexion of the Oriental to the pink and white of the Caucasian, it can't be helped."

In 1909, one year before the Times Building was bombed by unionists and twenty employees were killed, Otis identified "our twin plagues" as "race relations and strikes." After the bombing, Carey McWilliams wrote, "a reign of unmitigated political terror was unleashed in Los Angeles." Thugs, professional gunmen, private detectives, and a sympathetic city police force were enlisted in the war against unions and suspected communists. Morrow Mayo wrote of this period in his history of the city: "It is somehow absurd but nevertheless true that for forty years the smiling, booming sunshine City of the Angels has been the bloodiest arena in the Western World!"

The Yellow Peril returned in 1919 as race hatred again flared throughout the state. Harry Chandler's *Times* warned of a foreign takeover because of an increase in the Japanese birthrate. IT IS NO FALSE

ALARM, warned the newspaper with the only use of all capital letters in a lengthy editorial headlined ALARMING JAPANESE MULTIPLICATION IN CALIFORNIA. Picture brides were the danger. The editorial, reflecting the belief held seventy years later that Latinas were "breeders," read, in part:

> The fruitfulness of those brides is almost uncanny. A new baby arrives as often and as regularly as the springtime to the Japanese mother in California. They are a hardy lot, reproducing at a rate that threatens to become appalling.

On Sunday, July 25, 1920, the page-one headline declared YELLOW RACE OVERRUNNING STATE: HOW THE JAPANESE, IN DEFIANCE OF THE INTERNATIONAL AGREEMENT, ARE GRADUALLY ABSORBING CALIFORNIA SET FORTH AND PROVEN; ACTION DEMANDED. An accompanying article was titled THE JAP MENACE IN CALIFORNIA. Along with similar clarion calls in the McClatchy and de Young newspapers in Sacramento and San Francisco, action was taken against the Japanese in the Central Valley and elsewhere throughout the state. New, restrictive laws were passed in Sacramento and Washington, D.C.

The *Times*, having historically favored the more pliant Chinese, recommended importing more of the "Celestials" to break the Japanese "monopoly" on farm production. The newspaper opted for "selective immigration," an issue that would return in the 1990s.

For nearly thirty years, from 1922 to 1950, Carey McWilliams had a "ringside seat at the year-round circus" that was Southern California in its most formative period.

Born into a Colorado ranching family that went bust, and later kicked out of a Denver college for partying, McWilliams landed in Los Angeles like many of its other inhabitants—a migrant on the verge of failure. He worked in the credit department of the *Times* while attending the football factory and finishing school for the sons and daughters of Southern California gentry that was the University of Southern California (USC) in the 1920s. McWilliams got his undergraduate and law degrees at USC. He practiced law for a time in one of Los Angeles's most prestigious firms, handling some minor legal matters for Mrs. Harry Chandler and her son, Norman.

Outside of work he cultivated a literary career. McWilliams was part of the intellectual circle that gathered around Jake Zeitlin, a bookseller. He was a friend and protégé of H. L. Mencken, a friend of poet Robinson Jeffers and socialist Upton Sinclair, and corresponded with novelist Mary Austin. His first book was a biography of that independent-minded Californian Ambrose Bierce.

In the Depression decade of the 1930s, McWilliams turned to the subject of race relations. He was the first writer to document and comment upon the social fabric of California in a realistic manner. Almost single-handedly, he brought the long-buried issue of racism to the attention of his generation through a series of books and numerous newspaper and magazine articles. McWilliams, however, went beyond the passive act of writing and became actively involved in the major issues of his time. He served on and headed numerous committees, represented criminal defendants and labor unions, and headed the state agency that oversaw farm labor.

The author-lawyer-activist was at the center of such issues as migratory farm labor, labor unions (including a strike in the desert community of Trona), the prewar anti-Fascist movement, Los Angeles crime, the internment of Japanese-Americans, the Sleepy Lagoon murder case, the Zoot Suit riots, the emergence of Richard M. Nixon, the early McCarthy era, and the Hollywood Ten. A fast typist, McWilliams wrote ten books in California and contributed to local publications and such national journals as the newspaper *PM,* and *Harper's, The New Republic,* and *The Nation* magazines. From his writings would emerge the ideas for the movie *Chinatown* and the stage play and subsequent movie *Zoot Suit.*

At all those levels McWilliams tried to educate himself and others. He later wrote, "In one way or another the books came out of my California experience and were very much a part of the times and the place." A present-day California historian, Kevin Starr, wrote that "no one mastered his region more completely" than McWilliams, who left California in 1950 to go to work for *The Nation,* where he was editor from 1955 to 1975.

If the Chandlers and their newspaper were the leaders of the conservative Southern California faction, McWilliams and his writings represented the best of the liberal viewpoint. That both could coexist in Los Angeles was a reflection of the region, which, in McWilliams's words, was "a paradoxical land with a tricky environment." *Times* columnist Harry Carr, who also ran political errands for Harry Chandler, once asked

McWilliams, "Why do you like to pick on L.A.? It is a sap town: but so are the rest of them!" McWilliams saw the *Times* as the institution most responsible for the phenomenal growth of Southern California and his onetime boss, Harry Chandler, as "the region's most influential resident."

McWilliams later looked back and cited two events in his life that shaped his political philosophy: First, his father going broke when the cattle market collapsed in 1919 "left me with a distrust of the system." All was not as it appeared to be, and McWilliams decided to distance himself from the mainstream. Second, his experience of working in the credit department of the *Times* during the 1920s "gave me an extraordinary insight into the underside, the seamy side, of the boom in Southern California."

Near the end of his life, McWilliams was asked by an interviewer to explain his theory of race relations as it evolved from his California experience. He could have been referring to the Mexicans and the Japanese, among others, when he said:

> I was convinced there were two kinds of stereotypes, two kinds of minority groups. One is the working-class kind of stereotype. It's the group that comes in and lacks the skills, the education, the background to get up. So you exclude them. You give them the undesirable jobs, and you sort them out, residentially and otherwise. And, of course, you fashion a stereotype to justify what you've done and to perpetuate it. So you say of this group that they are lazy, shiftless; they have too many children; their sexual morals are terrible, etcetera, etcetera.
>
> And then there is another kind of stereotype. It involves another kind of minority. And you say of this kind of minority, "They're too smart. They're too clever. They're too resourceful. They're too clannish." It's a stereotype, but it's quite a different kind of stereotype because it has to rationalize a different kind of situation. They're both rationalizations, of course.

These were not the messages that Harry Chandler and his newspaper delivered to the citizens of Los Angeles. Otis and Chandler had been involved in the machinations that resulted in Owens Valley water being brought south to the San Fernando Valley; Chandler, in particular, had gone about amassing a huge real estate empire. Otis was the founder and ideologue; Chandler was the practical empire builder.

At the peak of acquisition, Chandler and his associates owned from one and a half to two million acres, a figure equivalent to between 3 percent and 4 percent of the private lands in the state. Their holdings were concentrated south of the Tehachapi Mountains in the San Fernando Valley, in downtown Los Angeles, along the southern coast, and through the interior mountains and valleys of Southern California. There were additional holdings elsewhere in the West and in Baja California. An historian of the *Times*, Robert Gottlieb, wrote that Chandler was "California's leading land baron."

In a democracy, there was something feudal about Chandler's operations. He made it a point of not leaving a paper trail, but scraps of his jottings could be found elsewhere, such as a note to H. J. Whitley, "The Great Developer." Whitley had written Chandler boasting of his plans in San Luis Obispo County. Chandler replied, boasting of his plans in Mexico. It was a case of one-upmanship; Chandler won easily. He wrote:

> We have a little more than 200,000 acres of land under cultivation now on our ranch in Lower California, with something over 100,000 acres in cotton this year. We are getting our railroad built through from Calexico to the Gulf of California and hope to have it in full operation within the next 12 months. Our harbor development (Port Otis) there is going to make a seaport for Arizona, Imperial Valley and all of our own immediate territory. The change that has occurred since you were down there nearly 20 years ago I am sure would interest you and some time when you can spend a day or two I hope you will come down and make the trip to the ranch with me again because we have been going ahead as best we could and constantly for just about an even 30 years since I first took the option on the property in December, 1898. We have a little over 3,500 miles of main and lateral canals with drain ditches and headgates. I believe it is the biggest irrigation system in the world in private ownership—at least we can find no record to the contrary.

The story differed for the have-nots. Great economic disparity, much of it very visible, was another Southern California characteristic. During the early years of the century, Mexicans returned to California in increasing numbers. Revolution and poverty in Mexico and the development of irrigated agriculture and industry in Southern California contributed to a rise in the Mexican population of Los Angeles from 5 percent in 1900 to 20 percent in 1930. Industrial developments surrounding the Plaza area pushed the Mexicans eastward across the Los Angeles River, which became the western boundary of the barrio.

The Mexicans were an invisible population, sealed off from the rest of the community by occupation, segregated housing, and segregated schools. They built and maintained the railroads and streetcar lines and "filled a labor vacuum created in numerous agricultural communities by the restrictions on Asian workers," according to historian Ricardo Romo.

The image of Los Angeles as the "White Spot of America," as the *Times* called it, peaked around 1930. *Saturday Evening Post* readers were assured that Los Angeles residents were "native Americans almost in total." The coastal city of El Segundo boasted it had "no negroes or Mexicans." Long Beach, which actually had a sizable Mexican population, claimed that 98 percent of its inhabitants "are of the Anglo-Saxon race." Joseph Lilly, a former New York City newspaperman, wrote in the *North American Review* that, despite a large Mexican population, Los Angeles was "peculiarly 'white' in the composition of its population." He added that the city is "an interesting experiment for the Anglo-Saxon in America."

Along with the influx of Mexicans—actually, more of a return home, as some regarded it—there were a number of "brown scares." These racial outbursts climaxed in the deportations of the Depression years. As usual, economics was a factor: White migrants from the Dust Bowl lessened the need for Mexican farm laborers. There were attempts to organize Hispanic farmworkers. The farm laborers struck. McWilliams wrote that "every strike in which Mexicans participated in the borderlands in the 1930s was broken by the use of violence and was followed by deportation."

The federal government was not immune to racial panic. In 1931, the Los Angeles coordinator for the federal unemployment relief agency wired Washington, D.C.:

> We note press notices this morning figure four hundred thousand deportable aliens United States. Stop. Estimate five per-

cent in this district. Stop. We can pick them all up through police and sheriff channels. Stop. United States Department of Emmigration incapacitated to handle. Stop. You advise please as to method of getting rid. Stop. We need their jobs for needy citizens.

Between three thousand and four thousand Mexicans, many of them U.S. citizens, were immediately picked up and held without access to legal counsel. The *Times* and the Los Angeles Chamber of Commerce opposed the raids. The *Times,* with its extensive agricultural holdings and dependence on advertisers who in turn depended on the Hispanic market, noted that "somebody has been throwing a scare into large numbers of Mexican laborers who have, in fact, a legal right to be here." The economy would be harmed by an agricultural labor shortage and the loss of business to merchants. *Economic suicide* was the term that the paper applied to the deportations in an editorial that went on to note, "The vast majority of Mexicans are gentle, kindly, hard-working people. They are easily intimidated and are very sensitive to affront."

The *Times* eventually fell silent on the issue, and the deportations— financed by federal and local tax monies—continued through 1939. A half-million Mexicans were deported nationwide, the majority of them being from Southern California. McWilliams wrote, "Still an important factor in the farm-labor groups in California, the Mexican, like his predecessors, was used for a purpose, and, when other developments intervened, was, like the Chinese and Japanese, discarded." By 1937 most farm laborers were white. OUR RACE PROBLEMS VANISH, declared a San Francisco newspaper.

The Los Angeles River underwent a dramatic change. Large property losses, drownings during floods, and war or the threat of war during and after World War II sold Congress on stabilizing the river. A great deal of help came from national taxpayers who funded the joint Los Angeles County Flood Control District and U.S. Army Corps of Engineers projects. During the middle years of this century, Los Angeles was the major beneficiary of the pork-barrel funds earmarked by Congress for national flood control projects. The benefit was mainly local; land that shouldn't have been developed in a floodplain was made safe, and the port was prevented from becoming a sandbar.

Drought in 1937 preceded the disaster that would prompt a remedy. A series of storms originating in Siberia in February 1938 circled southward over Midway Island and then swung eastward near Hawaii. They descended upon the Southern California coast with a fury that had not been felt for seventy years. Between February 27 and March 4, an average of 22.5 and a maximum of 32.2 inches of rain fell in the Los Angeles Basin. The storms came in waves. There was no constancy: Light, intense, and then continuous rain was followed by a lull. Then came the heaviest rain. It tapered off to light and then intermittent rain.

The floods peaked on the afternoon of March 2, when nearly three hundred thousand acres in Southern California were inundated. Eighty-seven people died, and the property damage was estimated at nearly $80 million, in prewar terms. No greater flows had ever been recorded on the Los Angeles River.

Something else was new. A report by the U.S. Geological Survey noted:

> The covering of absorptive cones and valley floors with houses and streets, the construction and other surficial changes in mountain areas, and the changes in land cover due to fires and other causes may all change, within a comparatively short time, the runoff characteristics of a drainage basin, especially in an area as young geologically as Southern California.

Little had been done before the flood of 1938 to control the river. Now the federal and county governments got busy, and in the next thirty years cemented the banks of the river and its tributaries. Property that was becoming quite valuable as real estate on the floodplain and Los Angeles's defense industries needed to be protected. Social historian Mike Davis wrote, "The city's original prime amenity—the lovely, meandering Los Angeles River—had been sacrificially transformed into an ugly concrete 'storm sewer' in the 1940s in order to protect adjoining real estate from the periodic flood menace."

In January and February 1969, heavy rains fell. In some areas the precipitation exceeded the 1938 totals. The system held. There was a close call in 1980: Six storms filled the system to capacity, and a seventh was on its way. It never arrived. In the summer of 1993, the Army Corps of Engineers and the Los Angeles County Department of Public Works prepared to raise the artificial banks on the lower one-third of the river, at a cost to exceed a half billion dollars. That year, the Los Angeles River was named one of the most environmentally degraded rivers in the country.

The racial hysteria that flowed just underneath the surface of the city's culture emerged again after Pearl Harbor. Once again the target was Japanese-Americans. The newspapers insistently beat the "tom-toms of hate," as one federal official put it. The head of the War Relocation Authority, Dillon S. Myer, thought the *Los Angeles Times* and the Hearst newspapers in San Francisco and Los Angeles were the greatest "fomenters of hate."

In early 1942, the *Times* editorialized, "A viper is nonetheless a viper wherever the egg is hatched—so a Japanese American, born of Japanese parents—grows up to be a Japanese, not an American." The Hearst newspapers had owned the phrase *yellow peril* for some years. One of their columnists wrote, "I am for immediate removal of every Japanese on the West Coast to a point deep in the interior. I don't mean a nice part of the interior either. Herd 'em up, pack 'em off and give 'em the inside room in the badlands. Let 'em be pinched, hurt, hungry and dead up against it."

The Japanese-Americans were shipped to relocation camps at Tule Lake and Manzanar in California and Heart Mountain in Wyoming, among other western locations. By October, the *Times* observed that for the first time in two generations Southern California was free of Japanese. Fewer than one hundred remained in various hospitals; more than 110,000 had been forcibly removed. During the intervening months there was a plethora of hate-inspiring stories: there were Jap spy rings, Jap plots to destroy the water system, and the general Jap peril. None proved to be true.

Nearly forty years later a *Times* archivist, a third-generation Japanese-American, wrote to a reporter in the Washington, D.C., bureau who had used the phrase *sneak attack* in a story about Pearl Harbor. The archivist was involved in preparing an exhibit on the newspaper's history for the lobby of the Times Building. He wrote:

> One thing we tried to avoid in preparing copy for the Lobby
> Exhibit, especially on sensitive stories, many of which arose

during the 40s, including the internment, the zoot suit riots and this newspaper's participation in them, was to describe those events with some perspective, some objectivity.

"Sneak attack" harkens back to the hysteria that put my family behind barbed wire. It seems odd to me to use the phrase in a contemporary news account; we called Pearl Harbor the "surprise attack" in our exhibit.

I really doubt if anyone else in our million-plus readership will take you to task on this—I just felt I had to—knowing of *Times'* history and their role during the war years and how this paper has come so far from those days; knowing of my parents' struggle to regain their respect and place in a society that forced them both out of college in 1941 and into Manzanar, and eventually the desolate Heart Mtn. relocation center. And of remembering how my mother cried while watching [the movie] "Go For Broke" as she remembered her younger brother who went off to war with the 442nd—and lost his life fighting German tanks in the Battle of the Bulge.

I think we've come too far to use a descriptive phrase nearly 40 years old a month from now. I've come to learn how history is determined by whomever happens to be describing it—which I feel you did well, except for the one phrase.

The need to fear and hate intensified during the war years. Without missing a beat—and sometimes pounding two drums at once—the *Times* and other Los Angeles newspapers and radio stations switched from the Japanese-Americans to the Mexican-Americans. The Latinos had returned during the economic boom times of World War II, along with a sizable number of African Americans for the first time.

McWilliams, who watched the violent events unfold, wrote, "The riots were not an unexpected rupture in Anglo-Hispano relations but the logical end-product of a hundred years of neglect and discrimination." He blamed the press and the police, the police and the press—indistinguishable during this time—for building up "within a short period of six months, sufficient anti-Mexican sentiment to prepare the community for a full-scale offensive against the Mexican minority."

The hate campaign centered on young Mexicans, who, along with blacks and some whites, had adopted a style of loose-fitting clothing which they called "drapes" and the press and police called "zoot suits." The word *Mexicans* was dropped from press accounts after federal authorities pointed out the damage it did to President Franklin D. Roosevelt's Good Neighbor Policy and the opportunities it presented to Axis propagandists. The code words *zoot suit, pachuco,* and *gang* were substituted. The gangs, McWilliams wrote, were loosely organized neighborhood groups: "Many of them are nothing more than, in effect, boys' clubs without a clubhouse."

There were stories about SIX ARRESTED IN GANG DEATH, GANG'S VICTIM TELLS OF STABBING, and JAIL 85 BOY GANG RIOTERS. (There was an echo of this overuse of the word *gang* during the 1980s and early 1990s in press accounts referring to Latino and black youths.) The different style of clothing and the long hair infuriated the authorities. Headlines declared: TEENAGERS BEATEN BY ZOOT SUITERS, ZOOT SUIT GANG CHIEF HELD, BAN GOON HAIRCUTS, and BYE, BYE DUCKTAILS, referring to policemen shaving the heads of youths, a not-uncommon practice of the later hippie years.

An entity within the Los Angeles County Sheriff's Department known as the "Foreign Relations Bureau" issued a report to the grand jury that had been impaneled to investigate Mexican youth activities in 1942. The report was entitled "Statistics," although it contained only opinions. The chief of the Los Angeles Police Department concurred in the report's findings. It thereby became official law enforcement dogma.

There was discrimination in Los Angeles, but there was good reason for it, the report implied: There was a biological difference between the races. The metaphors the report used were the wild cat and the domestic cat. They were of the same family, but their "biological characteristics" were different. So it was with the human race:

> The Malay [Filipino] is ever more vicious than the Mongolian, to which race the Japanese and Chinese of course belong. In fact, the Malay seems to have all the bad qualities of the Mongolian and none of the good qualities.
>
> As to the negro, we also have a biological aspect to which the contributing factors are the same as in respect to the Mexican, which only aggravates the condition as to the two races.

The report advocated jailing every member of a gang: "take them out of circulation until they realize that the authorities will not tolerate gangsterism." There were police sweeps and raids. One of the biggest police show-ups followed the arrest of three hundred "gang suspects," the *Times* reported. "The show-up was ordered by authorities to give hoodlum victims an opportunity to identify their assailants," said the newspaper. In January 1943 the grand jury issued its report, stating that there should be more vocational training in Los Angeles schools, and that there really had not been sufficient time to consider the problem.

In early June 1943, there began one week of what McWilliams referred to as a "private war." With police standing by and doing nothing at first, young sailors and marines from nearby bases who had been influenced by press accounts and the unified position of the chief law enforcement agencies roamed the barrios and beat and undressed Mexican youths.

The police, organized into a "Vengeance Squad," joined in the melee. Mexicans and some African Americans and Filipinos were yanked from movie theaters, streetcars, and the sidewalks of their neighborhoods and savagely beaten with nightsticks, rocks, bricks, clubs, fists, knees, and feet. Al Waxman, the editor of the *Eastside Journal,* wrote:

> At Twelfth and Central I came upon a scene that will long live in my memory. Police were swinging clubs and servicemen were fighting with civilians. Wholesale arrests were being made by the officers.
>
> Four boys came out of a pool hall. They were wearing the zoot-suits that have become the symbol of a fighting flag. Police ordered them into arrest cars. One refused. He asked: "Why am I being arrested?" The police officer answered with three swift blows of the night-stick across the boy's head and he went down. As he sprawled, he was kicked in the face. Police had difficulty loading his body into the vehicle because he was one-legged and wore a wooden limb. Maybe the officer didn't know he was attacking a cripple.

McWilliams made the following point:

> There is a deadly parallel between the pictures of naked Mexican boys lying on the streets of Los Angeles in pools of blood—with grinning mobs standing around—and the pic-

tures one began to see a few years ago of Jews being made to clean the streets of Vienna.

The *Times* cheered. ZOOT SUITERS LEARN LESSONS IN FIGHT WITH SERVICEMEN was one headline. The newspaper added that the riots were "having a cleansing effect." Their chief effect, however, was one of a contagion spreading, as race riots erupted across the country that summer in San Diego, Philadelphia, Chicago, Detroit, and New York. There were deaths elsewhere that summer, thirty-four in Detroit alone; but for some miraculous reason, there were none in Los Angeles.

There were other oddities. There was little property damage, no political manifestos, no martyrs or heroes, and youths were both the victims and the perpetrators. It was, as one historian pointed out, "a highly symbolic style of mass protest." Another noted, "It was a Mexican version of the Chinese Massacre of 1871, and it lost little in the translation."

There was a strange addendum, a last gasp. The *Times,* and the people and institutions it represented, did not give up easily. On Sunday, July 16, 1944, the newspaper ran a two-column, unsigned article on the first page of the second section headlined YOUTHFUL GANG SECRETS EXPOSED. The sources for the story were ambiguous. The article qualified as a blind item in a gossip column or a hatchet job called up by some person of authority within the newspaper but displayed with equivocation, since it did not run on the front page but was in the Sunday paper, which had the largest circulation. The story was the newspaper equivalent of the law enforcement agency's "statistics."

There were a number of accusations: Gangs were organized by "subversive elements" who wanted "the experience of handling mob groups." With the newspaper's history of labor- and red-baiting, there was no doubt what *subversive* meant. Gang members "speak a strange argot unintelligible to the uninitiated, perform sadistic mutilations upon unwilling neophytes, and smoke marihuana." Girl members were tattooed "with the secret cabalistic symbol of their gang." Their high hairdos concealed knives and nail files. "They are required to obey and yield to male members of their gangs in all things." *Yiska* was the term for marijuana, smoked by gang members in that most sacrosanct of Los Angeles places, the motion-picture theater.

The Council of Civic Liberty, an organization that McWilliams helped to found, called a meeting at the Biltmore Hotel to object to the story and urged people to write Harry Chandler and tell him that such articles divided the community and hampered the war effort. Two months after the story ran, Harry Chandler died at the age of eighty; his son, Norman, took over as publisher.

When I arrived at the *Times* in September 1964, the newspaper, the city, and the state were at the height of their promise. Otis Chandler had become publisher in 1960, and the emphasis was now on journalistic excellence, rather than the promotion of parochial concerns. The Music Center, made possible by the persistence of his mother's fund-raising, was about to open on a site that was the Los Angeles equivalent of the Parthenon; and the state had just surpassed New York in population. Within one year the firestorm of Watts would reduce these symbols of greatness—"One of the World's Great Newspapers" was the *Times's* motto in those years—to the equivalent of ashes.

As I write these words in the fall of 1993, Los Angeles is burning. I have seen it burn before; undoubtedly it will burn again. My first taste of fire, Los Angeles–style, was a week or two after I arrived. I was sent to Santa Barbara where I watched a wildfire burn through the stylish enclave of Montecito. I saw other such fires: one in the Angeles National Forest where ten firefighters were burned to death, the 1971 Malibu fire which was repeated twenty-two years later, and the grass that burned annually on our hillside in the Mount Washington area when youths torched it with matches.

The rich and middle classes—I was among the latter—were the victims of wildfires. We lived in the fire-prone hills. Poor people lived on the flat plain. Their structures were more likely to be consumed by fires spawned by racial warfare.

The *Times* was oblivious to the coming racial conflagration. The number of blacks in Los Angeles had increased from 25,000 before World War II to 650,000 in 1965. At most, African Americans would make up 18 percent of the population.

There were signs. The large majority of blacks were squeezed into the South Central part of the city, an area that included the former Mexican community of Watts. Among large cities, Los Angeles was the only one whose mayor was hostile to the federal antipoverty program. In 1964, white California voters passed an initiative measure that repealed a fair-housing bill enacted by the state legislature. The extremes in wealth among the races, always quite physically visible in Los Angeles, had increased. There were constant complaints regarding police brutality, about which little was done in the age before camcorders.

Like the Mexicans, the blacks were an invisible minority in the mid-1960s. Raphael J. Sonenshein, a professor of political science at California State University, Fullerton, wrote, "Blacks were invisible still, especially in such citywide media outlets as the *Los Angeles Times*." He added, "It took massive civil violence to unmistakably express Black concerns and command the full attention of the city." Another researcher counted only one hundred twenty inches of news space devoted to the black community since 1943, and those stories were either short, concerned only crime, or originated from city hall handouts about municipal projects.

At the *Times*, a white reporter by the name of Paul Weeks covered the antipoverty and civil rights beats. A Mexican-American reporter, Ruben Salazar, covered the Latino community, between other chores. Another Mexican-American reporter covered the courts, and there were two women reporters. The remainder of the large staff was white males.

In 1963, Weeks wrote a story that quoted civil rights leaders as concluding at a meeting that "racial strife such as California has never seen before will erupt" if the housing initiative qualified for the ballot, let alone pass, as it eventually did. In April 1964, Weeks wrote an article about how African Americans in the San Fernando Valley community of Pacoima felt after violence flared there when police officers stopped a black motorist—the same type of incident that resulted in widespread violence in 1965 and 1992.

Not long afterward the reporter was taken off the racial beat. "He got personally involved in ways observers shouldn't get involved," said one of Weeks's editors. "There was also no question that he was subjected to the remnants of bigotry, both within and outside the paper." Weeks also had written scathingly of the John Birch Society, and some Chandlers were Birchers.

At seven P.M. on a Sunday night in early January 1965, A. S. "Doc" Young, the editor of the *Los Angeles Sentinel*, an African American newspaper, called the *Times*'s city desk to object to the word *Negress* that was used in an early edition of the paper. He pointed out that the word had no legitimacy, was derogatory, and no one called a white woman a "Caucasianess."

The person who answered replied, "Well, a fucking nigger is a fucking nigger," and—"laughing like some maniac," according to Young—hung up the phone. Young called back and someone else said that the mistake would be fixed in the next edition. *Times* editor Nick B. Williams wrote Young a letter of apology.

The *Times*, under young Chandler, was in a period of growth and rebuilding. The paper's editorial goals for 1965 were to strengthen regional coverage by beefing up the entertainment section, hiring "a very superior reporter" for local news, and adding full-time correspondents in San Diego, Ventura, and San Bernardino. An article in the *Saturday Review* in June noted that the *Times*, perhaps the single most profitable paper in the nation, "has undergone the most complete editorial overhaul of its history, and the pace of change seems to be accelerating rather than diminishing."

At seven P.M. on Wednesday, August 11, 1965, a California Highway Patrol officer arrested a twenty-one-year-old black man on suspicion of drunk driving. A crowd gathered. Other officers arrived; someone spit on them. They grabbed a black woman who wore a smock that resembled a maternity dress. Rocks flew. The officers fled.

The *Times* rarely covered crime or other stories in African American or Hispanic neighborhoods, but this time white police and highway patrol officers were involved. I was working that night and was dispatched to the scene. I later wrote:

> The rock shattered the right wind-wing and sent a shower of glass over *Times* photographer Joe Kennedy and myself—the first newsmen to drive down Avalon Blvd. Wednesday night.
>
> We had been forewarned. A few seconds earlier screams and yells had erupted on our left. Then a shower of rocks, bottles and bricks hit the right side of the car.
>
> We sped out of the dimly lit area, guessing the car was stoned either because of our white faces or the mobile radio antenna, which perhaps made the station wagon look like a police car. The car in front of us, driven by a Negro, was not hit. Other nearby cars were.

We sought the comparative safety of the police command post at Avalon Boulevard and Imperial Highway. Other *Times* reporters and photographers arrived. The crowd grew. The tension increased. The police charged the rock-throwing crowd at 11:50 P.M. The book *Burn, Baby, Burn* took up the story:

> At that moment came a new eruption, exceeding anything that had gone before, and which, in newsman Fradkin's estimate, set the pattern for the bloody days and nights ahead. The fresh outburst left the tall, athletic-looking, thirty-year-old writer among the fallen.
>
> "For some reason," Fradkin said, "the police took after a group of troublemakers on one of the four corners. Then the whole scene seemed to disintegrate—with people running in all directions. It was a visual thing, almost impossible to describe. That was when the real riot began. Until then, the people had directed their venom specifically at the police. But just before midnight, the pattern of the next few days of senseless, brutal, non-discriminatory attacks emerged. The people began attacking, not just white policemen, but anyone who was white. After then, I'm not really sure what happened. I got hit.

A man with a brick in his hand swung at my head. I saw the arc of the descending object in my peripheral vision and managed to duck. I took the blow on the side of my head and collarbone. I remember slowly sinking to the ground, thinking, If I pass out, I've had it. I willed myself up and ran.

Don Cormier, a *Times* photographer, continued the account in the book:

> "Suddenly, I heard another cry of: 'Get him, get him.' I looked around and saw a running figure do a somersault in the air and then crash against a car. The figure disappeared from view and I knew he had hit hard. I could see several Negroes clustering around whoever was down. I trained my camera on them, and they saw me. They ran off. They didn't want to be photographed. It scared them away."

Cormier's photographic instinct may have saved Philip Fradkin's life, kept him from being the first to die in the Los Angeles riots of 1965. "The man who had been knocked

down," said Cormier, "slowly got to his feet. It was Phil Fradkin. He was shaken and bleeding and said to me: 'Let's get out
of here.' I got into my car with Bob Jackson. Fradkin got into
the car with Jack Gaunt. We left."

My injuries turned out to be minor. I was back on the streets of South
Central Los Angeles the next night and for the duration of the riot. But
my faith, acquired during the Eisenhower years, in the security that I
thought was guaranteed to middle-class, white Americans was irrevocably
shaken that night. (It was to be eventually destroyed by other events that
I witnessed as a journalist during that decade. Those years were my cru-
cible. Like McWilliams, my epiphany took place while I was employed by
the *Times*.)

Los Angeles burned during the five-day racial war that followed the
first night of rioting. When it was over, the city within was devastated. It
was clearly an armed conflict between blacks and whites. Thirty-one
African Americans were killed. Three Caucasians—a firefighter, a deputy
sheriff, and a policeman—were also killed.

The predominant view was that the African Americans lost; some
blacks thought they had won, since they had been noticed for the first
time. The *Times* won a Pulitzer Prize for its coverage of the conflagration.
It was the first such honor the newspaper was awarded under Otis Chandler's reign, and he was extremely proud of the accomplishment. *Time*
magazine subsequently wrote a laudatory article about the newspaper,
describing Chandler as a "fun-loving surfer and weight lifter." He sent a
copy of the article to his father with the notation: "When do I have the
time to be fun loving?" Two black reporters were hired immediately, but
neither lasted long. Others eventually took their places.

The governor formed a commission to study the cause of the riot and
propose remedies. One of the commission's members was Asa Call, a
close friend of the elder Chandlers. He said, "I did not rate the colored
community in the same class I do the white community for intelligence
or for integrity nor for ambition to work or progress." Civil rights activist
Bayard Rustin expressed a different view. He wrote, "The whole point of
the outbreak in Watts was that it marked the first major rebellion of
Negroes against their own masochism and was carried on with the
express purpose of asserting that they would no longer quietly submit to
the deprivation of slum life."

Riot, revolt, rebellion, and *revolution* were some of the words applied to those five days and to the racial disturbances of the next three years when violence spread eastward from California. In his book *Violence as Protest,* Robert M. Fogelson declared, "There is, however, substantial agreement that the 1960s confronted America with the greatest threat to public order since the dreadful industrial disputes of the late nineteenth and early twentieth centuries."

The violence stunned the man who was the model Southern Californian. Otis Chandler said ten years after the riot, "Certain things had an impact on me, certain trips we took, the Watts riot." He continued, "I was so shocked by Watts and what it signified that I wanted to personally involve myself to get a better feeling for the problem so I could have a little more intelligent input into our editorials and news coverage." Thereafter, black leaders were periodically invited to the *Times* for discussions.

Chandler was bred to his role, and he looked the part of the scion of Southern California's leading family. He was tall, blond, muscled, ruggedly handsome, and intelligent. His recreational interests and hobbies were similar to those of *Times* readers; at the very least, they were activities that Southern Californians could relate to: surfing, hunting, fishing, photography, tennis, waterskiing, weight lifting, classic and sports cars, and motorcycles. He belonged to surfing, hunting, and athletic clubs in Southern California and New York City and was a member of the California Club. Chandler owned a beautiful home in the exclusive suburb of San Marino, had an assured future, was married to a capable woman who, like him, had been educated at Stanford University, and had five handsome children.

He was not like his great-grandfather, grandfather, and father. Chandler once told a radio interviewer, "I have trouble with the word *power.* I really like *opportunity* better than *power.*" He shied from power, passing political supplicants on to his editors. Chandler was a registered Republican, but he felt more at ease with the Democratic presidents that he knew. His publishing philosophy was: "You have to have a very good editorial product before you can sell advertising and before you can sell readers on it. Community service is an integral part of it."

Chandler defined the *Times* as a "mass and class" newspaper, once telling a radio interviewer:

> You're aiming at two audiences. You're aiming at a mass audi-
> ence, which is the middle-income people in Los Angeles and
> Orange counties primarily. Then you are also aiming the
> *Times* and certain of its features and news at the opinion lead-
> ers, the opinion molders, the high demographic profile in
> Southern California and somewhat throughout the state.

That was why, he said, the newspaper contained such diverse features
as the comics and editorial pages. Then with typical Chandler bluntness,
which earned him public criticism after the broadcast, the publisher outlined who the paper was not for. Chandler was asked about minority coverage. He replied:

> I don't know how we get the readership. We could provide the
> news, even if it cost us millions of dollars, which it would
> because we couldn't get the advertising to support it. The
> mass Black and Chicano audiences in Southern California do
> not have the purchasing power that our stores require to
> spend additional money advertising in the *Times*. So, we can
> make the editorial commitment, the management commit-
> ment to cover those communities; but then how do we get
> them to read the *Times*? It's not their kind of newspaper, it's
> too big, it's too stuffy, and, if you will, it's too complicated.

Chandler relentlessly pursued excellence, and he also wanted recognition for it. He thought his newspaper was great, perhaps the greatest,
but he thought that easterners who passed judgment on such matters
were biased in favor of their own kind. However, when it came time to
quietly hand out kudos, Chandler rated Kay Graham, publisher of the
Washington Post, and Punch Sulzberger, publisher of the *New York
Times,* as the most influential persons in his field. Chandler regretted
that his close friendship with the latter prevented Chandler's approaching
him for a story on the *Times*. Chandler told a colleague, "I guess we will
have to wait and wait and wait."

We looked up to Otis Chandler as a demigod, which probably wasn't
fair to him. He was, after all, a mere mortal; that was proved in 1972
when another newspaper revealed that Chandler had acted as a shill for
a close friend from his Stanford days. He had accepted finder's fees, a car,

and credit cards in return for introducing his friend to prospective buyers of a questionable oil company stock. Why did a man who seemed to have it all accept those minor perks? we asked ourselves. We felt betrayed. When discovered, Chandler returned the money, the car, and the credit cards, but it was too late. Like the region, he had lost his luster.

Years later Chandler sent a newspaper clipping about a drug bust to a son who enjoyed partying, along with a note that read:

> It reminded me, once again, of the obligations that we all face as members of the Chandler family. Any time any of us stumble, it will be well reported in the media and all the other members of the family will be negatively affected, even though they had no involvement whatsoever. Just remember who you are and who you represent. It is not easy to be who you are and to say no to temptation.

The Family, meaning the web of relatives who, through private investment companies and trusts, held the controlling interest in Times Mirror Company, the parent company of the *Times*, were always of paramount concern to Otis Chandler. During Chandler's tenure as publisher, the Mormon Church held the second-largest block of stock in Times Mirror. The church once asked for a directorship, but was not given one. Chandler said, "We just don't believe in prorating boardships based on the ownership of stock." But a lot of family members and associates were directors.

Chandler once characterized The Family as "quite blunt, outspoken, and unemotional." A trusted attorney resolved family disputes before they became lawsuits and public knowledge, and Chandler thanked him for his services. When *Forbes* magazine in 1983 was about to publish The Family's net worth as being $640 million—a figure that Chandler said was incorrect—he was in a bit of a quandary. He was, after all, the chairman of the board of a media holding company that owned a slew of newspapers and magazines dependent on the free flow of information. Chandler, however, declined to set the record straight, stating, "On the other hand, I doubt that any members of the Chandler family wish to respond with any details to this request."

For years the *Times* was the cash cow for the media holding company, and there was pressure on the newspaper to produce earnings for the corporation and The Family.

The *Los Angeles Times* thrashed about during the inter-riot period attempting to resolve the dichotomy between the audience it needed to serve in order to attract and keep advertisers and the audience it should serve in order to fulfill its public service mandate. The situation was complicated by the white flight to the suburbs, leaving the central areas of the city devoid of the newspaper's historical readership.

The *Times,* not surprisingly, chose to follow its traditional subscribers, but success was not axiomatic. In the suburbs of Orange County and the San Fernando Valley, the local newspapers beefed up their product and successfully competed with the *Times.* With the newest flood of immigrants, the region and the potential readership were changing, becoming much more diverse; the newspaper needed to change with them. But how?

The dichotomy had first become visible to *Times* executives in the immediate post-riot period. Nick Williams, a perceptive editor, wrote a downtown attorney, "And despite the enormous size of the effort that would be involved, I think we've simply got to get down into the root causes of urban decay and racial exasperation or we won't much like the way we live." He defined the *Times*'s audience shortly after he retired in 1971 as being "upper income, tending in that direction. The more education they had, the more likely they were to be *Times* readers. And, we had no intention of letting go of that. That is the solid base on which a successful newspaper operates. You can have an awful lot of low income circulation and no advertising pull at all."

The newspaper produced a central-city-zone section, but it did not last long: There was little advertising to support it, the circulation department complained about the large percentage of bad debts, and distributors did not relish driving the streets of South Central Los Angeles in the early morning hours. But the *Times* was getting laudatory reviews from such liberal publications as *Harper's* and *The Nation,* the latter noting, "No other newspaper in the country has so thoroughly parted from an undistinguished past." In 1970, the *Los Angeles Times* ranked second behind the *New York Times* in a survey reported in the latter newspaper.

Midway through the decade the newspaper looked back at the ten years since the riot and concluded: "Failure. Futility. Frustration, angry frustration. Sick apathy. Despair." The editorial concluded:

The urgency is not the risk of another riot, although that could happen. It is a broader risk—the risk of the city's deluding itself into thinking that what happens to people in Watts is not relevant to the lives of every other citizen.

To let that happen, to abandon that part of the city, would be to accept a lethal corruption of our society.

Yet that part of the city was virtually abandoned by the newspaper, described in 1975 as "the media colossus of the West" by *Advertising Age* magazine. Certainly, Watts was not embedded very deeply in the consciousness of the types of people who read the *Times*. Coverage was sporadic, at best. There were occasional lengthy articles or entire sections of the paper, sometimes in a foreign language, devoted to the African American, Jewish, Korean, Vietnamese, Chinese, Armenian, Latino, and Caucasian communities, the latter featured as victims in a package of stories that ran one day on "marauders" from the inner city.

That story about the black and brown criminal underclass was accompanied by a graphic map in which the path taken by marauding inner-city criminals resembled the launching of atomic bombs upon the white suburbs. The article began: "One by one and in small bands, young men desperate for money are marauding out of the heart of Los Angeles in a growing wave to prey upon the suburban middle and upper classes, sometimes with senseless savagery."

Like the river in its natural state, the direction of the *Times* swung back and forth. William F. Thomas, who had succeeded Williams as editor, personally supervised the marauder story and insisted on its strong language. He had also begun to insert yearly requests for another central-city-zone section in the editorial budget. And those stories that ran at great length about minorities were superior efforts, the one on the Latinos winning a Pulitzer Prize.

In 1980, for the first time in the *Times*'s history, an outsider took over as publisher, and Chandler, who had largely accomplished what he had set out to do, moved up to the positions of editor in chief and chairman of Times Mirror. Tom Johnson, once a protégé of Lyndon B. Johnson, was Chandler's handpicked choice to succeed him, there being no young Chandler ready to assume the reins.

At the *Times*'s 1981 Centennial Dinner in the Harry Chandler Auditorium, Jack Smith, the newspaper's popular columnist who was regarded as the prose equivalent of the poet laureate of Los Angeles, set the tone of celebration in a speech entitled "Our City, Our *Times*." He said:

> Of course, the *Times* and the city have grown together. Their relationship has been symbiotic. It is hard to imagine one without the other. A city is as much a mirror of its newspaper, in a sense, as a newspaper is of its city.

With a Chandler in the publisher's corner office on the second floor of the Times Building at Second and Spring Streets, corporate encroachment into the operation of the newspaper was nonexistent. That changed during the 1980s, with more decisions made by consensus generated by paperwork, meetings, polls, and focus groups. "Keep the editorial budget down" was the most frequently heard message from the corporate types, who tended to pay more attention to the marketing end of the equation as the decade neared a close.

Going, going, gone was excellence as a primary goal; now it was, What do the readers want—meaning those persons variously defined by the marketing research department as belonging to "essential upscale target groups," "upscale audience," "top (or upper) demographic groups," "able-to-buy adults," and "Los Angeles's most affluent adults." These were the readers that the *Times* courted.

In an article on the *Times*, the *Columbia Journalism Review* pointed out:

> Newspapers play a risky game when they follow affluent readers out of the city and county. Not only do they bank on their staff's ability to compete with local newspapers on their home turf, but they also gamble that they will be able to attract upwardly mobile segments of the minority community as they move up the demographic scale—even if these newest members of the middle class came from families that had no newspaper-reading habit.

The newspaper, after all, was only a partial mirror of Los Angeles. There was a gap in the reflection that resembled the hole in a donut. The editorial component of the *Times*'s five-year plan for the years 1981–1985 read, in part:

> The South-Central and East Los Angeles areas are the only communities that are not served by our suburban sections. Thus, we carry virtually no truly community news from these areas. Both to round out our coverage, and to meet our community obligations, this effort is required of us.

The document went on to point out that the "so-called" minorities were on their way toward overtaking the majority. It quoted a member of a Latino focus group as stating, "The *Los Angeles Times* does not give a sense of representing the community." For most of the eighties, the request for such a section was made each year by the editorial department. It survived preliminary budget discussions, only to be shot down during the final rounds. Those responsible for marketing, advertising, circulation, and distribution opposed it. Where before Chandler could rule by fiat, consensus was now the order of the day.

Chandler looked on from above. He was close to Johnson, cheering him on and educating him about the region and the politics of the newspaper. Meanwhile, Chandler was getting hammered in interviews. He complained to another Times Mirror executive that, regardless of the ground rules, interviewers "always try to back me into a corner on the issue of minority coverage." As editor in chief, Chandler was mostly concerned with the weather page, sports, the television log, and turning the Sunday supplement that dealt with homes into a *New York Times*–type, general-interest magazine.

By the end of 1985, Chandler had surrendered all his titles except that of chairman of the executive committee of Times Mirror. He told fellow media mogul Walter H. Annenberg, "I want to have the opportunity to do some personal things which I have not had the chance to do up until now." Top *Times* executives, however, believed that The Family had a hand in easing out Chandler. Other assets had been sold off, family numbers were multiplying, and more members were dependent on income generated by Times Mirror stock, which meant greater pressure for higher dividends and less money plowed back into the product to maintain its level of excellence.

With Chandler gone, Johnson became more vulnerable. He was also responsible for some failures. Corporate officers took a stronger hold on the operations of the *Times*. The corporate side, with its emphasis on advertising and marketing, gained dominance in 1989 when Johnson was ousted and David Laventhol, first with the *Washington Post* and then *Newsday,* a Times Mirror holding in New York, was named publisher and chief executive officer; Laventhol already was president of Times Mirror. Shelby Coffey III, from the *Washington Post* via *U.S. News & World Report,* was named editor to replace the retiring William F. Thomas. Thomas Plate was imported from *Newsday* as editor of the editorial pages.

It was the triumph of corporate politics and the takeover of the *Times* by easterners. Change was in the wind. While the Sulzbergers of the *New York Times* and the Grahams of the *Washington Post* had young sons in place, no young Chandlers were in the line of succession. The dynasty was no longer. In 1993, Otis Chandler, who was now living on a twenty-acre ranch in Ojai, was suffering from prostate cancer and injuries sustained in a hunting accident.

The easterners took over with a five-year plan already in place for the years 1988–1992. Of course, there was no way to foresee the future. But when the plan was drafted in late 1987, there were again some clues. The makeup of Los Angeles's population was changing rapidly, and the boom of the eighties could not last forever.

The plan contained a lot of cheerleader-type statements, such as "Our mission is TO BECOME THE BEST NEWSPAPER IN THE UNITED STATES" by improving "our already world-class journalism" and controlling "the '*Times* Profile' demographic market." There was an ongoing study of a western/national edition, but no mention of a central city or East Los Angeles edition. It was predicted that circulation and profits would increase. The competition was no longer the *New York Times* or the *Washington Post,* but rather the *Register* of Orange County and the *Daily News* of the San Fernando Valley.

Buried deep within the document was a question that foretold the breakup of the region into archipelagoes. The question was in reference to naming the regional editions of the *Times.* It was:

> What name should we use?
> Early indications are that the "Los Angeles" is part of the problem. Are we willing to be "The Times"?

The answer, as it emerged with the naming of the separate editions, was yes. Ventura, the Valley, and Orange County were the geographic designations that were chosen to represent the newspaper. The name

"Los Angeles" had lost its magic. The newspaper, like the region, became Balkanized as it chased after its readers who'd fled in different directions. There were some who mourned the loss of the core newspaper, a single entity with zoned inserts rather than the regionally concocted mixture of news. For others the different faces of the *Times* on a given day reflected the splintered reality of the province.

A study by the marketing research department in 1989, the first year the Hispanic population made up one-third of the newspaper's market area (Los Angeles and Orange counties), pointed out that Los Angeles's Hispanics were the "largest single ethnic market in the United States." To tap this market, Times Mirror took the easy route, purchasing 50 percent of the successful Spanish-language daily *La Opinión* and publishing a tepid, bilingual insert called *Nuestro Tiempo* fifteen times a year. In contrast, the *Times,* seeking East Coast recognition, put together an expensive editing, satellite transmission, printing, and distribution system that delivered some two thousand copies of the newspaper daily to "opinion leaders" in Washington, D.C. For a time, there was even a Moscow edition.

As the new decade began, the reach and resources of the *Times* and the region appeared to be simply stupendous. *Multiculturalism, diversity,* and *Pacific Rim* were the buzzwords. A document put together by a broadly based community group that looked ahead to the year 2000 announced, "The vision of Los Angeles as a leader of 21st century citizens is within our reach." Historian Kevin Starr wrote in the epilogue that the 1984 Olympics and the Pope's recent visit "heralded the emergence of Los Angeles as a world crossroads city."

The newspaper was first in circulation among metropolitan dailies, led all newspapers since 1955 in total advertising, had the largest editorial department of any newspaper in the country, maintained twenty-seven foreign and twelve domestic bureaus, had been awarded eighteen Pulitzer Prizes, and was widely praised for its Gulf War coverage. There was another side to the story. Advertising revenues sank lower and lower. For the first time, Times Mirror recorded a net loss in 1992.

Southern California and the newspaper quickly sank to the depths of their joint existences. The speed and the extremes of change were once again remarkable. The Times Mirror annual report for 1992 noted, "The economic recession and the unusually slow recovery have continued to dampen our results, as they have for many others." Morale was poor;

management was nervous. When recovery came, the newspaper did not expect to bounce back to former levels of performance.

It all came crashing down that year in the devil wind of violence that was the nation's worst civil disturbance of the twentieth century. Appropriately enough, the region was undone by a videotape.

There were differences from past racial battles. The newspaper itself was a target of rioters this time, and there were bad feelings within the editorial department between reporters and editors of different ethnic backgrounds. Outside the stolid building—whose ground-floor windows had been smashed—it was also a multiracial conflict.

There were also similarities dating back to the 1850s. The hatred, fear, violence, deaths, injuries, and property losses were comparable. The *Times* was awarded a Pulitzer Prize for its riot coverage; within months after the violence ended, the newspaper reintroduced a central-city section that lost money. How long it would continue to exist was unknown.

This time I watched from the safety of Northern California as they torched palm trees. I saw the city burn as it had before and as it would undoubtedly burn again. Perhaps the final blaze would be extinguished eventually by a river loosened from its concrete bonds and once again in natural flood.

Past the downtown area I was on my own. South of the Atlantic Boulevard bridge in Vernon there was a bicycle path that led to the end of the river, not that there were many cyclists on the levee on this early summer evening. There was one jogger, two men walking on the path, and one man exercising his dog on the concrete riverbed by allowing it to chase a disabled duck. All the action was just off the riverbank in the teeming communities of Southeast Los Angeles.

A Bell police captain had told me that on occasion cyclists got knocked down and robbed, their bicycles stolen. It was a high crime area, he emphasized. The captain rattled off the figures: forty-three known gangs and seven thousand gang members. Recalling the overemphasis on gangs by police and Los Angeles newspapers during the 1940s, I asked him where he got his statistics. The captain was vague on his source.

But he was very precise on the facts of the first murder discovered in the brand-new City of Bell in 1929. He pointed to a picture on the wall showing the police chief of the day, clad in high-laced boots, and two uniformed patrolmen posing proudly beside a blackened skull that lay partially buried in the river mud. He had carefully researched the case of "the Torso Murder," as the newspapers dubbed it.

A woman's severed body parts were found in the river. Investigators traced the murder back to a vacant house where bloodstains were found. The Hearst newspaper described "the sinister house near the river" that was "the setting for the gruesome crime." It developed that the vivisection, done by the doctor-murderer, was the result of a love triangle. Now passionless crimes, where murderer and victim rarely knew each other, predominated. Guns were fired at a distance. Recently, one Bell High School student had ridden his bicycle up to another and killed him with a large-caliber pistol.

I thought I heard a shot and speeded up. I flashed past a man with no shirt, one foot in a cast, and crutches lying nearby. He sat motionless, like a specter, on the riverbank. I passed neat bungalows with children playing on the sidewalk, adults sitting in chairs in the shade, and countless vehicles parked in the crowded streets. The parks were teeming with Little Leaguers and basketball players. I rode past a wire company and a foundry, the remnants of heavy industry that had fled elsewhere. A small truck farm was sandwiched between machine-filled yards. On the river, graffiti was everywhere.

The path narrowed as it rose to meet Imperial Highway, where it merged with the roadway, crossed the bridge, and resumed on the east bank. It was the perfect spot for an ambush. Five shirtless youths blocked the way. They moved toward me. I braked, dismounted, and raised the tape recorder to my lips. I muttered police jargon into it. They hesitated. I spun about quickly and plunged down the angled concrete riverbank on my mountain bike.

There was a jolt where the bank met the flat riverbed, but I managed to keep my balance. I pedaled south under the bridge. I was in no-man's-land—a flat, white expanse of gleaming concrete with angled sides, above which I could see only drooping, high-voltage transmission lines. I heard the roar of freeway traffic all about me, and the *thwack, thwack, thwack* of helicopter blades. I felt terribly exposed in that empty concrete pit, whose barrenness and flatness reminded me of a desert playa. Except for the ravens, who fed on the organisms produced by the nutrient-rich water, there was only myself and my shadow.

A doll's severed head beside the bike path

I expected a shot to drop me at any moment. This was uninhabited terrain in the midst of a vast city. I left no tracks on the concrete. No one knew I was here. I carried just a few dollars and no wallet with identification. I, too, could wind up as a weathered skull.

A couple miles downstream I stopped and dipped my sweat-soaked T-shirt into the foul water that ran through the low-flow channel. I wrung it out, put it back on, and felt grateful for the sudden coolness.

This was the territory Luis J. Rodríguez inhabited during his youth. He knew about barriers, and later wrote about them. At the age of two, Rodríguez and his family left Ciudad Juárez for Los Angeles and wound up in the Mexican quarter of Watts called *La Colonia*. It was the oldest part of the community. Roosters crowed at daybreak, goats were tethered in backyards, there were weeds and junk lying about, gaunt dogs roamed the streets, and rats and cockroaches overran the house where the young child slept in a cobweb-infested attic. Rodríguez wrote in his memoir, *Always Running*:

> Our first exposure in America stays with me like a foul odor. It seemed a strange world, most of it spiteful to us, spitting and stepping on us, coughing us up, us immigrants, as if we were phlegm stuck in the collective throat of the country.

There were constant borders to cross. First the Rio Grande and then the Los Angeles River, a barrier that kept Latinos in their neighborhoods on the vast east side of the city for years, except for occasional forays downtown. At the age of six in 1960, the year that Otis Chandler became publisher of the *Times* and I entered California, Rodríguez and his older brother crossed another border—Alameda Street, a racial boundary that separated black and brown Watts from the "all-white, all-American" City of South Gate.

They were seeking groceries, but got a fight instead. The two boys were badly beaten up by Anglo kids. "My brother and I," Rodríguez wrote, "then picked ourselves up, saw the teenagers take off, still laughing, still talking about those stupid greasers who dared to cross over to South Gate."

As the early 1960s edged toward open racial warfare, the family moved from the lower floodplain to the San Fernando Valley and then the San Gabriel Valley, to the east of the river. Rodríguez later wrote a poem entitled "The Concrete River," which was published in a collection of his poems that bore the same title. It read, in part:

> *Our backs press up against*
> *A corrugated steel fence*
> *Along the dried banks*
> *Of a concrete river.*
> *Spray-painted outpourings*
> *On walls offer a chaos*
> *Of color for the eyes.*

To Rodríguez the Los Angeles River was not only a barrier but "an urban-spawned stream of muck" and "a steaming, bubbling snake of water, pouring over nightmares of wakefulness."

The racial wall that was Alameda Street eventually fell. Latinos flooded Southeast Los Angeles, and whites fled elsewhere. The factories that employed blue-collar workers in Los Angeles's industrial heartland departed, too. The litany of lost blue-chip industries was long: U.S. Steel, Bethlehem Steel, GM South Gate, Firestone Rubber, Uniroyal Rubber, American Can. Those plants had drawn their white workers from Bell, Maywood, and Cudahy, among other southeastern Los Angeles communities. The three tiny cities funneled their youths to Bell High School.

The history of these three communities was the history of rapid change. The Mexican owner of 4,239 acres lost his land to an American in 1864. A railroad siding was known as Patata, the name the Chinese laborers gave the tuber. A winery was built on the east bank of the river, but it was washed away in the December 1889 flood. Wine and brandy barrels floated downstream on Christmas Eve.

There was an oil boom. A granddaughter of the founder of Bell recalled playing tennis on her uncle's court in 1920. He was the Pacific Coast champion at the time. She said:

> We all sat down and cried because our tennis courts were gone and the adults were out in the back jumping up and down with glee because the oil had come in. But the oil went over the whole house and the tennis courts. They were drilling for water first, and they kept getting oil and my uncle went around to the oil companies and he said he spent days and weeks and months trying to persuade them that there was oil there but nobody would listen. They just laughed and said no, he's a tennis player, he doesn't know, and all this sort of thing.

In this manner was the Bell family fortune made. The 1933 Long Beach earthquake damaged oil wells in the area. Houses that survived the earthquake and those built in the 1930s and 1940s now qualify as historic structures in Cudahy. Named after a Chicago meatpacking family who set up shop in California, the one-mile-square City of Cudahy, just north of South Gate, was incorporated as Los Angeles County's smallest and seventy-first city in 1961. A Los Angeles newspaper predicted TINY CUDAHY, COUNTY'S NEWEST CITY, HAS BIG FUTURE.

And for a while, what was called the Tri-City Area did grow. The growth continued through the sixties. Ground was broken for the largest W. E. Grant store in California at one end of a $3 million shopping plaza. A poem, which conveniently overlooked the smog and heat that were endemic to the inland area, was read at a chamber of commerce meeting. It went:

> Oh, Cudahy delightful spot,
> With weather neither cold nor hot;
> Where scenic mountains are in view,
> And ocean breezes fan us too.
> Here golden sunsets close the day
> On those who live in Cudahy.

There was another side to the story, related by Mike Davis:

> During the Watts rebellion in August, 1965, black teenagers stoned the cars of white commuters, while Lynwood and South Gate police reciprocated with the beating and arrest of innocent black motorists. In the wake of the rioting on its borders, the skilled white working class began to abandon the Southeast.

The first wave of plant closings followed soon after the riot. The region was the center for rabid anti–school integration and anti–fair housing movements throughout the sixties. As the whites left, Latinos returned to supply the need for low-wage workers for the smaller manufacturing concerns, like clothing and furniture makers—the "new sweatshop economy," in Davis's words—that replaced the departed industrial giants.

To these workers were added the newest immigrants from Central America and Mexico during the 1980s. The three towns, on paper, were among the lowest in per capita income in the nation and the most overcrowded in Los Angeles County. Referring to 1990 census figures, the *Times* declared, "Overcrowded, low-income cities such as Bell and Maywood predicted a gloomy future for a large region of Los Angeles County."

But on the ground, especially when compared to the battered war zone of South Central Los Angeles just to the west, a flourishing cash economy that did not show up in census figures accounted for a feeling of vibrancy in the three communities. McDonald's came to Cudahy, the local Kmart was expanded, and the giant HMO Kaiser Permanente opened a state-of-the-art medical facility. In 1980, about 70 percent of Cudahy's population was Latino; in 1990, the figure was closing in on 90 percent and was bound to increase. The population of the three communities increased between 26 percent and 33 percent during the decade.

CONGRATULATIONS CLASS OF 1993 read the signs on businesses along Atlantic Boulevard, the north-south–running main drag that bisected the three small cities. Bell High School, one of the most crowded and largest high schools in the Los Angeles Unified School District, had a student enrollment of four thousand that was 96 percent Hispanic. The faculty was 67 percent white and 17 percent Hispanic. Because schools were so crowded in Southeast, some students were bused to Reed Junior High School in the San Fernando Valley.

Bell High School, founded in 1925, had tradition. The eagle was the school's symbol; its motto was "Honor lies in honest toil." The main buildings were completed in 1929, and then rebuilt after the 1933 Long Beach earthquake. A football team of that era that ranked first in the state defeated Torrance High School 102–0, tying the state record for high scoring. The game was 36–0 at the end of the first quarter. One Bell player scored ten touchdowns. The coach was known as "the little dictator." Eventually, All-American college players emerged from the ranks of Bell alumni, who tended to be sons of factory workers.

A principal wrote:

> Where now twenty thousand people reside, but a few short years ago cattle roamed. Old stock trails have given way to broad, paved highways; and clusters of adobe ranch houses have been torn down to make room for smoking factories.

On December 5, 1941, school enrollment was 2,200. Three days later 1,700 students lined up to enlist in the armed forces. Before and after World War II the high school served middle-class communities, mainly of eastern European descent. The homecoming parade in 1950 featured two students in blackface leading a donkey that pulled a cart with a southern motif. The names of homecoming queens began to change from Nancilee in the mid-1960s to Concetta in the mid-1980s. Now more students turned out for soccer than football.

Bell High School was located under the flight path of jets landing at Los Angeles International Airport, a fact that became quite evident outside during the midmorning recess. The demographics were visually lopsided. The principal, Melquiades Mares, Jr., told a blond-haired student that he was easy to identify when he ducked out of school early the day before. A Latino student standing nearby in the asphalt yard wore a T-shirt that declared WE DIDN'T CROSS THE BORDER. THE BORDER CROSSED US.

The gang members hung out at the south end of the West Quad. A youngster was discovered with a sawed-off shotgun hidden in his baggy pants just the other week. Two full-time employees were kept busy erasing the graffiti-splashed walls of the beige school buildings, which were also adorned with vibrantly colored murals sanctioned by school authorities. There was little active parent participation in the school, except when a safety issue was involved.

A middle-aged woman entered Mares's office after the break. She was carefully dressed and her long, graying hair was pulled back in a neat bun. She hugged her purse to her stomach. The mother and Mares spoke Spanish. The daughter entered, and Mares spoke English to her. She addressed Mares in English and her mother in Spanish. The mother spoke Spanish to both, but seemed to understand English. They were comfortable with the linguistic arrangement.

The daughter, a senior who lacked certain classes to graduate, wanted to go to summer school, which involved taking a school bus from Bell High School to another high school that offered the courses she needed. The mother was frightened about her daughter being so far from home. She mentioned the shootings. Her signature was needed on a document.

Mares: Don't you want her to graduate?
Mother: I don't want her to go so far away.
Mares: Here. Here is the pen. Help her.

Mares leaned forward and made thrusting motions with the pen, which was aimed straight at the mother. The mother drew back in her chair. The pantomime of thrust and retreat went on for several moments. The mother shrank back as far as the chair would allow. She reluctantly took the pen and signed.

There was another side to Bell High School. Nine of the top ten graduates were young Hispanic women. They had been heavily recruited by eastern Ivy League colleges and universities offering full scholarships. Two were going to Yale and one each to Harvard, Princeton, Brown, Dartmouth, and Columbia. A vice principal pushed them in this direction, believing that they needed to get away and see another world. Mares differed. He thought it was better if they went to equivalent schools in California, where they would not have to cope with cold weather and homesickness.

That afternoon Freddy Hernandez pecked away at the keyboard of an Apple computer in Glenda Wright's tenth-grade English class. The assignment was to write a story. He wrote:

> The vandalism and crime that occur on our streets every night and every day should be stopped. The policemen should not tolerate with young delinquents on the streets causing havoc. Of course there are some bad cops. But that doesn't mean that all policemen are bad. Also there are racial comments told by policemen when they stop people. I know. I was a victim of one of those cops who wouldn't stop saying something bad about my race. There are lots of races in our community. There are blacks, Hispanics, Samoans, Lebanese, Jews, Asians and others. We have to stop the hate and forget the past, but learn from it not to be fighting among each other. So we could share our thoughts, our cultures. So that everyone would know what other people like and dislike.

Then the bell rang, ending class.

A few days later the graduation ceremony was held in the East Los Angeles College stadium. The valedictorian, who was on her way to Princeton University, spoke first in English and then in Spanish. She said, "We are celebrating victory over all the obstacles that might have kept us from being here today. Statistics say only half of us will graduate."

A little farther down the Los Angeles River and still on the west bank is Lynwood, also part of Southeast. In 1930, at the peak of the region's reputation as the white spot of America, the City of Lynwood had reported to the Los Angeles Chamber of Commerce, "Lynwood, being restricted to the white race, can furnish ample labor of the better class."

I found myself in Lynwood with a *Los Angeles Times* reporter and photographer. I spent a day revisiting my alma mater and had asked to go out on assignment with one of the reporters. The assignment was the drive-by shooting of a ten-year-old Latina. The black youths were on bicycles when they sprayed Palm Avenue with .45-caliber slugs. They were hoping to hit a group of Hispanic youths, whom they thought had stolen and burned their car.

When we arrived, a Hispanic woman was picking up the remains of the yellow barrier tape, stamped with repeated "Caution" markings, that festooned the sidewalk in front of a small bungalow shaded by palm trees. The stucco was peeling off the front of the house, and the windows were covered with steel mesh. Before the family moved in less than one month ago, it had been a crack house.

A young black boy with a SANE T-shirt ("Sane kids say no to drugs and gangs") stood across the street. He pointed to the sidewalk in front of the bungalow and said:

> It was just like there, man. They were right there. First they came from over there; came right to here. The first shot, he missed. Then she got shot right there, and she fell on the ground. Then just started shooting all over the place; and then they rode off there.

A short distance away two black men unloaded lumber from a pickup truck, on whose rear bumper was the sticker HOLINESS IS THE WAY OF LIFE. ACTS II:38. They were remodeling a house. One said:

> Oh, you know whose car that was? They got it day before yesterday. I came home at one o'clock and saw the car burning. I said, "Oh-oh." Fire department came when the fire was going down. As they shoot, I ran out here. Everyone tell me not to

run out, because they look at me crazy. Those Hispanics look at me like they want to shoot me.

What the sad thing is, we have Hispanic, we have black here, and now what's going to happen?

It wasn't gangs. It was one or two dudes pissed off because someone burned their car. Now it is gangs. They all got their guns. They going to be all over this place.

A relative of the girl arrived to pick up some clothes. She recited the mother's version of the story in a flat voice:

She heard the shots. So she turned around, she seen her daughter on the ground, because kids know what to do when there are shots. She said, "Get up! Get up!" And when she didn't move, she kneeled down. She was so full of blood in her head. When she picked her up, she seen the bullet through her forehead. The doctor said, "It will take a miracle for her to live."

Another relative, who was standing nearby, added:

She had black, curly hair. She liked to dance, play, study. She studied a lot. She was a good girl. She was in elementary school. They moved in not even a month ago. They moved from El Segundo Boulevard to get away from the violence there. Cars pass by, and they shoot at anything. Part of her brain is damaged, they told us. There was an operation. She went into a coma about one o'clock this morning.

One block away on Euclid Avenue two mobile television vans, one serving "Eyewitness" and the other "Action" news, had just completed live feeds to their respective studios. The crews were packing up. African American and Hispanic neighbors stood about outside the house where the mother had sought refuge with a sister and her family.

We were ushered inside. The reporter and photographer, in what was journalism's most difficult situation, were extremely considerate. The family acted with grave dignity in the midst of great sorrow.

Two television sets, one atop the other, dominated one wall of the small living room. The other walls were festooned with family mementos, religious pictures, and photographs, mostly of children in school situations. Seven small children, one with a toy pistol in his hand, sat quietly on a sofa and watched intently. The mother, held by her oldest son, sat on another couch.

A relative translated, but the mother's keening needed no explanation. The reporter thanked the relative and asked her to tell the mother he was sorry that we had intruded. We left.

On the way back to the office, the reporter outlined his problem. Drive-by shootings were so common that they were no longer news. The governor had just been in Los Angeles and declared, in reference to such crimes, "Los Angeles is literally under siege." The reporter thought the no-escape angle was the best way to get the story in the newspaper. He asked my opinion. I agreed.

At the daily 2:30 P.M. editorial conference, the shooting story was quickly passed over for page-one consideration. The floods in the Midwest were the top story, and a picture of the partially flooded golden arches of a McDonald's reflected in the floodwater would accompany that front-page article.

The story of the shooting ran the next morning on the front page of Section B, the local news section, accompanied by a photograph of the grief-stricken mother and son on the sofa and a school picture of the smiling, curly-haired girl in a pink and black-lace dress. The article and two color photos, which made a powerful package, ran in the edition of the *Times* that circulated in the more central areas of the city under the headline NO ESCAPE.

The story was buried or nonexistent in the Ventura, Valley, and Orange County editions of the *Times*. The young girl died a few days later and became history, in the dismissive sense the term was used throughout the region.

Cooler and calmer, I pushed my bicycle up the embankment and rode it south to the end of my journey. I passed temporary shelters constructed of plywood and plastic tarpaulins. Presumably illegal aliens sat around portable radios and plastic water jugs. Farther on, four highway patrolmen, one carrying what looked like a confiscated rifle, led a handcuffed youth back to two patrol cars. The Dominguez Flood Control Channel intersected the river near where a tagger had inscribed KILLA CORY in giant letters on the concrete, which was tilted toward the sky.

I passed under the San Diego Freeway, from whose girdings a construction worker had been plucked a few weeks earlier during a flash flood. He was carried downriver to his death in the turbulent water, now flowing placidly in the low-flow channel. Below the Willow Street bridge the concrete gave way to crushed rock, a flexible material that could accommodate the subsidence caused by nearby oil drilling. A pelican dove into the tidal water. In the background were the nodding heads of oil wells and flames spouting from refineries.

I spent the last night in my van, camped at Shoreline RV Park. "On the water in Long Beach" was the motto; "24-hour security patrol" was the enticement. It was Sunday. The park at the mouth of the river was crowded. It attracted the inhabitants of the city that I had just passed through. People were picnicking and fishing. There was a Porsche owners' show on the grass. The tourists flocked to the shoppes and restaurants of the fake nautical village. The headline displayed in the newspaper racks read U.S. IMMIGRATION POLICY FAILS EVERYONE.

The end of the river

At the mouth of the Los Angeles River, the black hull of the *Queen Mary* ceaselessly plowed upstream. The river was bracketed by relics from different eras, the missile test facility at its beginning and the *Queen Mary* moored at its end. The giant ship was purchased by the City of Long Beach and brought to California to attract attention at about the same time a land developer bought London Bridge and plopped it down in the middle of the desert for the same reason.

The night I camped at the end of the river, a Christian singles' dance was held on board the aging tourist attraction. I watched the fog slip over the hump in the distance that was Point Fermin, where I had once lived, and roll across the water toward the twinkling lights of the ship.

Acknowledgments

Because this is a book about California written by a Californian, I am first going to thank my vehicle for her faithful performance over two hundred thousand miles. My 1981 ivory-colored Volkswagen camper was my home away from home and my traveling office for three book projects. The stove and refrigerator freed me from gut-wrenching café food, and the fold out bed was much more comfortable than the Lysol-scented motel rooms of the American West.

In many ways, it was the ideal California vehicle. The dented and rusted body did not attract attention, which meant that freeway shooters, carjackers, and stereo thieves were not a problem. The camper gave me incredible mobility; I sometimes caught myself counting the number of California provinces that I crossed in a given day without having to make advance reservations. When the Big Quake comes, it could serve as an emergency shelter.

There were three people who played a key role in making this book a reality. Working with Carl D. Brandt, the New York agent who represents me, is an exercise in patiently refining an idea, not through such modern conveniences as telephones and faxes, or even Express Mail or Federal Express for that matter, but rather through the regular mail that takes five days to a week to cross the continent. Ideas ripen during that length of time; for a person who sometimes wants to jump in five directions at once, like myself, this is a calming experience.

At first, I wanted to write a racial history of California; Carl thought this was not wise. I insisted. Reluctantly, Carl circulated my proposal. Jack Macrae at Henry Holt read it, along with other books of mine, and saw that much of what I had written was related to landscape. He suggested that I put the more timeless landscape in the forefront, and other issues, like changing demographics, would fall into place. I immediately sensed he was right. Jack was the first editor who worked with me to alter an idea rather than simply to reject it. What he proposed fit his needs and my interests.

Because I have run afoul of what I think of as the New York mentality—people back there just think differently—and because I need a first reader physically as well as culturally closer to home, I have employed Doris Ober, a tremendously competent freelance editor who lives a few miles down the road in Dogtown, to read my manuscripts. I trust her judgment. She is also a good friend, meaning, among other things, that she knows how to be critical in a gentle yet effective manner.

At the University of California at Berkeley, where I taught for a dozen years, Jim Gregory allowed me to monitor his history seminar on Los Angeles, and Doug Greenberg did the same for his California geography course. From listening to the teachers and students I reaffirmed some concepts, dropped others, and found new directions to probe. Each semester in my advanced nonfiction writing course in the Department of Rhetoric I gave the students a portion of the manuscript to read and comment upon, as I had done with their work. Their comments were incisive.

Gerald Haslam read the chapter on the Central Valley, and a person knowledgeable about the inner workings of the *Los Angeles Times,* who preferred that his name not be used, read the relevant portions of the last chapter. Both had helpful suggestions. Michael Mery and John Anderson built our new home, thus giving me a comfortable and productive place to live and work. Marty Knapp did his usual fine job developing the film and printing the negatives, using my directions. Carrie Smith caught errors in the manuscript that make me blush to think about. And Dianne, to whom this book is dedicated, was always there as a reader, companion, and wife.

Source Notes

Because I have chosen an unconventional way to access California, many of my sources were located in out-of-the-way places. I visited remote city and county libraries, local college and university libraries, museums, historical societies, the files or archives of government agencies and their outposts, and specialized bookstores, such as those found in national parks or the one in Lee Vining that carries only publications pertaining to Mono Lake.

I made prolonged and repeated stops at such better-known repositories of Californiana as the Bancroft Library at the University of California at Berkeley, the special collections section of the main library at the University of California at Los Angeles, and the Huntington Library. I frequently tapped into Melvyl, the on-line catalog that takes in not only the many libraries on the Berkeley campus, which is closest to my home, but also all other UC campus libraries and the California State University, Stanford University, and the California State Library systems. At Berkeley, besides the main and the Bancroft libraries, the anthropology, life sciences, biosciences, and environmental design branch libraries provided materials. There were less-frequent pauses at the engineering, transportation, and forestry libraries.

The incredible volume of information concerning California was a problem. Fortunately, by deciding to approach California through specific landscape features, the deluge was reduced to a manageable flow.

While working on this book I traveled nearly one hundred thousand miles over a three-year period, during which time I talked to an estimated one hundred fifty people. Some of these conversations were formal interviews conducted with a tape recorder or a notebook. Others were more informal. While backtracking over my life in California, I encountered many memories. Some were preserved in what I had written at the time; others were summoned to mind when I returned to previous haunts.

I do not research a place or a topic lightly. To describe a landscape I first go there and immerse myself in it—meaning to live, camp, drive, walk, climb, ski, paddle, or sail across that space. Then I will draw on what I have seen and recorded in my tape recorder, notebook, and camera; historical and contemporary still photographs and videotapes; topographical and geological maps; oral memories; and a wide variety of written materials.

To back up my impressions of California, I have relied throughout this book on what I regard as basic reference works: James D. Hart, *A Companion to California,* Berkeley: University of California Press, 1987; Walton Bean and James J. Rawls, *California,* New York: McGraw-Hill, fifth and sixth editions, 1988 and 1993 (The authors' names switched places on the last edition; I depended mainly on the 1988 edition.); Mary Hill, *California Landscape,* Berkeley: University of California Press, 1984; Elna Bakker, *An Island Called California,* Berkeley: University of California Press, 1984; Allen A. Schoenherr, *A Natural History of California,* Berkeley: University of California Press, 1992; and William H. Brewer, *Up and Down California in 1860–1864,* University of California Press, third edition, 1974.

The separateness, size, and richness of the state have resulted in a phenomenon of publishing that is more associated with advanced nations—its own almanac, atlases, and numerous guidebooks. The almanac and atlases are: James S. Fay, ed., *California Almanac,* Santa Barbara: Pacific Data Resources, sixth edition, 1993; Michael W. Donley, ed., *Atlas of California,* Culver City, Calif.: Pacific Book Center, 1979; Warren A. Beck, ed., *Historical Atlas of California,* Norman, Okl.: University of Oklahoma Press, 1974; William L. Kahrl, ed., *The California Water Atlas,* Sacramento: State of California, 1979. There is an atlas and gazetteer published separately for Northern and Southern California by DeLorme Mapping Company that show topography, roads, and outdoor recreation sites. Current Thomas Brothers maps are indispensable for urban areas.

There are numerous guidebooks, including *The WPA Guide to California,* New York: Pantheon, 1984. My two current favorites are aimed at the budget minded: Tom Stienstra's *California Camping,* San Francisco: Foghorn Press, and the Berkeley Guide's *On the Loose in California,* New York: Fodor's Travel Publications. (Both titles are updated regularly.)

Let me warn the literal-minded that, as a reader, I find the use of the ellipsis very distracting. I am always trying to guess what was left out. I have not used the ellipsis in direct quotations the few times that intervening words were dropped, nor have I deleted material that was substantive. The reader seeking exactness can always check the cited source for the full quotation.

The following sources are not all-inclusive. I had to cut off the citations at some point, so I chose what was most directly useful to me and what might be most accessible to the reader.

Preface

Josiah Royce's essay on geographical determinism can be found in Leonard Pitt, ed., *California Controversies,* San Rafael, Calif.: ETRI Publishing Co., 1985, p. 3. I also consulted Alfred Kazin, *A Writer's America,* New York: Knopf, 1988; David Wyatt, *The Fall into Eden,* New York: Cambridge University Press, 1986; Mary Austin, *Land of Little Rain,* Albuquerque: University of New Mexico Press, 1974, p. 3; Mary Austin, *The Land of Journey's Ending,* New York: The Century Co., 1924, p. 4; Tony Hiss, *The Experience of Place,* New York: Vintage, 1991, pp. xi–4; John Brinkerhoff Jackson, *Discovering the Vernacular Landscape,* New Haven, Conn.: Yale University Press, 1984, p. xi; Michael Hough, *Out of Place,* New Haven, Conn.: Yale University Press, 1990; D. W. Meinig, ed., *The Interpretation of Ordinary Landscapes,* New York: Oxford University Press, 1979; Jan O. M. Broek, *Geography,* Columbus, Ohio: Charles E. Merrill Books Inc., 1965, p. 18; and Ellen Churchill Semple, *Influences of Geographic Environment,* New York: Henry Holt, 1911, p. 2. Background for this controversy came from David L. Sills, ed., *International Encyclopedia of the Social Sciences,* New York: Macmillan and the Free Press, 1968, Volume 5, pp. 93–97. Besides a conversation with James Parsons, I consulted his article "The Uniqueness of California," *American Quarterly,* Spring 1955, pp. 45–55.

The Approach

The derivation of the name California is explained in Bean, pp. 11–12; Hart, p. 466; Erwin G. Gudde, "The Naming of California," *Names,* June 1954, pp. 121–133; Erwin G. Gudde, *1,000 California Place Names,* Berkeley: University of California Press, 1959, pp. 11–12; Carey McWilliams, *California,* Westport, Conn.: Greenwood Press, 1971, p. 3; Dora Beale Polk, *The Island of California,* Spokane, Wash.: Arthur H. Clarke Co., 1991, pp. 117–132. A large, fractious population is not new to California: Bean, pp. 4–5; Robert F. Heizer, *The Natural World of California Indians,* Berkeley: University of California Press, 1980, p. 11; James J. Rawls, ed., *New Directions in California History,* New York: McGraw-Hill, 1988, p. 7.

Deserts

On geological time and the changing environment see: U.S. Geological Survey, William L. Newman, *Geologic Time,* Washington, D.C., 1991; Stephen G. Wells, *Origin and Evolution of Deserts,* Albuquerque: University of New Mexico

Press, 1983, pp. 58, 133–157; E. C. Pielou, *After the Ice Age,* Chicago: University of Chicago Press, 1991, pp. 5–15, 251, 311; Emma Lou Davis, ed., *The Ancient Californians,* Los Angeles: Natural History Museum of Los Angeles County, 1978, pp. 18–20.

The chain of lakes is dealt with in Hill, 1984, pp. 187, 190; California Division of Mines: Mineral Information Service, Robert P. Blanc and George B. Cleveland, *Pleistocene Lakes of Southern California—I,* San Francisco, April 1961, pp. 1–8; Samuel G. Houghton, *A Trace of Desert Waters,* Salt Lake City: Howe Brothers, 1986, pp. 127–132.

Playas are explained in the following: Hill, 1984, pp. 192, 216–221; James T. Neal, "Recent Geomorphic Changes in Playas of Western United States," *The Journal of Geology,* September 1967, pp. 512–516, 522; Office of Aerospace Research, James T. Neal, *Geology, Mineralogy, and Hydrology of U.S. Playas,* Air Force Cambridge Research Laboratories, L. G. Hanscom Field, Bedford, Mass., April 1965, pp. vii, 99, 149–151; Ward S. Motts, ed., *Geology and Hydrology of Selected Playas in Western United States,* Geology Department, University of Massachusetts, Amherst, Mass., May 1970, pp. 239, 264, 266, 270; Robert P. Sharp, "Sliding Stones, Racetrack Playa, California," *Geological Society of America Bulletin,* December 1976, pp. 1704–1707, 1716.

On the Lone Pine earthquake see: W. A. Chalfant, *The Story of Inyo,* Bishop, Calif.: Chalfant Press, 1975, pp. 251–264; Mary Hill, *Geology of the Sierra Nevada,* Berkeley: University of California Press, 1975, pp. 165–168; Hart, p. 143; *Inyo Independent,* March 30, 1872; Houghton, p. 143; Schoenherr, p. 75.

On the Calico Early Man Site and the migration into North America, I consulted: James L. Bischoff, "Uranium-series and Soil-geomorphic Dating of the Calico Archeological Site, California," *Science,* December 1981, pp. 576–582; Thomas C. Blackburn, ed., *Woman, Poet, Scientist: Essays in New World Anthropology Honoring Dr. Emma Louise Davis,* Los Altos, Calif.: Ballina Press, 1985; "Calico Early Man," Bureau of Land Management, Riverside, Calif., undated pamphlet; Brian M. Fagan, *The Great Journey,* New York: Thames and Hudson, 1987, pp. 146–147.

The intaglios are explained in: Jo Anne Van Tilburg, ed., *Ancient Images on Stone,* Los Angeles: Institute of Archeology, University of California at Los Angeles, 1983, pp. 86–93; Evan Hadingham, *Lines to the Mountain Gods,* New York: Random House, 1987, p. 268; Frank M. Setzler, "Giant Effigies of the Southwest," *National Geographic,* September 1952, pp. 389–404; William H. Goetzmann, *Army Exploration in the American West, 1803–1863,* New Haven, Conn.: Yale University Press, 1959, pp. 244–245; Malcolm J. Rogers, "Early Lithic Industries of the Lower Basin of the Colorado River and Adjacent Desert Areas," *San Diego Museum Papers,* May 1939, pp. 9–22.

A number of publications put out by the Jet Propulsion Laboratory in Pasadena, Calif., deal with Goldstone: "Goldstone Deep Space Communications Complex," May 1989; "The Deep Space Network," January 1988; "Goldstone Solar Radar," November 1987; and "Search for Extraterrestrial Intelligence,"

June 1990. See also "Astronomers Launch New Hunt for Alien Life," *New York Times,* October 13, 1992. For basic information on China Lake, refer to "Vital Statistics, 1991," Naval Weapons Center, China Lake, Calif., January 1991.

A number of works supply general background on the desert. Among them are: Schoenherr, pp. 11–15, 406–415; Bureau of Land Management, *Final Environmental Impact Statement and Proposed Plan, California Desert Conservation Area,* Riverside, Calif., 1980; and Bureau of Land Management, *The California Desert Conservation Area Plan,* Riverside, Calif., 1980; California Division of Mines, Norman E. A. Hinds, *Evolution of the California Landscape,* Bulletin 158, San Francisco, 1952, pp. 63–89; Edmund C. Jaeger, *The California Deserts,* Stanford, Calif.: Stanford University Press, 1965; Reyner Banham, *Scenes in America Deserta,* Salt Lake City: Peregrine Smith, 1982; John C. Van Dyke, *The Desert,* Salt Lake City: Peregrine Smith Books, 1987.

Information on Mono Lake was drawn from the following sources: Kenneth Robert Lajoie, "Quaternary Stratigraphy and Geologic History of Mono Basin, Eastern California," Ph.D. thesis, Department of Geology, University of California at Berkeley, 1968, pp. 3–4; Hill, 1975, pp. 108–112; Roy A. Baily, "Volcanism, Structure and Geochronology of Long Valley Caldera, Mono County, California," *Journal of Geophysical Research,* February 1976, p. 741; National Academy of Sciences, *The Mono Basin Ecosystem,* National Academy Press, Washington, D.C., 1987, pp. 1–7, 15, 17, 132–134, 267, 270, 281, 390; U.S. Geological Survey, C. Dan Miller, *Potential Hazards from Future Volcanic Eruptions in the Long Valley–Mono Lake Area, East-Central California and Southwest Nevada—a Preliminary Assessment,* Geological Survey Circular 877, Washington, D.C., 1980; Inyo National Forest, *Long Valley Caldera–Mono Crater Contingency Plan,* Bishop, Calif., 1983, pp. 2–4; Brewer, pp. ix–xxiii, 415, 417, 421; Israel C. Russell, *Quaternary History of Mono Valley, California,* Lee Vining, Calif.: Artemisia Press, 1984, pp. 270, 273, 276, 278, 288, 290, 300, 371; U.S. Geological Survey, Israel C. Russell, *Quaternary History of Mono Valley, California,* Washington, D.C., 1889, unpaged preface, pp. 267–392; Roger D. McGrath, *Gunfighters, Highwaymen & Vigilantes,* Berkeley: University of California Press, 1984, pp. 258–259; William L. Kahrl, *Water and Power,* Berkeley: University of California Press, 1982, pp. 429–436; Norris Hundley, Jr., *The Great Thirst,* Berkeley: University of California Press, 1993, pp. 332–341; David Gaines, *Mono Lake Guidebook,* Lee Vining, Calif.: Kutsavi Books, 1989, pp. 14, 55–60, 80–82; Los Angeles Department of Water and Power, "Mono Lake Report," Los Angeles, Calif., undated; The Mono Lake Committee, "Lakewatch," Lee Vining, Calif., 1991.

For information on the birds of Mono Lake and the ravens of the desert see the following: Gaines, pp. 26–27, 36; *Ecosystem,* pp. 60–61, 101–105, 133; Anna Birgitta Rooth, *The Raven and the Carcass,* Helsinki: Academia Scientiarum Fennica, 1962, pp. 11–13; Tony Angell, *Ravens, Crows, Magpies, and Jays,* Seattle: University of Washington Press, 1978, p. 52; Bernd Heinrich, *Ravens in Winter,* New York: Summit Books, 1989, pp. 18–19, 23, 32, 76, 200, 247, 252; Edgar Allan Poe, *The Raven,* Boston: Northeastern University Press, 1986; Barry

Lopez, *Desert Notes,* New York: Avon Books, 1981, p. 18; Bureau of Land Management, *Raven Management Plan for the California Desert Conservation Area,* Riverside, Calif., 1990, pp. 7–9, 20, 29; Bureau of Land Management, *Environmental Impact Statement for the Management of the Common Raven in the California Desert Conservation Area,* Riverside, Calif., 1990; and Charles Bodin, *Blue Desert,* Tucson: University of Arizona Press, 1988, pp. 29–38.

For the view from Black Point I consulted: California Department of Parks and Recreation, "Mono Lake Tufa State Reserve," Sacramento, Calif., June 1987; Genny Smith, *Mammoth Lakes Sierra,* Mammoth Lakes, Calif.: Genny Smith Books, 1989, pp. 34, 47, 52, 115, 198; Russell, 1889, pp. 275, 378–379; Inyo National Forest, *Land and Resource Management Plan,* Bishop, Calif., 1988, p. 48; Brewer, p. 416; Francis P. Farquahar, *History of the Sierra Nevada,* Berkeley: University of California Press, 1965, pp. 79, 81; Thomas C. Fletcher, *Paiute, Prospector, Pioneer,* Lee Vining, Calif.: Artemisa Press, 1987, pp. 19–22, 96; and McGrath, pp. 137–138, 140–142.

On Mary Austin and the Owens Valley I drew information from: Austin, 1974; Wyatt, 1990, pp. 67–95; Augusta Fink, *I-Mary,* Tucson: University of Arizona Press, 1983, pp. 65–120; Esther Lanigan Stineman, *Mary Austin,* New Haven, Conn.: Yale University Press, 1989, pp. 49–68; Helen MacKnight Doyle, *Mary Austin,* New York: Gotham House, 1939, pp. 190, 202; Chalfant, pp. 342, 353, 409; Kahrl, pp. 106–107.

The goings-on around Owens Lake are covered in: G. Schumacher Smith, *Deepest Valley,* San Francisco: Sierra Club Books, 1978; *Mountains to Desert,* Independence, Calif.: Friends of the Eastern California Museum, 1988, pp. 75–95; Brewer, p. 537; Chalfant, pp. 180–181, 187–188, 223–225, 277–279; McGrath, pp. 36–37, 43–47, 60–61; Austin, 1974, p. 103; Richard E. Lingenfelter, "The Desert Steamers," *Journal of the West,* October 1962, pp. 149–160; "Digging Up a Mystery," *Los Angeles Times,* March 24, 1988; Donald Chaput, curator of history, Los Angeles County Natural History Museum, to State Land Commission, December 26, 1986; Kahrl, pp. 224, 273, 314; Pierre Saint-Amand, "Dust Storms from Owens and Mono Valleys, California," *China Lake Naval Weapons Center Technical Paper,* September 1986; "The Great Dust War," *Eastside Journal,* Spring 1990; "Mountain Lake Turns to Desert," *Sacramento Bee,* June 1991.

For the Pinto culture and sidewinder rattlesnake see: Amand, pp. 8, 15; William T. Eckhardt, archaeologist, Naval Weapons Center, China Lake, Calif., May 24, 1991; Emma Lou Davis, 1978, pp. 20, 164–166; Mark Raymond Harrington, *A Pinto Site at Little Lake, California,* Los Angeles: Southwest Museum, 1957, pp. 10, 82; Robert Jobson, Bureau of Land Management archaeologist, Ridgecrest, Calif., May 24, 1991; Laurence M. Klauber, *Rattlesnakes,* Berkeley: University of California Press, second edition, 1972, volumes I and II, pp. 33–120, 260, 370–379, 383, 388, 403, 450, 481–488, 821, 987, 1005–1009, 1238, 1291; Laurence M. Klauber, *Rattlesnakes,* Berkeley: University of California

Press (abridged edition), 1982, pp. 19–20; Raymond B. Cowes, *Desert Journal,* Berkeley: University of California Press, 1977, pp. 57–60, 126–127; "Pit Viper's Life," *New York Times,* October 15, 1991.

For China Lake and Emma Lou Davis I consulted: "Vital Statistics," 1991; "Notable Achievements of the Naval Weapons Center," Naval Weapons Center, China Lake, Calif., 1990, p. 8; Leonard Michaels, *West of the West,* San Francisco: North Point Press, 1989, p. 322; Emma Lou Davis, 1978, pp. 15–20, 30–31, 71–73, 175–177; Blackburn, pp. v–vi; "In the Shadow of Emma Lou Davis," *Newsletter of the Society for California Archeology,* March 1989; "Emma Lou Davis," *Mammoth Trumpet,* Center for the Study of First Americans, Orono, Maine: University of Maine, January 1989; "Emma Lou Davis," *San Diego Union,* October 21, 1989; and Emma Lou Davis, *vita,* undated.

Background on the Naval Weapons Center and its activities was obtained from the following: Albert B. Christman, *Sailors, Scientists, and Rockets,* Washington, D.C.: Naval History Division, 1971, pp. 86, 168; Mike Davis, *City of Quartz,* New York: Vintage, 1992, pp. 54–61; Donald W. Moore, *Ridgecrest, California,* Ridgecrest, Calif.: Maturango Museum, 1989; Brewer, p. 393; China Lake Naval Weapons Center, James R. Ouimette, *Survey and Evaluation of the Environmental Impact of Naval Weapons Center Activities,* Naval Weapons Center Technical Memorandum 2426, China Lake, June 1974, pp. 24, 31, 78; Gregory A. Davis, "Garlock Fault," *Geological Society of America Bulletin,* April 1973, pp. 1407–1422; Fremont E. Kast, *Science, Technology, and Management,* New York: McGraw-Hill, pp. 166–176; Ron Westrum, "Sidewinder," *Invention & Technology,* Fall 1989, pp. 57–58, 61–62; "To the Sea—'A Sidewinder,'" *Rocketeer,* China Lake, October 19, 1956; J. D. Gerrard-Gough and Albert B. Christman, *The Grand Experiment at Inyokern,* Washington, D.C.: Naval History Division, 1978, pp. 32, 47–49, 52, 63, 104, 106, 176, 183, 187, 289, 292; China Lake Naval Weapons Center, R. E. Kistler, *Notable Achievements of the Naval Weapons Center,* China Lake, August 1990, pp. 3, 19; Jerry Hartmann, Sidewinder Project Office, China Lake, May 21, 1991; "Sidewinder," Naval Weapons Center, China Lake, October 1988; David M. North, "Navy Center Expands Mission to Include Technology Management," *Aviation Week & Space Technology,* January 20, 1986; Leroy L. Doig III, "The Navy's 'Secret City' of China Lake," *Wings of Gold,* The Association of Naval Aviation, Winter 1985; Leroy L. Doig III, historian, China Lake, May 22, 1991; *Los Angeles Times,* May 1, 1986; undated videotape of flood, Naval Weapons Center, China Lake; Associated Press story on flood, August 16, 1984; Alan Alpers, head of Protocol Office, Naval Weapons Center, China Lake, May 22, 1991.

The information on Raymond Taylor came from Raymond G. Taylor, *Men, Medicine & Water,* Los Angeles: Friends of Los Angeles County Medical Association Library, 1982, pp. 14, 32, 56, 138, 159, 162.

On Searles Lake and Trona see: Bureau of Land Management, *Management Plan for the Trona Pinnacles Area of Critical Environmental Concern,* Ridgecrest, Calif., February 1989; David W. Scholl, "Pleistocene Algal Pinnacles at Searles Lake, California," *Journal of Sedimentary Petrology,* September 1960, pp. 414, 416, 428, 429; U.S. Geological Survey, George I. Smith, *Core KM-3, a Surface-to-Bedrock Record of Late Cenozoic Sedimentation in Searles Valley, California,* Geological Survey Professional Paper 1256, Washington, D.C., 1983, pp. 3, 11, 15, 18, 20; American Association of University Women, Searles Lake Branch, *Searles Valley Story,* 1975, pp. 12, 51, 81–84; John E. Teeple, "The Industrial Development of Searles Lake Brines," *American Chemical Society Monograph,* 1929, pp. 20–21; "Compendium of Searles Lake Operations," G. F. Moulton, Jr., Kerr-McGee Chemical Corporation, Trona, undated pamphlet; Dr. O. N. Cole, *Trona,* Trona, Calif.: High Desert Scribe, 1984; Dr. O. N. Cole, Trona, May 20, 1991; *Trona Argonaut,* March 18, May 5, June 3, July 1, July 8, and July 15, 1970; "3 Unions End 115-Day Strike at Trona Plant," *Los Angeles Times,* July 11, 1970; Jim Smith, director of community relations, North American Chemical Company, Trona, May 20, 1991; Gail F. Moulton, chief geologist, North American Chemical Company, Trona, May 20, 1991; U.S. Geological Survey, George I. Smith, *Surface Stratigraphy and Geochemistry of Late-Quaternary Evaporates, Searles Lake, California,* Geological Survey Professional Paper No. 1043, Washington, D.C., 1979; U.S. Geological Survey, George I. Smith, *Late-Quaternary Geologic and Climatic History of Searles Lake, Southeastern California,* Menlo Park, Calif., 1966.

For the Panamint Valley, Shoshone Indians, and creosote see: Richard E. Lingenfelter, *Death Valley and the Armagosa,* Berkeley: University of California Press, 1986, pp. 364–366; Frederick Vernon Coville, "The Panamint Indians of California," *The American Anthropologist,* October 1892; Julian H. Seward, "Basin-Plateau Aboriginal Sociopolitical Groups," *Bureau of American Ethnology Bulletin,* 1938; T. J. Mabry, ed., *Creosote Bush,* Stroudsburg, Penn.: Dowden, Hutchinson, and Ross Inc., 1977, pp. 116, 135, 209, 231; Gary Paul Nabhan, *Gathering the Desert,* Tucson: University of Arizona Press, 1985, pp. 11–12, 14–15; Walter Collins O'Kane, *The Intimate Desert,* Tucson: University of Arizona Press, 1985, pp. 121–122; Bakker, pp. 311–313; James A. MacMahon, *Deserts,* New York: Knopf, 1985, pp. 49–50, 498–499; Frank C. Vasek, "Creosote Bush," *American Journal of Botany,* February 1980, pp. 240, 250, 255; Janice C. Beatley, "Effects of Radioactive and Non-Radioactive Dust Upon *Larrea Divaricata* Cav., Nevada Test Site," *Health Physics,* December 1965, pp. 1621–1625; Janice C. Beatley, "Effects of Rainfall and Temperature on the Distribution and Behavior of *Larrea Tridentata* (Creosote Bush) in the Mojave Desert of Nevada," *Ecology,* Spring 1974, pp. 245–261.

The ghost town of Ballarat and Charles Manson are discussed in: Paul B. Hubbard, *Ballarat, 1897–1917,* Lancaster, Calif.: Paul B. Hubbard, 1965, pp. 5, 18, 20–23, 26, 35–37; Lingenfelter, 1986, pp. 23, 42, 47, 90, 94, 199–200; D. H. Walsh, Bureau of Land Management ranger, Ridgecrest, Calif., May 24, 1991; John Gilmore, *The Garbage People,* Los Angeles: Omega Press, 1971, p. 95; Bob Murphy, *Desert Shadows,* Billings, Mont.: Falcon Press, 1986, pp. 28–29, 33–36,

40, 91, 100; Ed Sanders, *The Family*, New York: E. P. Dutton, 1971, pp. 41, 277–278, 386; Vincent Bugliosi, *Helter Skelter*, New York: W.W. Norton and Co., 1974, pp. 42, 128, 130, 233, 476, 482.

For Death Valley and Lake Manly see: Lingenfelter, 1986; Houghton, pp. 150–160; U.S. Geological Survey, Charles B. Hunt, *Hydraulic Basin, Death Valley, California,* Geological Survey Professional Paper 494-B, Washington, D.C., 1966; C. B. Hunt, *Death Valley,* Berkeley: University of California Press, 1975; Eliot Blackwelder, "Lake Manly," *The Geographical Review,* July 1933, pp. 466–469; Edwin L. Rothfuss, superintendent, and Ross R. Hopkins, supervisory park ranger, Death Valley National Monument, May 17, 1991; National Park Service, *Draft General Management Plan, Draft Environmental Statement, Death Valley National Monument,* April 1989, pp. 114, 128, 135; National Park Service, Western Region, *102nd Congress Issues Briefing Statement, Drug Interdiction,* Death Valley National Monument, January 1991; National Park Service, *Summary of Monthly Narrative Report for July, 1960,* Death Valley National Monument, Calif., 1960.

The Sierra

For Donner Pass and the Sierra Nevada see: Semple, pp. 543–545; Hinds, pp. 13–29; R. W. Durrenberger, *The Geography of California in Essays and Readings,* Los Angeles: Brewster Publications, 1959, pp. 9–12; N. King Huber, *The Geologic Story of Yosemite National Park,* Yosemite National Park, Calif.: Yosemite Association, 1989; Oscar Lewis, *High Sierra Country,* New York: Duell, Sloan, and Pierce, 1955, pp. 3–7; Hill, 1984, pp. 111–125; Hill, 1975, pp. 1, 4, 41–42, 100; Schoenherr, pp. 1–4, 69–227; Farquhar, pp. 1–2, 23–26; Tracy I. Storer and Robert L. Usinger, *Sierra Nevada Natural History,* Berkeley: University of California Press, 1963, pp. 1–12; John McPhee, "Assembling California—I," *New Yorker,* September 7, 1991; Clarence King, *Mountaineering in the Sierra Nevada,* Lincoln, Neb.: University of Nebraska Press, 1970; Tim Palmer, *The Sierra Nevada,* Washington, D.C.: Island Press, 1988; "The Sierra in Peril," *Sacramento Bee,* a series of articles that ran in June 1991; "Treasures and Troubles of the Sierra Nevada," *Sunset,* May 1992.

See Jackson, pp. 21–27, for his ideas on highways. For incoming traffic refer to California Department of Food and Agriculture, *Fiscal Year Report for Truckee Station,* Truckee, Calif., 1991–1992. On the Washo Indians and history of the pass up to the coming of the Donner party consult S. A. Barrett, "The Washo Indians of California and Nevada," *No. 67, Anthropological Papers,* Department of Anthropology, University of Utah, August 1963, pp. 78, 82, 91; Susan Lindstrom, "Great Basin Fisherfolk," Ph.D. thesis, Department of Anthropology, University of California at Davis, 1992; James F. Downs, *The Two Worlds of the Washo,* New York: Holt, Rinehart and Winston, 1966, pp. 4–5, 9–10, 12, 36–37; *Wa She Shu: A Washo Tribal History,* Reno, Nev.: Inter-Tribal Council of Nevada, 1976, pp. 38, 43–44, 46; George R. Stewart, *Donner Pass,*

San Francisco: California Historical Society, 1960; George R. Stewart, *The Opening of the California Trail,* Berkeley: University of California Press, 1953, pp. 7, 9, 15, 23–24, 35, 39–42, 67, 98–99, 100; Sister Gabrielle Sullivan, "Martin Murphy Jr.," *Monograph Number Four,* Pacific Center for Western Historical Studies, University of the Pacific, Stockton, Calif., 1974, pp. ii–iv; W. T. Hamilton, *My Sixty Years on the Plains,* New York: Forest and Stream Publishing Co., 1905, pp. 159–161; Erwin G. Gudde, "Robinson Crusoe in the Sierra Nevada," *Sierra Club Bulletin,* May 1951, pp. 19–28; Edwin Bryant, *What I Saw in California,* Lincoln, Neb.: University of Nebraska Press, 1985, pp. 228–230; California Department of Natural Resources, Division of Beaches & Parks, *Report of Investigation on Location, Cost of Acquisition & Development of Overland Emigrant Trail,* Sacramento, 1949, p. 40; Department of Parks and Recreation, Paul E. Nesbitt, state historian, *Report on the California Emigrant Trail in Response to 1989 Legislation,* Sacramento, May 1990; Department of Parks and Recreation, *Feasibility Study, Truckee Trail,* Sacramento, June 1991; George Hinkle and Bliss Hinkle, *Sierra Nevada Lakes,* Reno, Nev.: University of Nevada Press, 1987, p. 62.

For the preliminaries on the Donner party and grizzly bears see C. F. McGlashan, *History of the Donner Party,* Stanford, Calif.: Stanford University Press, 1940; George R. Stewart, *Ordeal by Hunger,* Lincoln, Neb.: University of Nebraska Press, 1986, pp. 15, 50, 67–70, 87–88, 241–242; Joseph A. King, *Winter of Entrapment,* Toronto: P. D. Meany Publishers, 1992, pp. 25, 43–44, 149, 158, 346; Lewis, pp. 28–39; Tracy I. Storer, *California Grizzly,* Lincoln, Neb.: University of Nebraska Press, 1978; Thomas McNamee, *The Grizzly Bear,* New York: McGraw-Hill, 1986, pp. 32–40; Brewer, p. 523.

The Donner party at Donner Lake is described in: Thomas Sanchez, *Rabbit Boss,* New York: Vintage, 1989, p. 3; Joseph King, pp. 44, 49, 51–56, 58–59, 79, 106–118; Stewart, 1986, pp. 16, 103–104, 219–220, 228, 232; Frederick J. Teggart, ed., *Diary of Patrick Breen,* Olympic Valley, Calif.: Outbooks, 1978, pp. 2–16; Peter W. van der Pas, "Eastward Travel on the Truckee River Route," *Nevada County Historical Society Bulletin,* April 1988; Richard Rhodes, *The Ungodly,* New York: Charterhouse, 1973, p. 365; McGlashan; and Lewis, pp. 39–45.

The gold rush and gold fever are documented in: J. S. Holliday, *The World Rushed In,* New York: Touchstone, 1983, pp. 30, 53, 116, 297, 380, 400; Charles K. Graydon, *Trail of the First Wagons Over the Sierra Nevada,* St. Louis: Patrice Press, 1986, pp. 36–49; Beaches and Parks, 1949, pp. 3–4, 20, 25, 27, 32–34, 49; Bean, pp. 79, 82, 93, 101; Hinkle, pp. 89–109; Jerry McKevitt, " 'Gold Lake' Myth Brought Civilization to Plumas County," *Journal of the West,* October 1964, pp. 489–500; Farquhar, pp. 68–69.

References to Louisa Clapp are found in: Rodman Wilson Paul, "In Search of 'Dame Shirley,' " *Pacific Historical Review,* May 1964, pp. 127–146; and Louisa Clapp, *The Shirley Letters,* Salt Lake City: Peregrine Smith, 1991. For the symbol of the elephant see: Thomas N. Layton, "Stalking Elephants in Nevada," *Western*

Folklore, October 1976, pp. 250–257; George P. Hammond, "Who Saw the Elephant," California Historical Society, San Francisco, 1964, p. 7; Hart, p. 148.

Quitting the diggings and Theodore Judah are described in: Holliday, pp. 357, 452–455; John D. Galloway, "Theodore Dehone Judah—Railroad Pioneer," *Civil Engineering,* October and November 1941, pp. 586–588, 648–651; Carl I. Wheat, "A Sketch of the Life of Theodore D. Judah," *California Historical Society Quarterly,* September 1925, pp. 219–271; Bean, pp. 154–160; John R. Signor, *Donner Pass,* San Marino, Calif.: Golden West Books, 1985, pp. 12–19; Farquhar, pp. 107–110; Tahoe National Forest, *History of the Tahoe National Forest, 1840–1940,* Nevada City, Calif., May 1982, pp. 92–93.

The construction of the railroad over Donner Pass and the part the Chinese played in it are taken up by: Bean, pp. 126–128, 160–162; Ronald Takaki, *Strangers from a Different Shore,* New York: Penguin, 1990, pp. 84–87; John R. Gillis, "Tunnels of the Pacific Railroad," *Transaction of the American Society of Civil Engineers,* New York, 1872, pp. 153–163; Paul A. Lord, Jr., ed., *Fire and Ice,* Truckee, Calif.: Truckee Donner Historical Society, 1981, pp. 15–20; Signor, pp. 19, 21; Gunther Barth, "Bitter Strength," master's thesis, Harvard University, 1964, pp. 117–120, 179; Farquhar, pp. 110–114; Bill Yenne, *Southern Pacific,* New York: Bonanza, 1985, p. 19; Stuart Daggett, *Chapters on the History of the Southern Pacific,* New York: Ronald Press, 1920, pp. 13, 47–48, 67, 81; Thomas W. Chinn, ed., *A History of the Chinese in California,* San Francisco: Chinese Historical Society of America, 1969, p. 45; Wesley S. Griswold, *A Work of Giants,* New York: McGraw-Hill, 1962, pp. 97, 194; Paul M. Ong, "The Central Pacific Railroad and Exploitation of Chinese Labor," *Journal of Ethnic Studies,* Summer 1985, pp. 119–122.

On granite see: Hill, 1975, pp. 64–77; Huber, pp. 10–13; C. R. Twidale, *Granite Landforms,* Amsterdam: Elsevier Scientific Publishing Co., 1982, pp. 36, 39; Brewer, pp. 520–521; Judi Culbertson, *Permanent Californians,* Chelsea, Vt.: Chelsea Green Publishing Co., 1989, pp. 273–275.

Charles McGlashan and Truckee are taken up by: M. Nona McGlashan, *Give Me a Mountain Meadow,* Fresno: Pioneer Publishing Co., 1977; Joanne Meschery, *Truckee,* Truckee, Calif.: Rocking Stone Press, 1978, pp. 48, 71–74; Elmer Clarence Sandmeyer, *The Anti-Chinese Movement in California,* Urbana, Ill.: University of Illinois Press, 1973, pp. 45, 47–48; Carmena Freeman, "The Chinese in Nevada County a Century Ago," *Nevada County Historical Society Bulletin,* January 1979; McGlashan, p. xxix; Doris Foley, "The Pioneer (Donner) Monument," Searles Historical Library, Nevada City, Calif., 1982, p. 23; Pat Jones, "Truckee," *Nevada County Historical Bulletin,* July 1976; Tahoe National Forest, pp. 102–103.

Subsequent travel over Donner summit is handled by: Drake Hokanson, *The Lincoln Highway,* Iowa City: University of Iowa Press, 1988, pp. xvi, 12, 23, 36, 70–71, 85, 95, 131; *The Lincoln Highway,* New York: Dodd, Mead, 1935, pp. 6–7; Lincoln Highway Association, "A Picture of Progress on the Lincoln Way," Detroit, 1920, pp. 9, 12, 13–14, 34; Stewart, 1960, pp. 57–59; Thomas R. Vale and Geraldine R. Vale, *U.S. 40 Today,* Madison, Wisc.: The University of Wisconsin Press, 1935, p. 165; Eleanor Huggins, *Adventures On and Off Interstate 80,* Palo Alto, Calif.: Tioga Publishing Co., 1985, pp. 176–205.

For Truckee and the agricultural inspection station see "Like Gold Rush of '49, People Pour Into Hills," *Sacramento Bee,* June 1991; Pat Jones, p. 8; California Department of Agriculture, *History and Accomplishments of the California Department of Agriculture, 1919–1956,* Sacramento, 1956, pp. 18–19; California Department of Food and Agriculture, Division of Plant Industry, *1990 Annual Report,* Sacramento, 1990, pp. 1–3.

Land of Fire

For an overall look at Mount Shasta see the following: David St. Clair, *The Psychic World of California,* New York: Doubleday, 1972, pp. 1–5; "The Magic Mountain," *Newsweek,* July 30, 1973; Bruce Walton, ed., *Mount Shasta, Home of the Ancients,* Mokelumne Hill, Calif.: Health Research, 1985, pp. 128–130; Stephen L. Harris, *Fire Mountains of the West,* Missoula, Mont.: Mountain Press Publishing Company, 1988, pp. 322–324; A. F. Eichorn, *The Mt. Shasta Story,* Mount Shasta, Calif.: The Mount Shasta Herald, 1987, pp. 13–15; Rosemary Holsinger, *Shasta Indian Tales,* Happy Camp, Calif.: Naturegraph Publishers Inc., 1982, pp. 7–8; Joaquin Miller, *Life Among the Modocs,* San Jose: Urion Press, 1985, p. xxx; Charles Lockwood Stewart, "The Discovery and Exploration of Mount Shasta," master's thesis, Department of History, University of California, 1927; R. B. Dixon, "The Shasta," *Bulletin of the American Museum of Natural History,* July 1907, p. 470; Catherine Holt, "Shasta Ethnology," *Anthropological Records,* Berkeley, 1946, p. 326; Andy Selters and Michael Zanger, *The Mt. Shasta Book,* Berkeley: Wilderness Press, 1989.

The Lemurian legend can be found in: W. Scott-Elliot, *Legends of Atlantis and Lost Lemuria,* Wheaton, Ill.: The Theosophical Publishing House, 1990, pp. ix, xxv, xxxii, 119, 122, 138, 140; Lewis Spence, *The Problem of Lemuria,* London: Rider & Company, 1933, pp. 73, 104–109, 241; Eichorn, pp. 10–13, 18–20; Wishar S. Cerve, *Lemuria, the Lost Continent of the Pacific,* San Jose: Rosicrucian Press, 1931, pp. 123–124, 126, 209, 223, 250–258, 260; Walton, p. 12; "A People of Mystery," *Los Angeles Times,* May 22, 1932. The history of the town is from exhibits at the Sisson Museum, Mount Shasta; and Eichorn, pp. 10–13.

The physical presence and volcanic history of Shasta were derived from: "The Whitney Survey on Mount Shasta," Francis P. Farquhar, ed., California Historical Society, 1928, pp. 125, 131; Brewer, pp. 309–324; Stephen L. Harris, *Agents of Chaos,* Missoula, Mont.: Mountain Press Publishing Company, 1990, p. 177; Harris, 1988, pp. 3, 15–19, 309; U.S. Geological Survey, Dwight R. Crandell and Donald R. Nichols, *Volcanic Hazards at Mount Shasta, California,* Washington, D.C., 1989, p. 3; U.S. Geological Survey, C. Dan Miller, *Potential Hazards from Future Eruptions in the Vicinity of Mount Shasta Volcano, Northern California,*

Bulletin 1503, Washington, D.C., 1980; Robert Decker and Barbara Decker, *Volcanoes,* New York: W. H. Freeman, 1989, pp. 127–128; Hinds, pp. 122–130.

For Medicine Lake Volcano and the geology of war see: Charles A. Anderson, "Volcanoes of the Medicine Lake Highland, California," *Bulletin of the Department of Geological Sciences,* Berkeley, 1941; "Road Log: Field Trip to Medicine Lake Highland," Julie M. Donnelly, 1979; Julie M. Donnelly-Nolan, "Medicine Lake Volcano and Lava Beds National Monument, California," *Geological Society of America Centennial Field Guide—Cordilleran Section,* 1987; Julie M. Donnelly-Nolan, "A Magmatic Model of Medicine Lake Volcano, California," *Journal of Geophysical Research,* May 10, 1988; Julie M. Donnelly-Nolan, "Post-11,000-Year Volcanism at Medicine Lake Volcano, Cascade Range, Northern California," *Journal of Geophysical Research,* November 10, 1990; U.S. Geological Survey, Julie M. Donnelly-Nolan, *Geologic Map of Lava Beds National Monument, Northern California,* Miscellaneous Investigations Series, 1987; Julie M. Donnelly-Nolan, Tulelake, July 11, 1991; "Medicine Lake Volcano and Its Recent Earthquake Swarm," *Stronghold '89,* Lava Beds National Monument, 1989; U.S. Geological Survey, Aaron C. Waters, *Captain Jack's Stronghold,* Circular 838, Washington, D.C., 1981, pp. 151–161; U.S. Geological Survey, Aaron C. Waters, *Selected Caves and Lava-Tube Systems In and Near Lava Beds National Monument, California,* Bulletin 1673, Washington, D.C., 1990; National Park Service, *Statement for Management, Lava Beds,* April 1990.

For the human and natural history of Tule Lake up to the Modoc War see: David P. Adam, "Tulelake, California: The Last 3 Million Years," *Palaeogeography, Palaeoclimatology, Palaeoecology,* 1989, pp. 89–103; Samuel N. Dicken, "Pluvial Lake Modoc, Klamath County, Oregon, and Modoc and Siskiyou Counties, California," *Oregon Geology,* November 1980, pp. 179–187; Carrol B. Howe, *Ancient Modocs of California and Oregon,* Portland, Ore.: Binford & Mort Publishing, 1986; Odie B. Faulk, *The Modoc,* New York: Chelsea House Publishers, 1988, pp. 25–26; Verne F. Ray, *Primitive Pragmatists,* Seattle: University of Washington Press, 1963, pp. xii, xiv, 3–4, 134, 142, 144, 223; Holt, pp. 302, 312; John Charles Frémont, *Narrative of Exploration and Adventure,* New York: Longmans, Green & Co., 1956, pp. 489–491.

For the Modoc War see: Keith A. Murray, *The Modocs and Their War,* Norman, Okl.: University of Oklahoma Press, 1976; Richard H. Dillon, *Burnt-Out Fires,* Englewood Cliffs, N.J.: Prentice-Hall, 1973, pp. 57, 212–213; Oregon Institute of Technology, "The Modoc War—a Symposium," *The Journal of the Shaw Historical Library,* Klamath Falls, Ore., Fall 1988; Department of Indian Affairs, *Annual Report of the Commissioner of Indian Affairs to the Secretary of the Interior for the Year 1873,* Washington, D.C., 1874, p. 13; Max L. Heyman, *Prudent Soldier,* Glendale, Calif.: Arthur H. Clarke Co., 1959, pp. 349–358, 362–363, 373–374, 380; Richard H. Dillon, "Costs of the Modoc War," *California Historical Quarterly,* June 1949, p. 161; Erwin N. Thompson, *Modoc War,* Sacramento: Argus Books, 1971, pp. 5, 52, 193–194; Oliver Knight, *Following the Indian Wars,* Norman, Okl.: University of Oklahoma Press, 1960, pp. 108, 109, 114, 116,

122–125, 135–139, 144; Jeff C. Riddle, *The Indian History of the Modoc War*, San Jose: Urion Press, 1988, pp. 177–179, 215–221, 224–229, 238, 280, 282; Michael Dorris, "Modoc Sketches," *The Journal of the Modoc County Historical Society*, 1990, p. 161; Frances S. Landrum, ed., *Guardhouse, Gallows, and Graves*, Klamath Falls, Calif.: Klamath County Museum, 1988, pp. 16, 102–109, 128; *Yreka Weekly Union*, April 8 and June 14, 1873; *New York Herald*, April 13 and 15, 1873; *New York Times*, April 14 and 15, 1873; "Dyer's Account of the Massacre of April 11, 1873," typescript from papers of Lieutenant Parke, Lava Beds National Monument, Calif.; "A Composite of Versions on the Capture of Captain Jack," Francis S. Landrum, undated typescript, Lava Beds National Monument.

The trial, hanging, and aftermath is covered by: *San Francisco Chronicle*, July 13 and October 5, 12, 18, 1873; Landrum, pp. 56, 58, 63, 70–75, 79–80, 85–135; *New York Herald*, October 4, 1873; *Yreka Weekly Union*, September 27, 1873; *New York Times*, October 5, 1873; Dillon, p. 333; Knight, pp. 154–155; *Los Angeles Times*, March 21, 1977; Gary Hathaway, interpretive ranger, Lava Beds National Monument, July 19, 1991; Jack Fincher, "The Grisly Drama of the Modoc War and Captain Jack," *Smithsonian*, February 1985; Robert E. Smith, ed., *Oklahoma's Forgotten Indians*, Oklahoma City: Oklahoma Historical Society, 1981, pp. 86–107; Odie B. Faulk, *The Modoc People*, Phoenix: Indian Tribal Series, 1976, pp. 92–99.

For bald eagles see: Mark V. Stalmaster, *The Bald Eagle*, New York: Universe Books, 1987, pp. 5, 17, 25, 28, 97; J. C. Cooper, *An Illustrated Encyclopedia of Traditional Symbols*, New York: Thames and Hudson, 1988, p. 58; Jon M. Gerrard, *The Bald Eagle*, Washington, D.C.: Smithsonian Institution Press, 1988, pp. 10, 34, 117; William Leon Dawson, *The Birds of California*, San Diego: South Moulton Co., 1923, p. 1712; U.S. Fish and Wildlife Service, *Pacific Bald Eagle Recovery Plan*, Portland, Ore., 1986; U.S. Fish and Wildlife Service, *Klamath Basin National Wildlife Refuges, California–Oregon*, Washington, D.C., 1988; "Wintering of the Migrant Bald Eagle in the Lower 48 States," National Agricultural Chemicals Association, Washington, D.C., 1976; George P. Keister, Jr., "Characteristics of Winter Roosts and Population of Bald Eagles in the Klamath Basin," master's thesis, Oregon State University, 1981; Roger Johnson, refuge manager, Tule Lake, July 10, 1991; Jim Hamline, wildlife biologist, Tule Lake, July 7, 1991; Rachel Carson, *Silent Spring*, Boston: Houghton Mifflin Company, 1987 (reissue), p. 45.

William Finley and Tule Lake are documented by: Worth Mathewson, *William L. Finley*, Corvallis, Ore.: Oregon State University, 1986, foreword and pp. 1–4, 8, 57, 63; Bureau of Reclamation, *A Half Century of Progress on the Klamath Federal Reclamation Project*, Washington, D.C., 1957; Bureau of Reclamation, *Project History, Klamath Project, Oregon–California*, Vol. 69, 1980; Stanton B. Turner, "Reclamation—a New Look for the Tule Lake and Klamath Basins," *The Journal of the Shaw Historical Library*, Spring 1988, pp. 31–33.

For the internment of Japanese-Americans and the Tule Lake War Relocation Camp see: Stanton B. Turner, "Japanese-American Internment at Tule Lake, 1942–1946," *The Journal of the Shaw Historical Library*, Fall 1987; Maisie

Conrat, *Executive Order 9066,* San Francisco: California Historical Society, 1972, p. 85; Daisuke Kitagawa, *Issei and Nisei,* New York: Seabury Press, 1974, pp. 74, 99; *Kinenhi: Reflections on Tule Lake,* San Francisco: Tule Lake Committee, 1980; Jeane Wakatsuki, *Farewell to Manzanar,* New York: Bantam, 1974, p. 71; "Oral History Project, Tule Lake Segregation Project," California State University, Fullerton, 1976; Stan Turner, *The Years of Harvest,* Eugene, Ore.: 49th Avenue Press, 1988, p. 231; Richard Drinnon, *Keeper of Concentration Camps,* Berkeley: University of California Press, 1987, p. 155; Yuji Ichioka, *Views From Within,* Los Angeles: Asian American Studies Center, UCLA, 1989, pp. 132–135, 147; Michi Weglyn, *Years of Infamy,* New York: William Morrow, 1976, pp. 134, 158–159, 308; Dorothy Swaine Thomas, *The Spoilage,* Berkeley: University of California Press, 1946, pp. 84–89, 95–96; War Relocation Authority, *WRA,* Washington, D.C., 1946, p. 65; Roger Daniels, *Concentration Camps,* Malabar, Fla.: Robert E. Krieger Publishing, 1981, pp. 114–116.

Material on the segregation center, George Kuratomi, the disturbances, and the postscript can be found: throughout Weglyn, Thomas, and Drinnon; Dillon S. Myer, *Uprooted Americans,* Tucson: University of Arizona Press, 1971, pp. 76–80; War Relocation Authority, *Semi-annual Report, July 1 to December 31, 1943,* pp. 11, 13, 22–23; "Tule Lake Segregation Center," War Relocation Authority, September 1944; California State Senate, *Partial Report of Senate Fact-Finding Committee on Japanese Settlement,* 1944, pp. 5–21; *Kinenhi,* p. 32; *WRA,* p. 116; Turner, 1988, pp. 27, 31; *Klamath Falls Herald and News,* January 10 and March 22, 1952; *Modoc County Record,* January 25, 1979; California Department of Parks and Recreation, news release, March 16, 1977; Western Region, National Park Service, *Photographs: Historical and Descriptive Data, Tule Lake Project,* Historic American Buildings Survey, San Francisco, undated; "Tule Lake Relocation/Segregation Center, Newell, California," TeCom Publications, 1987; "Tulelake Reunion III," Sacramento, May 17–30, 1988.

For Tulelake following World War II see: Turner, "Reclamation," pp. 32–34; Turner, 1988, pp. 62, 65–66, 319–326; the two previously cited Bureau of Reclamation publications on the Klamath Project; Bureau of Reclamation, Ten Broeck Williamson, *History 1946 Land Openings,* 1947; Statement of Frederick W. Fahner, manager of the Tulelake Irrigation District, at the public hearing on the acreage limitation law, Klamath Falls, November 8, 1977; *Tulelake Reporter,* October 11, 1984; Bill Quinn, Tulelake historian, Tulelake, July 17, 1991; Eleanor Jane Bolestra, Tulelake, July 17, 1991; *Klamath Falls Herald-News,* March 26, 1939; "The Causes and Significance of the Modoc War," Hugh Wilson, Jr., Tulelake High School P.T.A., 1953, p. 23; *Lost River Star,* July 10, 1991. For Donnelly-Nolan see previously cited works.

Land of Water

For Cape Mendocino see: U.S. Geological Survey, Robert E. Wallace, ed., *The San Andreas Fault System, California,* Professional Paper 1515, Washington, D.C., 1990, pp. 252–255; "Humboldt's Complicated Fault System," *San Francisco Chronicle*, April 27, 1992; National Oceanic and Atmospheric Administration, *United States Coast Pilot, Pacific Coast*, Washington, D.C., 1991, p. 207; Ray Raphael, *An Everyday History of Somewhere*, Covelo, Calif.: Island Press, 1974, p. 36; Del Wilcox, *Voyages to California*, Elk, Calif.: Sea Rock Press, 1991, pp. 55–61; William L. Schurz, "The Manila Galleon and California," *Southwestern Historical Quarterly*, October 1917, pp. 107–113, 124; Henry R. Wagner, *Spanish Voyages to the Northwest Coast of America in the Sixteenth Century*, San Francisco: California Historical Society, 1929, pp. 155–157, 181; U.S. Coast and Geodetic Survey, *Voyages of Discovery and Exploration on the Northwest Coast of America, 1539–1603*, Washington, D.C., 1887; Owen C. Coy, *The Humboldt Bay Region, 1850–1875*, Eureka, Calif.: Humboldt County Historical Society, 1982, pp. 20–21, 180–181, 252–255.

Humboldt Bay and northwestern California are described in: Coy, pp. 28–29; *Coast Pilot*, p. 208; Herbert Michael Eder, "The Geographical Uniqueness of California's North Coast Counties," Ph.D. thesis, Department of Geology, UCLA, 1963; David Rains Wallace, *The Klamath Knot*, San Francisco: Sierra Club Books, 1978; Hill, 1984, p. 227; Schoenherr, pp. 228, 231; Hinds, pp. 139–142; Brewer, pp. 480–481, 491.

The movement of water and the maritime history are delineated by: Humboldt Bay Development Commission, untitled report, 1969; Humboldt Bay Symposium Committee, *Proceedings of the Humboldt Bay Symposium*, Eureka, Calif., 1982, pp. 2–31; Willard Bascom, *Waves and Beaches*, Garden City, NY: Anchor Books, 1980, p. 311; Ken Conlin, "Tsunami '64," *The Humboldt Historian*, January–February 1991, pp. 4–10; National Park Service, Edwin C. Bearss, *History Basic Data: Redwood National Park*, 1982, pp. 210–212; "A Symposium on Humboldt Bay," The Center for Community Development and the University of California Extension, 1966, p. 11; Jack McNairn, *Ships of the Redwood Coast*, Stanford, Calif.: Stanford University Press, 1945; Richard Batman, *The Outer Coast*, New York: Harcourt Brace Jovanovich, 1986.

For John Hittel and natural resources and the early history of Humboldt Bay see: Claude R. Petty, "Gold Rush Intellectual," Ph.D. thesis, Department of History, University of California at Berkeley, 1951, pp. 27, 57, 143, 158, 183; John S. Hittel, *The Resources of California*, San Francisco: A. Roman & Co., 1863, pp. v, 1, 306–311, 359, 376; Coy, pp. 45–47; Susan J. Pritchard, "The Shifting Sands of Humboldt Bar," master's thesis, Department of History, University of California at Santa Barbara, 1987, pp. 3, 121, 135; Wallace E. Martin, *Sail & Steam on the Northern California Coast, 1850–1900*, San Francisco: National Maritime Museum Association, 1983, p. 56; Bearss, p. 156.

Hans Buhne and Ulysses Grant are covered by: Coy, pp. 45–47; *Memorial and Biographical History of Northern California*, Chicago: The Lewis Publishing Co., 1891, p. 639; Susie Baker Fountain papers, Humboldt Historical Society, Eureka, Calif., vol. I, p. 299; Albert D. Richardson, *A Personal History of Ulysses S. Grant*, Washington, D.C.: The National Tribune, 1885, pp. 132–133;

Ulysses S. Grant, *Personal Memories of U. S. Grant, Vol. I*, New York: Charles L. Webster & Co., 1885, pp. 206–210; Stuart Nixon, *Redwood Empire*, New York: Galahad Books, 1960, p. 171; "Fort Humboldt," California Department of Parks and Recreation, Sacramento, 1989; Raphael, p. 46.

For the Indian massacre and the fate of the Chinese in Eureka see: Andrew Genzoli, "Indian Island," *Humboldt Historian*, January–February 1982, pp. 3–6; Andrew Genzoli, "Robert Gunther's Account of Indian Island," *Humboldt Historian*, July–August 1982, pp. 6–7; A. J. Bledsoe, *Indian Wars of the Northwest*, San Francisco: Bacon & Co., 1885, pp. 302–303, 312–313; Jack Norton, *Genocide in Northwestern California*, San Francisco: The Indian Historical Press, 1979, pp. 81, 86, 88–89; Robert F. Heizer, *The Destruction of California Indians*, Lincoln, Neb.: University of Nebraska Press, 1993, pp. 154–157, 211, 255–265; Nixon, pp. 186–187; Brewer, p. 493; Lynwood Carranco, *Redwood Country*, Belmont, Calif.: Star Publishing Co., 1986, pp. 35–67, 112–113; Takaki, 1990, pp. 104, 111.

Details on the construction of the breakwaters, logging, and shipwrecks up to 1925 are found in: Coy, pp. 223–224; Pritchard, pp. 8–154; Martin, pp. 44, 61; W. E. Dennison, "Humboldt Bay and Its Jetty System," *Overland Monthly*, October 1896, pp. 381–390; *History of Humboldt County, California*, San Francisco: Wallace W. Elliott & Co., 1881, pp. 129–131, 199–201; James A. Gibbs, *Shipwrecks of the Pacific Coast*, Portland, Ore.: Binford & Mort, 1981, pp. 195, 199–201; Lynwood Carranco, *Redwood Lumber Industry*, San Marino, Calif.: Golden West Books, 1982; J. M. Eddy, ed., *In the Redwoods' Realm*, San Francisco: D. S. Stanley & Co., 1893, pp. 1–2, 11, 21, 47, 72, 114; Carranco, 1986, pp. 40, 55, 94–96, 189–218; Karl Kortum, "Sailing Days on the Redwood Coast," California Historical Society, San Francisco, 1971; "Scary Crossing of Bar Related by Witnesses," *Humboldt Historian*, July–August 1984, pp. 17–18; Virginia M. Fields, *100 Years of Humboldt County Culture and Artistry*, Eureka, Calif.: Humboldt Cultural Center, 1986, pp. 54–55; Melvin A. Krei, "Ghost of Corona Emerges from Sand," *Humboldt Historian*, March–April 1988, pp. 9–11, 18; Nixon, p. 190; Philip L. Fradkin, *California, the Golden Coast*, New York: Viking Press, 1974, pp. 39–40.

The condition of the jetties between 1910 and 1930 is covered in Pritchard, pp. 99, 102–105, 108–112, 118–120, 148, 150, 153–154. For the current logging situation see: "Bitterness in Timber Country," *San Francisco Chronicle*, May 19, 1992; "Timber Country Tries to Cut Losses," *San Francisco Chronicle*, March 15, 1993; "Forest Service Expected to Deal Blow to State Loggers," *Los Angeles Times*, January 11, 1993.

For salmon and Richard Hume see: Cooper, p. 144; Edward Gifford, ed., *California Indian Nights*, Lincoln, Neb.: University of Nebraska Press, 1990, pp. 171–177; Malcolm Margolin, *The Ohlone Way*, Berkeley: Heyday Books, 1978, p. 136; Wallace, pp. 49–59; Alan Lufkin, ed., *California's Salmon and Steelhead*, Berkeley: University of California Press, 1991, pp. x, 7–8, 14, 37–45, 80, 117, 142–143, 165–167, 170; Ian Dore, *Salmon*, New York: Van Nostrand Reinhold, 1990, pp. 1, 3–4, 6–7, 196; Frances Turner McBeth, *Lower Klamath Country*,

Berkeley: Anchor Press, 1950, pp. 47–51; R. D. Hume, *A Pygmy Monopolist,* Madison, Wisc.: State Historical Society of Wisconsin, 1961, pp. v, 18, 28, 61–63, 83; Gordon B. Dobbs, *The Salmon King of Oregon,* Chapel Hill, N.C.: University of North Carolina Press, 1959, pp. 5, 48, 71, 174–176, 185, 195, 201, 212–213, 225; Bureau of Fisheries, John N. Cobb, *Pacific Salmon Fisheries,* Document No. 1092, Washington, D.C., 1930, pp. 438–439; State Board of Fish Commissioners, David Starr Jordan, *Salmon and Trout of the Pacific Coast,* Bulletin No. 4, Sacramento, 1892; California Advisory Committee on Salmon and Steelhead Trout, *An Environmental Tragedy,* Sacramento, 1971; California Advisory Committee on Salmon and Steelhead Trout, *The Tragedy Continues,* Sacramento, 1986; R. D. Hume, "Salmon of the Pacific Coast," 1893, pp. 10, 19, 23, 34; a series of California Department of Fish and Game press releases, dating from April 15, 1993, to June 7, 1994.

The lumber strike is described in: Frank Onstine, *The Great Lumber Strike of Humboldt County, 1935,* Arcata, Calif.: Mercurial Enterprises, 1979, pp. 8–23; Daniel A. Cornford, *Workers and Dissent in the Redwood Empire,* Philadelphia: Temple University Press, 1987, pp. 6, 166–167, 213, 258.

For the floods see: U.S. Geological Survey, A. O. Waananen, *Floods of December 1964 and January 1965 in the Far Western States,* Washington, D.C., 1971; California Disaster Office, *The Big Flood,* Sacramento, 1965; Brewer, p. 495.

Shoreline erosion within the bay and its consequences are covered by: Army Corps of Engineers, *Environmental Assessment, Fiscal Year 1991, Maintenance Dredging of the Humboldt Harbor Bar and Entrance and North Bay Channels,* San Francisco, 1991; Pritchard, pp. 126–130, 135; Donald C. Tuttle, "Problems of the Sea at Buhne Point," *Humboldt Historian,* July–August 1982, pp. 11–13; Donald C. Tuttle, environmental services manager, Humboldt County Department of Public Works, August 13, 1991; the two symposia on Humboldt Bay cited previously; various publications of the Humboldt Bay Harbor Recreation and Conservation District; *San Francisco Chronicle,* December 3, 1984; *Eureka Times-Standard,* July 2, 1980, and October 17, 1983; *Arcata Union,* May 15, 1980, and June 30, 1983; "Humboldt Bay Power Plant, Nuclear Unit," Pacific Gas & Electric Co., 1965; "Humboldt Bay Power Plant Update," Pacific Gas & Electric Co., 1989; U.S. Nuclear Regulatory Commission, *Reconnaissance Report: Effects of November 8, 1980, Earthquake on Humboldt Bay Power Plant in Eureka, California, Area,* Washington, D.C., 1981; Bascom, pp. 230–239, 328–331.

For more recent activities in Humboldt Bay see: Carranco, 1986, p. 101; Les Westfall, Eureka, August 14, 1991; Burt Bessellieu, August 13, 1991; Jon Humboldt Gates, *Night Crossings,* Trinidad, Calif.: Moonstone Publishing Co., 1990; *Coast Pilot,* p. 208; *Humboldt Times-Standard,* February 23, 1987; Mary Siler Anderson, *Whatever Happened to the Hippies,* San Pedro, Calif.: R & E Miles, 1990, pp. 1–6, 100–105; Ray Raphael, *Cash Crop,* Mendocino, Calif.: Ridge Times Press, 1985, pp. 4, 133, 137, 157; California Senate, Committee on the Judiciary, *Interim Hearing on Marijuana Cultivation,* Eureka, Calif., 1983, pp.

14, 50; "Campaign Against Marijuana Planting (CAMP)," final reports, 1985 and 1989; Ray Raphael, *Tree Talk,* Covelo, Calif.: Island Press, 1981; Ray Raphael, *Edges,* Lincoln, Neb.: University of Nebraska Press, 1986.

The Great Valley

The overall view of the valley was gained from: Gerald Haslam, *The Other California,* Santa Barbara: Capra Press, 1990; Gerald Haslam, *The Great Central Valley,* Berkeley: University of California Press, 1993; Laurence J. Jelinek, *Harvest Empire,* San Francisco: Boyd & Fraser Publishing Co., 1982; Donald Worster, *Rivers of Empire,* New York: Pantheon, 1985, p. 10; John Brinkerhoff Jackson, *American Spaces,* New York: Norton, 1972, p. 194; Hill, 1984, pp. 28–29; Durrenberger, pp. 37–39; Hinds, pp. 145–152; Arthur D. Howard, *Geologic History of Middle California,* Berkeley: University of California Press, 1979, pp. 31, 37, 57; "Urban Growth, Drought Threaten Central Valley," *San Francisco Chronicle,* March 4, 1991; Scott Stine, "Extreme and Persistent Drought in California and Patagonia During Medieval Times," *Nature,* June 16, 1994.

The early natural and human history is described in: California Division of Forestry, L. T. Burcham, *California Rangeland,* Sacramento, 1957, pp. 7, 55, 68, 96–98, 125, 138–139, 146–147, 186, 198; Bakker, pp. 146–147; "San Joaquin Primeval," *Stanislaus Stepping Stones,* Stanislaus County Historical Society, March 1978; Charles Preuss, *Exploring with Frémont,* Norman, Okl.: University of Oklahoma Press, 1958, pp. 120–121; Donald Jackson, ed., *The Expeditions of John Charles Frémont, Vol. I,* Urbana, Ill.: University of Illinois Press, 1970, p. 659; Haslam, 1993, pp. 31, 34, 130–131; Brewer, p. 511; Haslam, 1990, pp. 33–34; David Hornbeck, *California Patterns,* Mountain View, Calif.: Mayfield Publishing Co., 1983, p. 55; Schoenherr, pp. 516, 520, 538–539; Raymond F. Dasman, *The Destruction of California,* New York: Macmillan, 1965, p. 60.

For the encounters between Hispanics and Indians see: Jelinek, pp. 15–16; S. F. Cook, "The Population of the San Joaquin Valley in Approximately 1850," *University of California Publications, Anthropological Records,* 1961, pp. 33–35; A. L. Kroeber, *Handbook of the California Indians,* Berkeley: California Book Co., 1953, pp. 474–491; Frank F. Latta, *Handbook of Yokut Indians,* Santa Cruz, Calif.: Bear State Books, 1977, pp. xix, 35–36; S. F. Cook, "Expeditions to the Interior of California," *University of California Publications, Anthropological Records,* Vol. 20, No. 5, 1962, pp. 165–166, 168–169, 170–172, 176; Ruth P. Nelson, "Archeology of the Hoods Creek Site," master's thesis, Stanislaus State College, 1984, p. 31; Sherburne F. Cook, *The Conflict Between the California Indians and White Civilization,* Berkeley: University of California Press, 1976, pp. 192–193, 200–201, 212, 247.

The material on the arrival of the Anglos was derived from: Bryant, pp. 365–369; California Department of Food and Agriculture, *Statistical Review, 1990,* Sacramento, 1991; Jelinek, pp. 24, 29; Burcham, pp. 144–145, 153, 192–193; Brewer, p. 509; John Steinbeck, *East of Eden,* New York: Penguin

Books, 1986, p. 7; Donald J. Pisani, *From Family Farm to Agribusiness,* Berkeley: University of California Press, 1984, pp. 3–4; Helen Hohenthal, *Streams in a Thirsty Land,* Turlock, Calif.: City of Turlock, 1972, pp. 7–9; Hittel, pp. 151–181.

John Mitchell, the Chinese in Turlock, and the early history of that community are dealt with in: "A Millionaire Dead," *San Francisco Chronicle,* November 30, 1893; L. C. Branch, *History of Stanislaus County,* San Francisco: Elliott & Moore, 1881; *Turlock Journal,* January 21, 1948; *Stanislaus County News,* January 5, 1872; George H. Tinkham, *History of Stanislaus County,* Los Angeles: Historic Record Co., 1921, pp. 489–490; Hohenthal, pp. 9–19, 21, 35, 45, 57–58, 170; Sucheng Chan, *This Bittersweet Soil,* Berkeley: University of California Press, 1986; William L. Preston, *Vanishing Landscapes,* Berkeley: University of California Press, 1981, pp. 177–183, 211; Sarah Jackson, "Turlock Founder Was Man of Foresight," *Stanislaus Stepping Stones,* December 1977; Wallace Smith, *Garden of the Sun,* Los Angeles: Lyman House, 1939, pp. 176, 378; Alan M. Paterson, *Land, Water and Power,* Spokane, Wash.: Arthur H. Clarke Co., 1989, pp. 25–26; Pisani, pp. 5–15, 103; Takaki, 1990, p. 91; Frank Norris, *The Octopus,* New York: Penguin, 1986, p. 130; Dio Lewis, *Gypsies, or Why We Went Gypsying in the Sierras,* Boston: Eastern Book Co., 1881, pp. 110–111; Carey McWilliams, *Factories in the Field,* Boston: Little, Brown and Co., 1939, p. 77; Preston, pp. 76–79; Jelinek, pp. 33–35, 39.

For the early irrigation history see: Paterson, pp. 58, 63, 66–67, 76–88, 288; "Quest For Deep Gold," Thorne Gray, LaGrange, Calif., 1973, pp. 10–18; Pisani, pp. 64–66, 69, 73–74, 252–266, 269–271; Hundley, pp. 98–102; Jelinek, p. 57; Donald Worster, *Under Western Skies,* New York: Oxford University Press, 1992, p. 60; E. J. Wickson, *Rural California,* New York: Macmillan, 1923, pp. 79–83, 320–323; Claude B. Hutchinson, ed., *California Agriculture,* Berkeley: University of California Press, 1946, pp. 1–3, 34–37; California Department of Engineering, Frank Adams, *Irrigation Districts in California, 1887–1915,* Bulletin No. 2, Sacramento, 1917, p. 16; Carla G. Emig, "Changing Agricultural Patterns in the Turlock Irrigation District in California, 1887–1986," master's thesis, Stanislaus State University, 1986, pp. 20–22; various minutes of the Turlock Irrigation District, 1887–1905; Sol P. Elias, *Stories of the Stanislaus,* Modesto, Calif.: 1924, p. 67; Mildred D. Lucas, *From Amber Grain to Fruited Plain,* Ceres, Calif.: Ceres Bicentennial Committee, 1976, pp. 95, 103.

The early history of the Crowell property and the surrounding region is portrayed in: various land sale documents and farm records in the possession of Verne Crowell; Verne Crowell, Turlock, May 13, 26–27, 1992; U.S. Department of Agriculture, Office of Experiment Stations, Frank Adams, *The Distribution and Use of Water in Modesto and Turlock Irrigation Districts, California,* Washington, D.C., 1905, p. 123; U.S. Department of Agriculture, Office of Experiment Stations, Elwood Mead, *Annual Report of Irrigation and Drainage Investigations, 1904,* Washington, D.C., 1905, pp. 120–126; Paterson, pp. 91, 102, 109, 111–115, 117–120, 127–132; Emig, p. 44; Adams, 1905, p. 113; Robert Santos, "A Decade of Change in Stanislaus County Agriculture, 1900–1910,"

Department of History, Stanislaus State College, 1974, pp. 35–40; Robert Santos, "Wheat Ranches Gave Way Under Irrigation," *Stanislaus Stepping Stones,* December 1977, pp. 55–57; Agricultural Extension Service, Victor P. Osterli, *Alfalfa Production in Stanislaus County,* Modesto, Calif., 1946; Hohenthal, pp. 72–80, 238–244; Santos, pp. 14–27; Wallace Smith, pp. 430–435; Esther Crowell, "Early Turlock History," unpublished typescript, 1962; Phebe Fjellstrom, "Swedish-American Colonization in the San Joaquin Valley in California," *Studia Ethnographica Upsaliensia,* Monograph Series No. 33, 1970, pp. 60, 63, 83; Ron Harrelson, "Why Turlock's Many Churches," graduate seminar in American history, Stanislaus State College, 1969.

The portrait of Roy Meikle was derived from: Paterson, pp. 136, 194, 278–279, 351; his son Jim Meikle, Modesto, Calif., June 11, 1992; R. V. Meikle, memo to employees, February 23, 1935; R. V. Meikle, comments on *Los Angeles Times* article on La Grange Dam, May 24, 1953; R. V. Meikle, memo on drainage problems, October 1962; R. V. Meikle, "Irrigation in the Turlock District and the Four Acre Feet Per Acre Problem," December 29, 1950; R. V. Meikle, "Turlock Irrigation District," 1959.

For a discussion of class and ethnicity in the valley see: Haslam, 1990, pp. 45, 165; Haslam, 1993, pp. 11, 73–76; McWilliams, 1939, p. 134; Arian Beit Ishaya, "Class and Ethnicity in Rural California," Ph.D. thesis, anthropology, UCLA, 1985, pp. 20–21, 27–28; Kevin Starr, *Inventing the Dream,* New York: Oxford University Press, 1985, pp. 169–173; Hohenthal, pp. 187–189; Paterson, pp. 264–267; "Food First" and "Rural Democracy at Delhi," State Land Settlement Board, Berkeley, 1920; Elwood Mead, *Helping Men Own Farms,* New York: Macmillan, 1920, p. 5; *Farmer's Daily Journal,* Turlock, Calif., July 19–26, 1921, and September 2–3, 1921; *Modesto Evening News,* July 20–23, 1921; *San Francisco Examiner,* July 20–23, 1921; *Los Angeles Daily Times,* July 21–22, 1921; *New York Times,* July 21–22, 1921; Takaki, 1990, pp. 179–229, 324–335, 388; Charles Wollenberg, ed., *Ethnic Conflict in California History,* Los Angeles: Timmon-Brown, 1970, pp. 104–111; Kesa Noda, *Yamato Colony,* Livingston, Calif.: Livingston-Merced JACL Chapter, 1981; Scottie H. Hagedorn, "Visayan Filipinos of the Central Valley of California," master's thesis, Stanislaus State College, 1981, pp. 9–10; Carey McWilliams, *Prejudice,* Boston: Little, Brown & Co., 1944, pp. 58–59, 62, 67; House Committee of Immigration and Naturalization, *Japanese in California,* June 20, 1919; "Our Racial Problem," Japanese Exclusion League of California, 1921; Bean, pp. 256–259, 340.

The saga of the Crowells, their land, and Turlock continues in: Jelinek, pp. 93–95; Pisani, p. 81; Emig, p. 51; Hornbeck, pp. 81–82; *WPA,* pp. 68–69, 443–444; Farm Housing Survey, *California Human Interest Stories, Stanislaus County,* Modesto, Calif., 1934; Hohenthal, p. 300; *Turlock Journal,* November 8, 1961; Tony Jerome, *Memories of a Portuguese Immigrant to the San Joaquin Valley,* Stevinson, Calif.: San Joaquin Publishing, 1991; Paterson, pp. 318, 332, 335, 340, 354–363; Roy Meikle memorandums of October 1962, March 1, March 21, and December 12, 1963, and June 15, 1964; Turlock and Modesto Irrigation Districts, report in support of settlement agreement with the California Department of Fish and Game, Vol. I and II, Turlock, May 1, 1992, pp. 5–7; Technical Committee Notebook, Leroy Kennedy, *Turlock Area Drinking Water Supply Study,* Turlock, July 5, 1988; Turlock Irrigation District, *Environmental Studies, Phase I Report, Turlock Area Drinking Water Supply Study,* Turlock, April 1988, pp. 3–15; annual reports of the Turlock Irrigation District for 1979–1991; EIP Associates, *Draft Environmental Impact Report, Turlock Area Drinking Water Supply Study,* Turlock, June 1, 1992, pp. 1–3, 4.1–7; Leroy Kennedy, Turlock Irrigation District, Turlock, June 12, 1992; Haslam, 1993, pp. 126, 129, 181–182, 217; City of Turlock, planning commission files for Rolling Hill Estates and North View Meadows II, 1989; sales literature for the two subdivisions; State Census Data Center, *1990 Census of Population and Housing, Census Tracts 003602 and 003903,* Sacramento, 1990.

For valley oaks I consulted: Bakker, pp. 127, 147; Bruce M. Pavlik, *Oaks of California,* Los Olivos, Calif.: Cachuma Press, 1991, pp. 3, 5, 10, 11; Cooper, p. 121; Brewer, p. 109; Sylvia Sun Minnick, *Sanfow,* Fresno, Calif.: Panorama West Books, 1988, p. 74; Schoenherr, p. 533; Haslam, 1993, pp. 95–96; California Department of Parks and Recreation, various undated pamphlets.

The Fractured Province

For the transition between provinces and an overall look at the San Andreas Fault and the region see: Stephen Vincent, ed., *O California,* San Francisco: Bedford Arts, 1990, p. 46; Howard, pp. 2–3, 12, 26; "Geo Watch," *San Francisco Chronicle,* December 3, 1993; Robert Wallace, pp. 3–4, 77, 120–121, 152–153, 173; Bruce A. Bolt, *Earthquakes,* New York: W. H. Freeman and Co., 1993, pp. 102, 124, 202–206; "Earthquakes," Louis C. Pakiser, U.S. Geological Survey, Washington, D.C., 1991; David Starr Jordan, ed., *The California Earthquake of 1906,* San Francisco: A. M. Robertson, 1907, pp. 57–60; Andrew C. Lawson, *The California Earthquake of April 18, 1906,* Washington, D.C.: Carnegie Institution of Washington, 1908, p. vii; Brewer, pp. 118, 127, 358; Hinds, pp. 157–176; "Lick Observatory," University of California, 1984; "Golden Gate Area Losing Its Glow," *San Francisco Chronicle,* May 18, 1992.

The history of Mission San Juan Bautista and the activity on Fremont Peak are dealt with in: "Mission San Juan Bautista," John M. Martin, undated pamphlet; "San Juan Bautista State Historic Park," Department of Parks and Recreation, Sacramento, 1979; "Mission San Juan Bautista," Mary Null Boulé, Vashon, Wash.: Merryant Publishing, 1988; Malcolm Margolin, *Monterey in 1786,* Berkeley: Heyday Books, 1989; California Department of Parks and Recreation, Glenn J. Farris, *Archeological Testing in the Neophyte Family Housing Area at Mission San Juan Bautista, California,* Sacramento, 1991, p. 7; "Former Mission Building Foundations on the Taix Lot at San Juan Bautista SHP," Glenn Farris, California Department of Parks and Recreation, Sacramento, 1991, p. 2; "Docents Manual," San Juan Bautista State Park Volunteers Association, undated; *Hollis-*

ter Advance, a series of undated articles, possibly in 1917, by Father P. Triana in the files of the State Historic Park; *History of San Benito County, California,* San Francisco: Elliott & Moore, 1881, p. 91; Haslam, 1993, pp. 30–31; Bryant, pp. 364–365; Francis J. Weber, ed., *The Precursor's Mission,* Los Angeles: Archdiocese of Los Angeles, 1982; Isaac L. Mylar, *Early Days at the Mission San Juan Bautista,* Fresno: Panorama West Books, 1985, p. 22; David Lavender, *California,* Lincoln, Neb.: University of Nebraska Press, 1987, pp. 128–130; Nevins, pp. 466–472; Frémont, 1956, pp. 466–472; Jackson, pp. 78–85; Andrew Rolle, *John Charles Frémont,* Norman, Okl.: University of Oklahoma Press, 1991, pp. 71–74; John Steinbeck, *Travels with Charley,* New York: Bantam Books, 1963, pp. 205–206; *San Jose Mercury News,* August 7, 1992.

The serpentine landscape is discussed in: Arthur R. Kruckeberg, *California Serpentines,* Berkeley: University of California Press, 1984, pp. xiii, 6, 8, 14, 18, 64, 83, 85; R. R. Brooks, *Serpentine and Its Vegetation,* Portland, Ore.: Dioscorides Press, 1987; Brewer, pp. 139–140; Arthur Quinn, *Broken Shore,* Inverness, Calif.: Redwood Press, 1987, p. 4; "Serpentine Plant Display," University of California Botanical Garden, undated pamphlet; Schoenherr, pp. 278–281; Wallace Stegner, *Angle of Repose,* New York: Ballantine Books, 1991, pp. 73–74.

The Carrizo Plain is described in: Myron Angel, *History of San Luis Obispo County,* Fresno, Calif.: Valley Publishers, 1979, pp. 240, 363; Paul Van Zuyle, "The Historical Geography of Land Management on the Carrizo Plain," California Polytechnic State University, 1989, pp. 4, 13–17; "A Tour of the Carrizo Plain Natural Area," Bureau of Land Management, Bakersfield, Calif., 1990; Thomas C. Blackburn, *December's Child,* Berkeley: University of California Press, 1975, pp. 30, 32, 91–93; Myron Angel, *The Painted Rock,* San Luis Obispo, Calif.: Padre Productions, 1979; Marijean H. Eichel, "The Carrizo Plain: A Geographic Study of Settlement, Land Use, and Change," master's thesis, geography, San Jose State College, 1971; Marijean Eichel, "Changing Ecology of the Carrizo Plain," *La Vista,* San Luis Obispo County Historical Society, June 1972; Robert Wallace, p. 16; California Division of Mines and Geology, *San Andreas Fault in Southern California,* Special Report 118, San Francisco, 1975, pp. 241–242, 246; Pauline Bradley Dubin, "Nellie Bromley of Painted Rock," *La Vista,* June 1980, pp. 9–14; "A History of Carrisa Plain," compiled by Ted R. Fisher and the students of the Carrisa Plain Elementary School, undated; Craig Woods, "An Inventory of the Terrestrial Vertebrates of the Carrizo Plain, San Luis Obispo County," master's thesis, biology, California Polytechnic State University, 1982; Ernest C. Twisselman, "Flora of the Temblor Range," *Wasmann Journal of Biology,* University of San Francisco, 1956; *San Luis County Telegram-Tribune,* March 6, 8, and 17, 1960, July 30, 1977, May 31, 1979, and November 4, 1983.

For information on the California condor I referred to: Sanford R. Wilbur, *Vulture Biology and Management,* Berkeley: University of California Press, 1983, pp. 474–477, 481–482, 489; Sanford R. Wilbur, "The Condor and the Native Americans," *Outdoor California,* September–October 1983, pp. 7–8;

Schoenherr, pp. 307–310, 544; Lloyd Kiff, "To the Brink and Back," *Terra,* Natural History Museum of Los Angeles County, Summer 1990, pp. 7, 11; Brewer, p. 130; Harry Harris, "The Annals of Gymnogyps to 1990," *The Condor,* Cooper Ornithological Club, January–February 1941, pp. 32, 39–40; Alden H. Miller, "The Current Status and Welfare of the California Condor," *Research Report No. 6: The National Audubon Society,* 1965; David Phillips, ed., *The Condor Question,* San Francisco: Friends of the Earth, 1981, p. 47; Philip Fradkin, "Poison: Killing Pests Versus Saving Wildlife," *Los Angeles Times,* December 6, 1970; David Darlington, *In Condor Country,* New York: Henry Holt, 1991; *Los Angeles Times,* August 5, 1981; Lloyd Kiff, "A New Beginning for the Condor," *Western Tanager,* Los Angeles Chapter of the National Audubon Society, December 1991, pp. 1–5; Carl B. Koford, "The California Condor," *Research Report No. 4: The National Audubon Society,* 1953; Faith McNulty, "Last Days of the Condor," *Audubon,* March and May 1978; Caroline Arnold, *On the Brink of Extinction,* New York: Harcourt, Brace, Jovanovich, 1993, pp. 7–8, 18, 31–34; *Los Angeles Times,* January 15, 1992, and February 3, June 2 and 15, and July 21, 1993.

For Grove Karl Gilbert and the earthquake of 1906 see: Jordan, 1907, pp. 5, 18–19, 215–216, 350; Gladys Hansen, "The San Francisco Numbers Game," *California Geology,* December 1987, pp. 271–273; U.S. Geological Survey, *The San Francisco Earthquake and Fire of April 18, 1906,* Bulletin No. 324, 1907, pp. 8–15; Stephen J. Payne, *Grove Karl Gilbert,* Austin: University of Texas Press, 1980, pp. 174, 228–230; Brewer, p. 248; Mason L. Hill, "The San Andreas Fault: History of Concepts," *Geological Society of America Bulletin,* March 1981, pp. 112–119; U.S. Geological Survey, Robert D. Brown, Jr., *Map Showing Recently Active Breaks Along the San Andreas Fault Between Point Delgada and Bolinas Bay, California,* Washington, D.C., 1972; California Division of Mines and Geology, Alan J. Galloway, *Geology of the Point Reyes Peninsula, Marin County, California,* Bulletin 202, Sacramento, 1977, pp. 2–3, 5, 15–16, 48, 50, 59; Ron Redfern, *The Making of a Continent,* New York: Times Books, 1986, p. 112; Quinn, p. 3; John McPhee, "Assembling California—III," *New Yorker,* September 21, 1992, p. 58; California Division of Mines and Geology, Robert Streitz, *Studies of the San Andreas Fault Zone in Northern California,* Special Report 140, Sacramento, 1980, pp. 79, 86; Lawson, pp. 71, 76–80, 94, 192–193, 197–198, 372, 378, 380–383; Jack Mason, "Edgar's Quake," *Pt. Reyes Historian,* Spring 1981.

The seismicity of Parkfield is discussed in: Robert Wallace, pp. 68–70, 125–126, 161, 172; "History of Significant Earthquakes in the Parkfield Area" and "Historical Vignettes of the 1881, 1901, 1922, 1934, and 1966 Parkfield Earthquakes," *Earthquakes and Volcanoes,* Vol. 20, No. 2, U.S. Geological Survey, Reston, Va., 1988, pp. 2, 45–46, 52–54; Coast and Geodetic Survey, Jerry L. Coffman, ed., *The Parkfield, California, Earthquake of June 27, 1966,* Washington, D.C., 1966, pp. 1–2, 17–23, 29; William D. Stuart, "Parkfield Slowing Down," *Nature,* May 31, 1990, pp. 383–384; M. Myss, "Seismic Quiescence at Parkfield," *Nature,* May 31, 1990, pp. 426–431; "Quake Experts Still Supporting

Forecasting Despite Recent Miss," *San Francisco Chronicle,* November 22, 1993; "Is 1993 the Year of the Big One?" *New York Times,* January 3, 1993.

Consulted for the James Dean segment were: Venable Herndon, *James Dean: A Short Life,* Garden City, N.Y.: Doubleday, 1974, pp. 221–234; *Atascadero News,* October 6, 1955; *San Luis Obispo Times-Tribune,* October 1 and 3, 1955, and October 20, 1977; Brock Yates, "Far From Eden," *Car & Driver,* October 1985; Joe Hyams, *James Dean, Little Boy Lost,* New York: Warner Books, 1992, pp. 23, 45, 265–281, 309; "Monument to a Rebel," *San Francisco Examiner,* March 13, 1988; "Obsessed with James Dean," *People,* August 7, 1989.

For the recent natural and human history of the Point Reyes Peninsula see: Jules G. Evens, *The Natural History of Point Reyes Peninsula,* Point Reyes Station, Calif.: Point Reyes National Seashore Association, 1988; Hill, 1984, pp. 119–124; Robert Wallace, pp. 61–62; Harold Gilliam, *Island in Time,* San Francisco: Sierra Club Books, 1962; Jack Mason, *The Solemn Land,* Inverness, Calif.: North Shore Books, 1970; National Park Service, D. S. Livingston, *Ranching on the Point Reyes Peninsula,* Point Reyes Station, Calif., 1993; Dave Mitchell, *The Light on Synanon,* New York: Wideview Books, 1982; Robert Graysmith, *The Sleeping Lady,* New York: Dutton, 1990, pp. 49–54, 70–73, 101–105, 159, 161, 173, 194; Brian Lane and Wilfred Gregg, *The Encyclopedia of Serial Killers,* New York: Diamond Books, 1994, pp. 112, 114, 401–408; *Point Reyes Light,* December 4 and 11, 1980.

On California prisons I used material from: Mike Davis, pp. 288–289; California Department of Corrections, *California Prisoners & Parolees, 1990,* Sacramento, 1991, pp. 3–4; Brewer, p. 495; California Department of Corrections, *California State Prison, Del Norte County, Environmental Assessment Study,* Sacramento, November 1986; *Eureka Times-Standard,* January 6, February 1, and June 20, 1991, and August 13 and September 19, 1992; *San Francisco Chronicle,* June 15, 1990, July 5, 1991, and January 3 and February 3, 1993; *Los Angeles Times,* May 1, 1990, June 24 and September 7, 1993; "Institutional Profile," California Department of Corrections Communications Office, March 1992; "Racial-Ethnic Distribution of the Institution Population by Institution," California Department of Corrections, January 10, 1993.

For the Point Reyes flood and most recent Northern California earthquake see: "Prelude with Storm: Point Reyes," Michael Sykes, January 1982; U.S. Geological Survey, Stephen D. Ellen, ed., *Landslides, Floods, and Marine Effects of the Storm of January 3–5, 1982, in the San Francisco Bay Region, California,* Professional Paper 1434, Washington, D.C., 1988, pp. 7, 129–130; Evens, p. 21; *Point Reyes Light,* January 7, 1982; "The Loma Prieta Earthquake of October 17, 1989," U.S. Geological Survey, 1990, p. 7; U.S. Geological Survey, *Lessons Learned from the Loma Prieta, California, Earthquake of October 17, 1989,* Circular 1045, Washington, D.C., 1989; California Division of Mines and Geology, *The Loma Prieta (Santa Cruz Mountains), California, Earthquake of 17 October 1989,* Special Publication 104, 1990; *Fifteen Seconds,* Covelo, Calif.: Island Press, 1989, p. 82; *Point Reyes Light,* October 19, 1989; U.S. Geological Survey,

Probabilities of Large Earthquakes in the San Francisco Bay Region, California, Circular 1053, Washington, D.C., 1990, pp. 1–3, 23; Risa Palm, *After a California Earthquake,* Chicago: University of Chicago Press, 1992, p. 110.

The Profligate Province

For Portolá's *entrada* across Los Angeles I consulted: Herbert E. Bolton, *Historical Memoirs of New California,* vol. 2, Berkeley: University of California Press, 1926, pp. 130–134, 138, 256–257; Bolton, *Fray Juan Crespi,* Berkeley: University of California Press, 1927, pp. xiii, xvi, xxiv.

The physical description of the region was derived from: Schoenherr, pp. 313–405; Brewer, pp. 46–47; Durrenberger, pp. 23–27; Hinds, pp. 185–215, 325; Los Angeles County Committee of Menaced and Flooded Areas, J. W. Reagan, *Flood History of Los Angeles County,* 1914 (hereinafter referred to as the Reagan Report); Paul F. Starrs, "The Navel of California and Other Oranges," *California Geographer,* vol. xxviii, 1988, pp. 1–41; "City Will No Longer Trap Coyotes for Any Reason," *Los Angeles Times,* June 29, 1993; Ken Nunn, *Pomona Queen,* New York: Pocket Books, 1992, pp. 147–148.

For an overall sense of recent human history and the rapid turnover of peoples see: Mike Davis, *City of Quartz,* New York: Vintage, 1992; David Rieff, *Los Angeles: Capital of the Third World,* New York: Simon & Schuster, 1991; Edward W. Soja, *Postmodern Geographies,* New York: Verso, 1989, pp. 190–248; Robert M. Fogelson, *The Fragmented Metropolis,* Cambridge, Mass.: Harvard University Press, 1967; Carey McWilliams, *Southern California,* Salt Lake City: Peregrine Smith, 1973; "Race and Ethnicity, Southern California Counties, U.S. Census," *Los Angeles Times,* undated data; "Land of Languages," *Los Angeles Times,* July 5, 1993; "Rich-Poor Gulf Widens in State," *Los Angeles Times,* May 11, 1992; David E. Simcox, ed., *U.S. Immigration in the 1980s,* Boulder: Westview Press, 1988, p. 175; "Race/Ethnicity in Los Angeles County—U.S. Census, 1990," Marketing Research Department, *Los Angeles Times,* 1991; John Miller, ed., *Los Angeles Stories,* San Francisco: Chronicle Books, 1991, p. 153.

The events of the summer of 1993 were documented in the following: "Jobless Rate Bad for U.S., Worse for California," *Los Angeles Times,* July 2, 1993; "Hard Times: Hardest on the Homeless," *Los Angeles Times,* July 5, 1993; "Riordan Sees Renewal for L.A.," *Los Angeles Times,* July 2, 1993; "Assault Weapons," *Los Angeles Times,* July 2, 1993; "A New Tide of Immigration Brings Hostility to the Surface, Poll Finds," *New York Times,* June 27, 1993; "America: Still a Melting Pot?" *Newsweek,* August 9, 1993; "7 Ships Being Tracked as Migrant Smugglers," *Los Angeles Times,* July 10, 1993; "Detention of Ships Continues," *Los Angeles Times,* July 12, 1993; "Wilson's Plan to Curb Illegal Immigration," *San Francisco Chronicle,* August 10, 1993; "Grand Jury Urges 3-Year Ban on U.S. Immigration," *Los Angeles Times,* June 17, 1993; "Southland is Ripe Turf for White Hate Groups," *Los Angeles Times,* July 25, 1993; "Which Way L.A.," KCRW-FM radio broadcast, July 20, 1993; "Is the City of Angels Going to Hell?" *Time,* April 19, 1993; Joan

Didion, "Trouble in Lakewood," *New Yorker,* July 26, 1993, p. 60; "Nation's Land of Promise Enters an Era of Limits" (the first of a three-part series titled "Reinventing California"), *New York Times,* August 24, 1993; "Can L.A. Renew Its Love Affair with Itself?" *Los Angeles Times,* July 16, 1993.

For an overall view of the Los Angeles River I consulted: Anthony F. Turhollow, *A History of the Los Angeles District, U.S. Army Corps of Engineers, 1898–1965,* Los Angeles: U.S. Army Engineer District, 1975; U.S. Army Corps of Engineers, *Los Angeles County Drainage Area Review, Los Angeles,* 1991, pp. 13, 45; Stephen R. Van Wormer, "A History of Flood Control in the Los Angeles County Drainage Area," *Southern California Quarterly,* Spring 1991; Natural History Museum of Los Angeles County, Kimball L. Garrett, *The Biota of the Los Angeles River, Los Angeles,* 1993, pp. 4–5, 9, D-2, F-2, 68, G-2; "Los Angeles River," *Coast & Ocean,* State Coastal Conservancy, Summer 1993; "Can the River of Concrete Live Again?," *American Way,* June 15, 1992; "Watching the River Flow," *California Living,* November 4, 1984; "In Search of the L.A. River," *Los Angeles Times,* a series of articles running from October 20, 1985, to January 30, 1986.

For information on the start of the river, see: National Park Service, *Cheeseboro Canyon, Palo Comado Canyon Draft EIS,* Santa Monica National Recreation Area, Agoura Hills, 1990, pp. 65, 69, 85; "Cheeseboro Canyon Site," Santa Monica National Recreation Area, Agoura, 1991; Steven R. Ovendale, environmental engineer, Rocketdyne Division, Santa Susana Field Laboratory, July 28, 1993; "Thirty-five Years in Power for America: A History of the Rocketdyne Division of Rockwell International," Rocketdyne Publications Service, Canoga Park, 1990; "Fact Sheet, RCRA Facility Assessment on the Santa Susana Field Laboratory," Environmental Protection Agency, February 5, 1991; "Rockwell Will Pay $280,000 to Settle Suit," *Los Angeles Times,* November 28, 1990; "SSFL Water Release to Bell Creek," Video Services, Rocketdyne Division, March 22, 1993; "Bell Canyon Bulletin," various issues of homeowners' association newsletter; Bell Canyon Association memorandum to homeowners, guidelines for sale signs, May 1993.

Flatlands and palm tree information were derived from: Reyner Banham, *Los Angeles: the Architecture of Four Ecologies,* London: Pelican Books, 1973, p. 161; Victoria Padilla, *Southern California Gardens,* Berkeley: University of California Press, 1961, pp. 223–226, 236; Kevin Starr, *Material Dreams,* New York: Oxford University Press, 1990, p. 185; Charles Henry Rowan, "Ornamental Plants as a Factor in the Cultural Development of Southern California," master's thesis, UCLA, 1957, p. 145; Desmond Muirhead, *Palms,* Globe, Az.: Dale Stuart King, 1961, pp. 2, 36–37; William Hertich, *Palms and Cycads,* San Marino, Calif.: The Henry E. Huntington Library, 1951, pp. 3, 7, 97; John Fante, *Ask the Dust,* New York: Stackpole Sons, 1930, pp. 15, 60.

For H. J. Whitley and the San Fernando Valley see: Jackson Mayers, *The San Fernando Valley,* Walnut, Calif.: John D. McIntyre, 1976; "The Changing Landscape of the San Fernando Valley Between 1930 and 1964," Richard E. Preston,

Center for Urban Studies, San Fernando Valley State College, Northridge, 1965; Interpretive Exhibit, El Encino State Park, Encino, 1993; W. W. Robinson, *The Story of San Fernando Valley,* Los Angeles: Title Insurance and Trust Company, 1961, pp. 4–5, 21; Bruce W. Miller, *The Gabrielino,* Los Osos, Calif.: Sand River Press, 1991, pp. 25, 29; Francis J. Weber, *The Mission in the Valley,* Santa Barbara: Kimberly Press Inc., 1987, pp. 5, 13; *Who's Who in the Pacific Southwest,* Los Angeles: Times Mirror Printing & Binding House, 1913, p. 391; various documents, Boxes 2–4, 15, 16, the Whitley Papers, UCLA Special Collections; Virginia Whitley, undated manuscript, Box 3, Whitley Papers, UCLA Special Collections, pp. 52–54, 58; Catherine Mulholland, *The Owensmouth Baby,* Northridge: Santa Susana Press, 1987, pp. 4–9, 11, 116–123, 126, 128–129.

For the events in the Sepulveda Basin, besides the tape-recorded conversation, see: "Officers Engage in Gun Battle With Alleged Gang Members," *Los Angeles Times,* July 18, 1993; Van Wormer, p. 88; Stuart H. Brehm III, U.S. Army Corps of Engineers, July 9, 1993.

For the background on the eastern end of the Valley and Los Angeles schools I consulted: Mayers, various pages; Preston, p. 16; "Valley Crime Rate Low, Fear High, Poll Shows," *Los Angeles Times,* May 9, 1993; "Economy Still Troubles Many Valley Residents," *Los Angeles Times,* May 11, 1993; "Valley Residents Give Low Grade to Public Schools," *Los Angeles Times,* May 10, 1993; Los Angeles Unified School District, *School Accountability Report,* Los Angeles, 1992; Judith Rosenberg Raferty, *Land of Fair Promise,* Stanford: Stanford University Press, 1992, pp. 111–114, 117–118, 153, 187; Mulholland, p. 121; W. Henry Cooke, "The Segregation of Mexican-American School Children in Southern California," *School and Society,* June 5, 1948, p. 417; John Caughey and LaRee Caughey, *School Segregation on Our Doorstep,* Los Angeles: Quail Books, 1966, pp. 3–5; John Caughey, *To Kill a Child's Spirit,* Itasca, Ill.: F. E. Peacock Publishers, 1973, pp. 6–7, 10–17, 131–140; Bean, pp. 387–388; "Executive Summary: Ethnic Survey Report," Information Technology Division, Los Angeles Unified School District, 1993; "L.A. School District Enrollment Growth Slows," Office of Communications, Los Angeles Unified School District, October 15, 1992; various *Daily News* and *Los Angeles Times* stories in the spring and early summer of 1993, including the following in the *Times*: "A Period of Unprecedented Crisis for the Schools," May 1, 1993; "Ethnic Minorities Now School Board Majority," July 2, 1993; "For L.A. Class of '93, Pomp and Tough Circumstances," July 5, 1993; "Expelled for Having Guns, Youths Must Reclaim Selves," July 12, 1993.

On Reed Junior High School, see: "Choices," Los Angeles Unified School District, 1993; "Reed Middle School," Los Angeles Unified School District School Accountability Report, 1992; "Skyline, Commemorative Edition," Walter Reed Jr. High, 1939–1989; Larry J. Tash, principal, July 14–15, 1993; *Los Angeles Times,* March 14–15, 1993, series on high birthrates of Latina teenagers; "Way Ahead of the Class," *Los Angeles Times,* June 27, 1993; "Reed Review," various issues of school newsletter for parents.

Universal's CityWalk attraction was the subject of numerous releases in an MCA press packet, Summer 1993, and the following newspaper stories that were not included in the packet: "A Glittery Bit of Urban Make-Believe," *Los Angeles Times,* July 18, 1993; "Theater Chain's Action Triggers Cries of Racism," *Los Angeles Times,* July 25, 1993; "Who Should Define a City," *New York Times,* August 15, 1993.

For the downtown section of the Los Angeles River, check: Mike Davis, p. 254; "A River in Need of Friends," *Downtown News,* January 2, 1989; "Even Buildings Turn Their Faces Away from L.A.'s River," *Wall Street Journal,* August 1, 1989; "Can the L.A. River Be Saved," *L.A. Weekly,* September 7, 1989; "Charting a New Path for the L.A. River," *Los Angeles Times,* April 2, 1990; Dave Weaver, U.S. Army Corps of Engineers, July 29, 1993.

The history of early flooding is contained in: "Report of the Board of Engineers Flood Control to the Board of Supervisors," Los Angeles County, 1915; "The Los Angeles River Is:," *Aqueduct,* Southern California Metropolitan Water District, Spring 1978; Richard Bigger, *Flood Control in Metropolitan Los Angeles,* Berkeley: University of California Press, 1959, pp. 2–3; Marc Blake, "A Study of the Los Angeles River," master's thesis, architecture, UCLA, 1990; "Life of a Rancher," *The Quarterly,* Historical Society of Southern California, September 1950, pp. 190–191; Reagan Report, various pages; J. M. Guinn, "Exceptional Years: A History of California Floods and Drought," Historical Society of Southern California, 1890, pp. 33–39; J. J. Warner, *An Historical Sketch of Los Angeles County,* Los Angeles: O. W. Smith, 1936, p. 18; Brewer, pp. 25, 30; Harris Newmark, *Sixty Years in Southern California, 1853–1913,* New York: The Knickerbocker Press, 1926, p. 258; Robert Glass Cleland, *The Cattle on a Thousand Hills,* San Marino, Calif.: The Huntington Library, 1951, p. 128; Ludwig Louis Salvator, *Los Angeles in the Sunny Seventies,* Los Angeles: McCallister & Zeitlin, 1929, pp. 39, 41, 141–142.

The early racial history of Los Angeles can be traced in: Horace Bell, *Reminiscences of a Ranger,* Santa Barbara: Hebberd Publishers, 1927, pp. 2, 13, 35, 155, 159; Cleland, pp. 35, 58–59, 90; Robert F. Heizer and Alan J. Almquist, *The Other Californians,* Berkeley: University of California Press, 1977, pp. 49–50; Heizer, 1993, pp. 30–32; Richard H. Peterson, "Anti-Mexican Nativism in California, 1848–1853," *Southern California Quarterly,* Winter 1980, pp. 309–327; Robert W. Blew, "Vigilantism in Los Angeles, 1835–1874," *Southern California Quarterly,* Spring 1972, pp. 11, 13, 26–27; Leonard Pitt, *The Decline of the Californios,* Berkeley: University of California Press, 1968, pp. 154, 166, 273–274; Richard Griswold del Castillo, *The Los Angeles Barrio, 1850–1890,* Berkeley: University of California Press, 1979, pp. 106, 115, 119; "Los Angeles in 1854–5: the Diary of Rev. James Woods," *The Quarterly,* March 1941, pp. 70, 74, 79, 83; Brewer, p. 14; Paul M. De Falla, "Lantern in the Western Sky, Part I," *Southern California Quarterly,* March 1960, pp. 57–88; Paul M. De Falla, "Lantern in the Western Sky, Part II," *Southern California Quarterly,* June 1960, pp. 161–185; William R. Locklear, "The Celestials and the Angels," *Southern California Quar-*

terly, September 1960, pp. 239–256; Newmark, p. 433; Cleland, p. 227; Ricardo Romo, *History of a Barrio,* Austin: University of Texas Press, 1983, pp. 5, 10–11; Kevin, 1990, pp. 62, 120; Starr, 1985, pp. 40–41; Carey McWilliams, *North From Mexico,* Philadelphia: J. B. Lippincott, 1949, pp. 128–131; McWilliams, 1973, p. 157.

For racism and the Harrison Gray Otis era of the *Los Angeles Times* see: McWilliams, 1973, p. 274; "1781–1908, Los Angeles and the *Times,* a Brief Historical Sketch of Both," *Los Angeles Times,* 1908; "This Above All—To Thine Own Self Be True," *Among Ourselves,* March 1931; "Hearst Turned Down," *Los Angeles Times,* December 14, 1905; David Halberstam, *The Powers That Be,* New York: Dell Publishing Co., 1986, pp. 136, 169; G. William Dornhoff, *Who Rules America Now?,* New York: Prentice-Hall, 1983, pp. 107–109; Terry Christensen and Larry Gerston, "The Rise of the McClatchys and Other California Newspaper Dynasties," *The Californians,* September–October 1983, p. 8; James Bassett interview of William F. Knowland, Oakland, June 12, 1973; Robert Gottlieb and Irene Wolt, *Thinking Big,* New York: G. P. Putnam's Sons, 1977, pp. 23, 192, 271; Marshal Berges, *The Life and Times of Los Angeles,* New York: Atheneum, 1984, p. 13; "A Chronological History of the Los Angeles Times," *Los Angeles Times,* 1993, p. 1; "Racial Attitudes of the *Times,* 1881–[1943]," working paper compiled for Bassett history, undated; "Heathen Chinee," *Los Angeles Daily Times,* December 8, 1881; Locklear, p. 253; McWilliams, 1973, pp. 277, 281; *San Francisco Call,* December 18, 1906; various articles in the *Los Angeles Times,* dating from January 22, 1885, to October 13, 1923.

McWilliams is covered in: Carey McWilliams, *The Education of Carey McWilliams,* New York: Simon & Schuster, 1978, pp. 43–44, 46, 81–82, 119–120; Starr, 1990, pp. 153–155, 305, 310–313; Carey McWilliams to James Bassett, September 8, 1975, with undated Carr letter and clipping attached; Joel Gardner, ed., "Honorable in All Things," oral interview with Carey McWilliams, UCLA special collections, 1982, pp. 43–44, 79, 82, 84, 262; McWilliams, 1973, p. 376; McWilliams, 1949, pp. 289–290.

For the Harry Chandler era of the *Times* and racism I consulted the following: Berges, p. 35; Gottlieb, pp. 150–151, 227; "Tejon Ranch," Tejon Ranch Company, undated; Mulholland, p. 128; Romo, 1983; Rodolfo Acuña, *Occupied America,* New York: Harper & Row, 1988, pp. 172–173, 200–205; George E. Frakes, *Minorities in California History,* New York: Random House, 1971, pp. 62–68; McWilliams, 1949, pp. 167–169, 185–187, 194, 281; Carey McWilliams, *Brothers Under the Skin,* Boston: Little, Brown, 1951, p. 113; Joseph Lilly, "Metropolis of the West," *North American Review,* September 1931, pp. 242–243; Raphael J. Sonenshein, *Politics in Black and White,* Princeton: Princeton University Press, 1933, p. 26; "The Mexican Migration," *Los Angeles Times,* April 18, 1931; "Immigration Abuses," *Los Angeles Times,* July 22, 1931; McWilliams, 1939, pp. 130, 305.

Paving the river is covered in: Bigger, pp. 13, 39, 134, 136–137; U.S. Geological Survey, Harold C. Troxell, *Floods of March 1938 in Southern California,*

Water-Supply Paper 844, Washington, D.C., 1942, pp. 1–14; Mike Davis, p. 199; Van Wormer, pp. 88–89; *Daily News,* April 19, 1993; "Disastrous Flooding Could Return to Los Angeles County," *LACDA Update,* U.S. Army Corps of Engineers, September 1987; "$300 Million Solution to Curb '100-Year' Flood Waters," *Los Angeles Times,* March 9, 1989.

For the *Times* during the World War II era and racism see: Dillon S. Myer, 1971, pp. 18, 84, 282; Roger Daniels, 1981, pp. 31–34; Takaki, 1990, pp. 387–388; *Los Angeles Times,* January 23 and February 2, 6, and 19, 1942; Darrell Kunitomi to Lee May, November 4, 1981 (copies of other internal correspondence and many documents were found in the *Times* historical archives); McWilliams, 1949, pp. 227, 237–238, 242, 245, 248–249, 251, 253, 255–257; Gottlieb, pp. 298–299; *The New Republic,* June 21, 1943; various *Los Angeles Times* stories contained in files of the McWilliams Collection, Special Collections, UCLA; Los Angeles County Sheriff's Department, Lt. Edward Duran Ayres, *Statistics,* Los Angeles, August 20, 1942, pp. 2–4 (Box 29, McWilliams Collection); *Los Angeles Times,* August 11, 1942; *Los Angeles Herald & Express,* January 5, 1943; Gottlieb, p. 299; *PM,* June 11, 1943; Romo, pp. 166–167; Acuña, pp. 256–259; McWilliams, 1951, p. 3; Maurico Mazón, *The Zoot-Suit Riots,* Austin: University of Texas Press, 1984, pp. 78–79; Beatrice Griffith, *American Me,* Westport, Conn.: Greenwood Press, 1977, pp. 15–28; John Caughey and LaRee Caughey, *Los Angeles,* Berkeley: University of California Press, 1977, p. 381; "Youthful Gang Secrets Exposed," *Los Angeles Times,* July 16, 1944.

The Otis Chandler era and racism were documented by the following: Bean, pp. 386–387; Sonenshein, pp. 7, 30–33, 66, 71, 77; Richard G. Lillard, *Eden in Jeopardy,* New York: Knopf, 1966; Jack R. Hart, *The Information Empire,* Washington, D.C., University Press of America, 1981, p. 312; Roger Daniels and Spencer C. Olin, *Racism in California,* New York: Macmillan, 1972, pp. 281–288; *Los Angeles Times,* November 11, 1963, and April 24, 1964; James Bassett interview with William F. Thomas, *Times* editor, Los Angeles, May 21, 1975; Ellis Cose, *The Press,* New York: William Morrow, 1989, p. 126; A. S. Young to Nick B. Williams, January 3, 1965; Nick B. Williams to A. S. Young, January 8, 1965; "Editorial Goals for 1965," Nick B. Williams, undated memo; "The New Look at the Times," *Saturday Review,* June 12, 1965; "Reporter Tells Violence, Fear During Riot," *Los Angeles Times,* August 13, 1965; Jerry Cohen and William S. Murphy, *Burn, Baby, Burn,* New York: E. P. Dutton, 1966, pp. 75–78; Sonenshein, p. 77; James Bassett interview with Asa Call, April 2, 1973; Bayard Rustin, "The Watts Manifesto and the McCone Report," *Commentary,* March 1966, p. 30; Frakes, pp. 117–127; Robert M. Fogelson, *Violence as Protest,* Garden City, N.Y.: Doubleday, 1971, p. 5; "Enterprise in Los Angeles," *Time,* May 13, 1966; Nick B. Williams to Frank P. Doherty, August 4, 1967; James Bassett interview with Otis Chandler, Los Angeles, July 29, 1975; Jack R. Hart interview with Otis Chandler, Los Angeles, August 4, 1975; résumé of Otis Chandler, 1974; KNXT radio interview with Otis Chandler, October 14, 1975; Otis Chandler's answers to the *U.S. News & World Report* survey "Who Runs America," March

17, 1983; Michael Jackson interview with Otis Chandler, Los Angeles, March 1978; Otis Chandler to F. Daniel Frost, March 3, May 25, and June 9, 1983; Otis Chandler to Robert F. Eburu, March 15, 1983; James Bassett interview with Nick B. Williams, Los Angeles, November 1, 1972; KNXT radio interview with Otis Chandler, November 20, 1981; Phil Kerby, "Los Angeles Times: Most Likely to Succeed," *Nation,* January 15, 1968, pp. 80–81; John Corry, "The Los Angeles Times," *Harper's,* December 1969; *New York Times,* April 27, 1970; "Watts Plus 10," *Los Angeles Times,* March 24, 1975; Bob Gottlieb and Irene Wolt, "The Changing of the Guard," *Los Angeles,* December 1978; David Halberstam, "The California Dynasty," *The Atlantic,* April 1979; "Fat Times in Los Angeles," *Newsweek,* September 22, 1980; "The Los Angeles Times Reaches for the Top," *Washington Journalism Review,* July/August 1982; "Marauders From Inner City Prey on L.A.'s Suburbs," *Los Angeles Times,* July 12, 1981; Jack Smith, "Our City, Our *Times,*" December 4, 1981; market profiles for the years 1983, 1987, 1989, Los Angeles Times Marketing Research; Bob Gottlieb, "The Gray Lady of Spring Street Hits 100," *Los Angeles,* November 1981, p. 404; Féliz Gutiérrez and Clint C. Wilson III, "The Demographic Dilemma," *Columbia Journalism Review,* January–February 1979; "Five Year Plan," *Los Angeles Times,* 1980; Otis Chandler to Tom Johnson, February 24, 1981; Otis Chandler to Robert F. Eburu, January 1, 1981.

(I should point out that I interviewed a number of former and current *Times* editors and reporters for information about the Otis Chandler and corporate eras. Many requested that their names not be used; some spoke on the record. I have decided not to list any names in fairness to all. The two top *Times* executives in the summer of 1993, publisher David Laventhol and editor Shelby Coffey III, said they either did not have the time to see me or made the conditions for the interview so restrictive that I did not deem it productive.)

The corporate era of the *Times* and racism is covered by: Otis Chandler to Walter H. Annenberg, July 31, 1985; various internal memos; Joe Domanick, "The *Times* After the Chandlers," *Los Angeles,* September 1987; "Los Angeles Times, A Publishing Power, Is Beset by Small Rivals," *Wall Street Journal,* February 7, 1989; "Los Angeles's Changing *Times,*" *Newsweek,* September 11, 1989; Joan Didion, *After Henry,* New York: Simon & Schuster, 1992, pp. 220–250; "Five-Year Plan, 1988–1992," *Los Angeles Times,* 1987; "Los Angeles Paper's Rival Visions," *New York Times,* June 18, 1990; "Hello, Sweetheart! Get Me Remake," *Time,* April 15, 1991; "Los Angeles Hispanic Market," Los Angeles Times Marketing Research, 1989; "L.A. 2000," Los Angeles 2000 Committee, 1988; "Facts About the Los Angeles Times," Los Angeles Times Public Relations Department, August 1991; "Times Mirror 1992 Annual Report," Times Mirror, 1992, p. 5; "Suddenly, Costs, Morale Are Problems," *USA Today,* March 13, 1992; "Update," newsletter issued by the publisher's office, *Los Angeles Times,* September 18, 1992, February 3, May 18, and June 10, 1993; "Covering the War at Home," *Washington Journalism Review,* July–August 1992; "Report from the Committee on Diversity," Shelby Coffey III memo to editorial staff, July 16,

1993; "Aftermath of L.A. Riots," *Editor & Publisher,* May 23, 1992; "The L.A. Riots and Media Preparedness," *Editor & Publisher,* May 23, 1992.

Information on Southeast Los Angeles and the lower river were derived from the following: Capt. Mike Trevis, Bell Police Department, June 26, 1993; various *Los Angeles Examiner* stories in 1929; Luis J. Rodríguez, *Always Running,* Willimantic, Conn.: Curbstone Press, 1993, pp. 17–30; Luis J. Rodríguez, *The Concrete River,* Willimantic, Conn.: Curbstone Press, 1991, pp. 38–39; Cudahay General Plan Update, City of Cudahay, March 20, 1992; "The History of Bell and the James George Bell Family," City of Bell, undated; untitled pamphlet celebrating Cudahay's 25th anniversary, 1960–1985; *Los Angeles Times,* November 29, 1963; *Industrial Post,* December 14, 1967; various issues of the *Industrial Post* and *Daily Signal*; David Reid, ed., *Sex, Death, and God in L.A.,* New York: Pantheon, 1992, pp. 54–71; "Census Figures Paint a Checkered Portrait of L.A. County," *Los Angeles Times,* August 29, 1990; *Newsweek,* June 26, 1989; Los Angeles County Department of Regional Planning Bulletin, special census edition, January 1992; City of Cudahay newsletters, December 1991 and Summer 1992; "Pursuing Excellence," Bell High School accreditation report, 1992–1993; "Bell High School," Los Angeles Unified School District Accountability Report, 1991–1992; Mel Mares, principal, and Tony Solorzano, vice principal, July 21, 1993; various issues of the "Eagle," Bell High School yearbook; various mimeographed histories, school-year highlights, and reviews of certain years at Bell High School; "Golden Days at Dear Bell High," 50th-anniversary celebration booklet; "Stiff Penalties in Drive-By Shooting Pushed," *Los Angeles Times,* July 10, 1993; "10-Year-Old Girl Shot in the Head by Bicyclists," *Los Angeles Times,* July 21, 1993; Metro Budget for Thursday, July 22, 1993, *Los Angeles Times*; "No Escape," *Los Angeles Times,* July 22, 1993; "2 Gang Members Held in Shooting of Girl, 10," *Los Angeles Times,* July 24, 1993; "10-Year-Old Lynwood Shooting Victim Dies," *Los Angeles Times,* July 25, 1993; "Queen Mary," David F. Hutchings, Kingfisher Publications, Southampton, England, 1986.

Index

455

Index